J. Simon · J.-J. André

Molecular Semiconductors

Photoelectrical Properties and Solar Cells

Editors: J. M. Lehn, Ch. W. Rees

With 166 Figures and 41 Tables

Springer-Verlag
Berlin Heidelberg New York Tokyo

PHYSICS

525 7 3 7 9 5

Prof. Jacques Simon
Ecole Supérieure de Physique
et Chimie Industrielles
de la Ville de Paris
10, rue Vauquelin
F-75231 Paris Cedex 05

Dr. Jean-Jacques André
Centre de Recherches
sur les Macromolécules (C.N.R.S.)
6, rue Boussingault
F-67083 Strasbourg Cedex

ISBN 3-540-13754-8 Springer-Verlag Berlin Heidelberg New York Tokyo
ISBN 0-387-13754-8 Springer-Verlag New York Heidelberg Berlin Tokyo

Library of Congress Cataloging in Publication Data
Simon, J. (Jacques), 1947– Molecular semiconductors.
Bibliography: p. Includes index.
1. Semiconductors. 2. Molecular crystals. 3. Photoelectricity. 4. Solar batteries. I. André,
J.-J. (Jean-Jacques), 1940–. II. Lehn, J.-M. (Jean-Marie). III. Rees, Charles W. (Charles
Wayne).
IV. Title. QC611.S583 1984 537.6'22 84-20197

This work is subject to copyright. All rights are reserved, whether the whole or part of the
material is concerned, specifically those of translation, reprinting, re-use of illustrations,
broadcasting, reproduction by photocopying machine or similar means, and storage in
data banks. Under § 54 of the German Copyright Law where copies are made for other
than private use, a fee is payable to "Verwertungsgesellschaft Wort", Munich.

© by Springer-Verlag Berlin Heidelberg 1985
Printed in Germany

The use of registered names, trademarks, etc. in this publication does not imply, even
in the absence of a specific statement, that such names are exempt from the relevant
protective laws and regulations and therefore free for general use.

Typesetting: Th. Müntzer, GDR; Offsetprinting: Sala-Druck, Berlin;
Bookbinding: Lüderitz & Bauer, Berlin

2152/3020-543210

QC611
S583
1985
PHYS

A Elisabeth
"Sans elle tout ce qui est ne serait pas"
                    (C. de Gaulle, à propos de Mme de Gaulle)

A la mémoire de Pierre Vitali

        A Madeleine

De grands physiciens ont fort bien trouvé pourquoi les lieux souterrains sont chauds en hiver et froids en été. De plus grands physiciens ont trouvé depuis peu que cela n'était pas.

(Fontenelle)

Soleil, je t'adore comme les souvages . . .

(Jean Cocteau)

Preface

During the past thirty years considerable efforts have been made to design the synthesis and the study of molecular semiconductors. Molecular semiconductors — and more generally molecular materials — involve interactions between individual subunits which can be separately synthesized. Organic and metallo-organic derivatives are the basis of most of the molecular materials. A survey of the literature on molecular semiconductors leaves one rather confused. It does seem to be very difficult to correlate the molecular structure of these semiconductors with their experimental electrical properties. For inorganic materials a simple definition delimits a fairly homogeneous family. If an inorganic material has a conductivity intermediate between that of an insulator $(<10^{-12}\ \Omega^{-1}\ cm^{-1})$ and that of a metal $(>10^3\ \Omega^{-1}\ cm^{-1})$, then it is a semiconductor and will exhibit the characteristic properties of this family, such as junction formation, photoconductivity, and the photovoltaic effect. For molecular compounds, such simplicity is certainly not the case. A huge number of molecular and macromolecular systems have been described which possess an intermediate conductivity. However, the various attempts which have been made to rationalize their properties have, more often than not, failed. Even very basic electrical properties such as the mechanism of the charge carrier formation or the nature and the density of the dopants are not known in detail. The study of molecular semiconductor junctions is very probably the most powerful approach to shed light on these problems. The physico-chemical characteristics of such devices depend on the molecular structure of the semiconductor and on the presence of minute amounts of impurities. Most of the fundamental transport parameters (mobility and nature of the charge carriers, density and distribution of the dopants, nature of the conduction mechanism, etc.) may be determined from junction studies in the dark or under illumination. In the latter case, we have the transformation of light into electricity, the photovoltaic effect. The term "solar cell" is often preferred to designate these devices. Although the terms "photovoltaic effect" and "solar cell" cover exactly the same type of scientific interests, they must not be confused; pure scientists and industrialists must each have a language of their own. Junctions studies are, in any event, a very powerful and promising way of studying molecular semiconductors.

Is the term "molecular semiconductor" a misnomer? It is certainly true that the mobilities normally measured are low. Would "doped insulator" be more appropriate? The first chapter of the book will go over a few basic notions of solid state physics and try to solve this semantic problem. As far as it it possible a chemical interpretation of notions such as mobilities, Fermi levels, or band schemes will be given. Such interpretations will of course necessitate some approximations. It is hoped that what the chemist gains in intuitive understanding will more than compensate for what the physicist loses in precision. Of all the possible modes of conduction, emphasis will be placed on those most characteristic of molecular materials. In all cases the importance of the nature and the concentration of the impurities on the electrical properties will be emphasized. The relative contributions of structural defects and chemical impurities on the various trapping processes will be estimated.

The second chapter is devoted to a general discussion of the various photoelectric phenomena present in molecular semiconductors. Basic photochemical properties of molecular compounds are first presented and the factors affecting the extent of energy transfer in molecular materials are briefly recalled. This then leads to a discussion of the various pathways for photogenerating charge carriers. After outlining the mechanisms of formation of semiconductor p/n junctions and semiconductor/metal junctions, the photovoltaic properties of these devices will be discussed. At this point most of the scientific ideas underlying the operations of organic solar cells will have been hopefully mastered.

A theme running through the first two chapters will be the distinction between two classes of molecular semiconductors: molecular crystals and polymers. The formers are available in highly purified form through the use of chromatography, distillation, sublimation, or crystallization. The mobility is limited by the usually small π overlap. Macroscopic migration of charge then involves electron hopping from molecule to molecule. Problems thus arise in applying band theory. The second class, the polymers, provide a strong contrast. Once formed they cannot be purified to the standards of molecular crystals. The intrinsic mobility is now determined by the strong, covalent intrachain interactions and is thus large. A band theory then becomes less inappropriate. However, the impurities present in these systems may now completely dominate the mobility through trapping processes. Any particular case will show properties somewhere between the extremes of the pure molecular crystal and the impure polymer, and a clear understanding would necessitate a detailed knowledge of its chemical and physico-chemical properties. So rather than attempt a necessarily superficial coverage of all known molecular semiconductors, a "typical" example of each of the two types will be discussed. Chapter 3 will discuss metall-ophthalocyanines, "typical" molecular crystals. Chapter 4 focuses on polyacetylene as the "typical" polymer. In both cases the synthesis and physico-chemical properties of the two systems will be discussed first. A review is provided to the literature up to approximately July 1982. Once the materials have been defined their electrical dark properties can be discussed. Each chapter will end with a section devoted to the photoconduction processes and the photovoltaic effect. At the light of the previous considerations, an overview of all the main other molecular conductors or semiconductors is given in Chapter 5.

To design and elaborate new molecular semiconductors will require an unusually close collaboration between chemists and physicists. It is hoped that this book will help the chemist to appreciate the importance of the physics involved, and will help the physicist appreciate the problems facing the chemist. By this means we hope to bridge the gulf between them.

Acknowledgments

Many people made possible the realization of this book, but we want to particularly acknowledge Professors J.-M. Lehn, H. Benoit, and C. Wippler. Professors M. Schott and G. Weill are thanked for helpful and stimulating discussions. We very gratefully acknowledge the help provided by everyone working at Centre de Recherches sur les Macromolécules of Strasbourg (C.R.M.). Thanks are due to Mr. Ph. Gramain for offering us the opportunity of developing this subject at C.R.M. Mr. M. Martin, Mr. C. Piechocki, and Miss D. Markovitsi are thanked for their collaboration. C.N.R.S., P.I.R.S.E.M., and ELF are acknowledged for their financial support.

This whole book has been rewritten and often re-written by Mr. S. Abbott.

Table of Contents

List of Symbols

A

a: constant

a: ratio of photoinjection quantum efficiency to dark injection rate

A: organic material

A: electron acceptor molecule

A*: Richardson constant

a_c: lattice constant (centered cubic lattice)

A_c: crystal electron affinity

A_g: gas-phase electron affinity

a_l: lattice constant

A_o: perfection factor of a diode

A_s: semiconductor electron affinity

B

b: constant

b: ratio of photoinjection quantum efficiency to dark injection rate (other electrode as a)

C

c: constant

c: velocity of light

C: concentration

C: capacitance

C_{bulk}: bulk capacitance

C_{cont}: contact capacitance

C_j: junction capacitance

C_p: parallel equivalent capacitance

C_s: series equivalent capacitance

c': lattice parameter

D

d: organic material thickness

D: electron donor molecule

D: diffusion coefficient

$\mathscr{D}_h$: hopping diffusion constant (in cm^2/s)

D_n: electron diffusion coefficient

D_p: hole diffusion coefficient

E

e: charge of electron

E: energy

E_{act}: activation energy

E_{BE}: binding energy of photoemitted electrons

E_c: energy at the bottom of the conduction band (conduction band edge)

E_{coul}: Coulomb energy

E_{c-i}: charge-induced dipole energy

E_{c-d}: charge-dipole energy

E_{c-q}: charge-quadrupole energy

E_{c-iq}: charge-induced quadrupole energy

E_{el}: pure electronic energy

E_{exc}: exchange energy

$E_F(n)$; $E_F(p)$: Fermi level in the neutral region of n and p-semiconductors

$E_F(s)$: Fermi level in semiconductor

E_F: Fermi level

E_{i-i}: induced dipole-induced dipole energy

E_{ip}: ion pair interaction energy

$E_i = E_F$: Fermi level in the intrinsic case

E_{in}: inner sphere reorganization energy

E_{KE}: kinetic energy of photoemitted electrons

E_{o-o}: octopole-octopole energy

E_{out}: outer sphere reorganization energy

E_{pe}: polarization energy

E_{reo}: reorganization energy

E_t: charge carrier trap depth

E_{tu}: height of the tunneling potential barrier

E_v: energy at the top of the valence band (valence band edge)

E_{vac}: vacuum energy

E_W: higher terms of the multipolar energy

E_t^{ex}: trap depth for excitons

E_t^e: trap depth for electrons

E_t^h: trap depth for holes

$E_{1/2}^{ox}$: half-wave oxidation potential

$E_{1/2}^{red}$: half-wave reduction potential

ΔE_{ac}: energy to create activated complex

ΔE_{cc}: energy to create charge carrier

ΔE_{MS}: energy barrier metal-semiconductor

ΔE_{rel}: vibronic bandwidth

ΔE_{SM}: energy barrier semiconductor-metal

F

$\mathbf{F}$: flux of particles

F: correction factor related to the refractive index of the medium

FF: fill factor

$F_D(\nu)$: spectral distribution of the donor emission

$\tilde{f}_{mn}$: oscillator strength

G

$\mathfrak{g}$: statistical distribution

g: Lande factor

$\mathcal{G}$: gain factor

$G_A(\nu)$: spectral distribution of the acceptor absorption

g_n: degeneracy of the nth state

G_p: $1/R_p$

H

$h(E)$: trap distribution

I

I: intensity of current

I_A: impurity (acceptor)

I_c: ionization potential (crystalline state)

I_D: impurity (donor)

I_{dark}: dark current

I_g: ionization potential (gas phase)

I_{max}: intensity at maximum power

I_{out}: output current

I_{ph}: photocurrent

I_S: surface-ionization potential

I_{SC}: short-circuit current

I_{SCL}: space-charge-limited current

I_Ω: ohmic current

J

J: current density

J: exchange integral

J_{ov}: spectral overlap

$\mathbf{J}_{n_1}$: diffusion current (electrons)

$\mathbf{J}_{n_2}$: generation or drift current (electrons)

$\mathbf{J}_{p_1}$: diffusion current (holes)

$\mathbf{J}_{p_2}$: generation or drift current (holes)

$\mathbf{J}_{sat}$: saturation current

K

$\mathbf{k}$: wavenumber vector

k: Boltzmann's constant

k_c: rate constant of production of free carriers

k_{CT}: charge-transfer-rate constant

k_{ET}: rate constant for electron transfer

k_n: rate constant for non-radiative transitions

k_Q: quenching processes (1st order)

k_T: rate constant for intersystem crossing to triplet state

L

$\mathcal{L}$: effective average orbital radius

l_a: light penetration depth

l_{ex}: exciton migration length

L_n: diffusion length of electrons

L_p: diffusion length of holes

l_{sc}: width of the space-charge region

l_{tu}: width of the tunneling barrier

M

m: mass of electron

m^*: effective mass of electrons

$\mathbf{M}$: dipole-moment operator

$\mathbf{m}_t$: transition-moment integral

N

n: density of free charge carriers

n: number of electrons

$\mathscr{N}$: Avogadro's number

N_a: density of ionized acceptors

N_c: effective density of states in the conduction band

N_{cd}: number of collisions with the barrier in a tunneling process

N_d: density of ionized donors

N_D: bulk donor density

N(E): density of energy states

n_i: number of intrinsic charge carriers

n_n: concentration of electrons in the n region

n_p: concentration of electrons in the p region

N_{surf}: density of surface acceptors

n_t: density of trapped carriers

N_t: density of traps

N_v: effective density of states in the valence band

N_t^e: density of traps for electrons

N_t^h: density of traps for holes

P

$\mathbf{p}$: momentum

p: number of holes

P: pressure

P_{ex}: probability of energy hopping

$\mathscr{P}_h$: overall probability of hopping

P_h: probability of hopping within the activated complex

P_{mn}: probability of light absorption between states m and n

p_n: concentration of holes in the n-doped region

P_{O_2}: partial pressure of O_2

p_p: concentration of holes in the p-doped region

$\mathscr{P}_{st}$: sticking probability

P_{tu}: probability of tunneling

Q

Q: charge of a capacitor

Q_B: quadrupole moment of B

$Q_{v'v''}$: Franck-Condon factor

R

R: resistance

R_{bulk}: bulk resistance

R_{cont}: interparticle contact resistance

R_j: junction resistance

R_p: equivalent parallel resistance

r_{rf}: rectification ratio

R_s: series resistance

R_{sh}: shunt resistance

S

s: second

$\mathscr{S}_d$: dipole strength

T

t: time

t_{ab}: absorption time of a monolayer of gas

T_{cm}: thermoelectric power in the charge-migration process

$T_{D^*\to A}$: period of transfer from D* to A

t_t: transit time

U

U_0: number of sites

V

$\mathbf{v}$: velocity

V: potential

V: volt

V_a: applied potential

$\mathbf{V}_{bi}$: built-in potential

V_{max}: voltage at maximum power

V_n: potential in the neutral region in an n-type semiconductor

V_{oc}: open-circuit voltage

V_{out}: output voltage

V_p: potential in the neutral region in a p-type semiconductor

W

W: bandwidth

ΔW: effective bandwidth

X

x_n, x_p: distance of the space-charge region extending in the n or p part

Z
Z_c: coordination number
$Z.e$: net charge of carriers

α
α: orbital overlap
α_a: absorption coefficient

β
β: coulombic integral

γ
γ: resonance integral
γ_{ex}: exciton splitting

ε
$\varepsilon(\mathbf{k})$: energy in $\mathbf{k}$ space

ζ
ζ_B: polarizability of molecule B
$\Delta\zeta$: deviation of molecular polarizability

η
η: primary quantum yield of charge carrier formation
η_{CC}: quantum yield of charge carrier formation
η_{CT}: probability of carrier formation from exciton states

ϑ
ϑ: fraction of free carriers

ϰ
$\varkappa$: dielectric constant

λ
λ: wavelength

μ
μ: mobility
μ_A, μ_B: dipole moments of A & B
μ_d: drift mobility
μ_{eff}: effective mobility
μ_n: electron mobility
μ_p: hole mobility

ν
ν_{ex}: exciton hopping frequency
ν_h: hopping frequency in the activated complex

ξ
ξ: electric field
ξ_{bi}: built-in electric field

ϱ
ϱ_f: free charge density
ϱ_{tot}: total charge density

σ
σ: conductivity
σ_o: experimental conductivity pre-exponential factor

τ
τ: lifetime of the charge carriers
τ_c: time between two collisions (Drude model)
τ_e: electronic polarization time
τ_f: fluorescence lifetime
τ_h: hopping time
τ_p: phosphorescence lifetime
τ_{rel}: vibrational relaxation time
τ_t: time of energy transfer
τ_v: intramolecular vibration relaxation time
τ_{vib}: vibration time
τ_∞: intrinsic exciton lifetime

φ
φ_{CT}: probability that the charge carrier created contributes to the photocurrent
φ_m, φ_n: one-electron wave functions
φ_{pe}: charge-carrier photogeneration efficiency

χ
χ_{pc}: power conversion efficiency

Γ
$\Gamma_{||}$: peak-to-peak ESR linewidth (parallel component)

$\Gamma_\perp$: idem (perpendicular component)
$\Gamma(r)$: potential profile

Θ

$\Theta_a(\Theta)$: angular dependence of the dipole-dipole interaction

Λ

Λ: distribution function

Ψ

Ψ: electronic wave function

Φ

Φ_F: fluorescence quantum yield
Φ_o: incident photon flux
Φ_M: metal work function
Φ_P: phosphorescence quantum yield
Φ_S: semiconductor work function
Φ_{sp}: spectrometer work function
Φ_{surf}: surface-state potential
Φ_T: triplet yield

Abbreviations

DSC: Differential Scanning Calorimetry
ESR: Electron Spin Resonance
IR: Infra-Red
MCD: Magnetic Circular Dichroism
NHE: Normalized Hydrogen Electrode
SCE: Saturated Calomel Electrode

SCLC: Space Charge Limited Current
TSC: Thermally Stimulated Current
UPS: Ultra-Violet Photoelectron Spectroscopy
XPS: X-ray Photoelectron Spectroscopy

I Basic Notions of Solid State Physics

I.1 Dark Conductivity: Generalities

The transport of electrical charges in a material is dependent upon the number of free electrons or holes (charge carriers) and their velocity. The conductance equation is the simplest way to express a relationship between these quantities.

$$\sigma = Ze \cdot n \cdot \mu \tag{I.1}$$

σ: conductivity in $\Omega^{-1} \, cm^{-1}$
Ze: net charge of the carriers
n: concentration of charge carriers
μ: mobility in $cm^2/V \cdot s$

The conductivity represents the charge transported across a unit cross-sectional area per second per unit electric field applied. The conductivity is proportional to the concentration of free charge carriers and to their velocity μ in a unit field. When more than one charge-carrier type is present, the overall conductivity is the sum of the respective contributions as long as there is a negligible interaction between the various charge carriers. In chemical terms, the transport of electrons in an organic material involves two factors: the creation of free charges and their ability to migrate.

$$AAA \xrightarrow{n} A^- AA \xrightarrow{\mu} AA^- A \tag{I.2}$$

The overall conductivity defines the classical frontiers between metals, semi-conductors, and insulators (Fig. I.1). Inorganic compounds span the entire scale. Undoped organic derivatives, on the contrary, form two distinct classes belonging either to the metals or to the insulators. No intrinsic organic semiconductors have ever been described if the charge-transfer-type complexes are excluded. The conductance equation shows that a good conductivity may be obtained when a large quantity of charge carriers is generated. This is the role of the dopants. Charges have to be created from neutral molecules and several chemical pathways are possible. If the free electrons or holes are thermally generated, the semiconductor is defined as "intrinsic":

$$AAA \overset{\Delta}{\rightleftharpoons} A^+ A^- A \rightleftharpoons h^+ \quad or \quad e^- \text{ (free)} \tag{I.3}$$

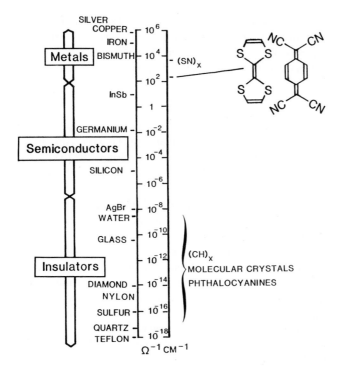

Fig. I.1. Conductivity domain of metals, semiconductors, and isulators. A few inorganic and organic compounds have been reported on the scale

An "extrinsic" semiconductor contains impurities capable of accepting or donating electrons:

$$AAAI_A \rightleftarrows AAA^+I_A^- \rightarrow h^+ \quad \text{(free)} \tag{I.4}$$

$$AAAI_D \rightleftarrows AAA^-I_D^+ \rightarrow e^- \quad \text{(free)} \tag{I.5}$$

The ionization step must be distinguished from the creation of free charge carriers. When the ion pair is formed, a very strong electrostatic interaction is operating and this must be overcome to create free carriers. This effect is particularly important for organic compounds which possess very low dielectric constants.

$$AAA^-A^+AA \rightarrow A^-AAAA^+A \quad \text{separation of charges} \tag{I.6}$$

When the free charges have been generated their mobility will be limited by various mechanisms. Chemical or structural "traps" may capture the free charges.

$$AAA^-I_AA \rightleftarrows AAAI_A^-A \quad \text{chemical trapping} \tag{I.7}$$

Many other trapping processes (scattering events in physical terms) may occur. Mobility is intimately linked to the nature of the molecular and lattice vibrations.

Electronic motion is much faster than nuclear motion; thus, within the limits of the Born-Oppenheimer approximation, electron transfer must occur with no exchange of energy with the external medium during the electronic hop. Consequently, vibrational excitations must take place before isoenergetic electron transfer can occur:

$$AAA^-A \xrightleftharpoons{E_{vib}} AA''A^{-''}A \xrightarrow{\text{transfer}} AAAA^- \tag{I.8}$$

Thus, relaxation of molecular or lattice vibrations may lead to an immobilization of the charges. More subtle is the effect of molecular overlap between contiguous molecules. The rate of transfer is directly correlated to this overlap; the overlap in turn determines the nature of the lattice vibrations which most affect the charge transfer.

To make more quantitative predictions it is necessary to use a model. Synthetic chemists need to be able to relate the density of free charges n and the mobility μ to the molecular structure of the isolated organic molecules and the way they stack in the solid state (lattice parameters and morphology of the solid). Almost unavoidably, a band theory is used to rationalize the properties of semiconductors. More and more elaborated models have been successfully applied to monocrystalline compounds. However, most of the organic semiconductors are in the form of amorphous or polycrystalline films. The band models under these conditions are far more difficult to handle. Rather than considering collective wave functions for the electronic levels of the solid, electronically "localized" energy levels may be used. This duality will be found all through the next chapters. In either case, the properties of the individual molecules can be a good starting point for elucidating the properties of the molecular solid state. It is necessary to explain a few terms and concepts arising from band theory. The next section is devoted to this task. It must be taken more as a glossary of terms than as an attempt to introduce solid state physics.

I.2 Conduction in Metals: Drude and Sommerfeld Models

In 1900, Drude introduced a model for explaining the conductivity of metals [1-3]. Electrons are separated into mobile conduction electrons and fixed valence electrons. The nucleons and the valence electrons form a lattice of net positively charged spheres against which conduction electrons collide (Fig. I.2). The free electrons are considered as a gas. The collision time is assumed to be negligible and no other forces than those acting during the collisions are operating. Two other important approximations are made:

— the electron-electron interactions are neglected (independent electron approximation).

— the electron-cation interactions between two consecutive collisions are neglected (free electron approximation).

Electrons experience collisions with a mean free time τ_c between collisions. If an electric field ξ is applied, the electrons will acquire an average velocity equal to:

$$\mathbf{v} = \frac{e \cdot \tau_c}{m} \xi \tag{I.9}$$

m: mass of free electron

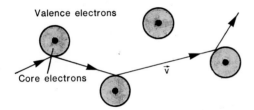

Valence electrons

Core electrons

$\vec{v}$

Fig. I.2. Schematic representation of the Drude model of electronic conduction in metals

The current density $\mathbf{J}$ is given by Ohm's law:

$$\mathbf{J} = -n \cdot e \cdot \mathbf{v} = \sigma \cdot \xi \tag{I.10}$$

$$\sigma = e^2 \frac{n \cdot \tau_c}{m} \tag{I.11}$$

$$\mu = e \frac{\tau_c}{m} \tag{I.12}$$

This last expression relates a macroscopic parameter, the mobility, which can be determined experimentally, to microscopic parameters, τ_c and m, which may be calculated, in principle, from the chemical structure and stacking properties of the material. Typical values for copper at room temperature are $\tau_c = 2.7 \cdot 10^{-14}$ s, $n = 8.47 \cdot 10^{22}$ cm^{-3}, $\sigma = 0.64 \cdot 10^6 \, \Omega^{-1}$ cm^{-1}, $\mu = 40$ cm^2/V $\cdot$ s. In a field of 1 V/cm, this corresponds to a velocity of 40 cm/s, quite small when compared to thermal agitation (10^7 cm/s).

The electronic velocity distribution in the Drude model is taken to follow the classical Maxwell-Boltzmann distribution:

$$g = a \cdot \exp \left(\frac{E_F - E}{kT} \right) \tag{I.13}$$

k: Boltzmann's constant
T: absolute temperature
a: constant

Sommerfeld retained the independent electron approximation and the free electron approximation but changed the form of the distribution function to take account of the Pauli exclusion principle which allows only two electrons in each energy level. He thus used a Fermi-Dirac distribution function of the form:

$$g = \frac{1}{1 + \exp (E - E_F)/kT} \tag{I.14}$$

At temperatures high relative to $E_F(kT \gg E_F)$, the Fermi distribution becomes very close to the Boltzmann distribution (Fig. I.3). In either case E_F is called the Fermi energy and at $T = 0$ K all levels below E_F will be occupied while the others remain empty. At this temperature, E_F is identical to the free energy change of the

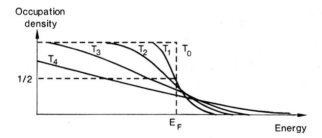

Fig. I.3. The normalized Fermi-Dirac function for increasing temperatures. At high temperatures the Fermi-Dirac function becomes the classical Boltzmann distribution

system per electron or, in other words, its chemical potential. Consequently, it must remain constant across any physical interface whenever two solids are joined together.

The concentration of electrons of energy E, n(E), is obtained by multiplying the distribution function g(E) by the density of the energy states N(E) available:

$$n(E) = g(E) \cdot N(E) \, dE \tag{I.15}$$

I.3 Band Model for Conduction

In the Drude and Sommerfeld theories, the metallic cations, composed of the nucleons and of the valence electrons, play only a minor role, since they have no effect on the motion of the electrons between two collisions.

According to de Broglie's equation, a particle of mass m and of velocity v has associated a wave of wavelength:

$$\lambda = \frac{h}{mv} \tag{I.16}$$

to the allowed wavelengths correspond the allowed energy levels for the particle. The lattice cations impose, however, a periodic potential (Fig. I.4). The electronic wavelength is therefore related to the lattice spacing for a given position of the atoms considered as immobile. In the one electron approximation developed by Bloch, the wave function of a single electron is:

$$\Psi_k = \Phi_k(\mathbf{r}) \exp (i \cdot \mathbf{k} \cdot \mathbf{r}) \tag{I.17}$$

$\mathbf{r}$ is the position vector of the electron, $\mathbf{k}$ the wavenumber vector $|\mathbf{k}| = \dfrac{2\pi}{\lambda}$, and Φ_k a function whose periodicity is the same as the crystal lattice.

Properties of electrons are usually described in the $\mathbf{k}$ space. If $\Phi_k(\mathbf{r})$ is constant throughout the crystal, the free electron approximation is valid and consequently also the Sommerfeld theory. As the values of $\mathbf{k}$ are closely spaced, the corresponding energy function $E(\mathbf{k})$ is almost a continuous function of $\mathbf{k}$. If $\Phi_k(\mathbf{r})$ is not constant,

potential

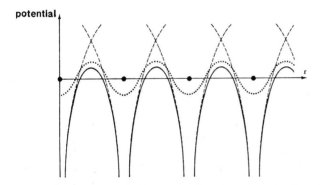

Fig. I.4. Periodic potentials found in a crystalline array of positive ions.
Solid curves: potential along the line of ions
Dotted curve: potential between plane of ions
Dashed curves: potential of single isolated ions. (After ref. 2)

zones of allowed and forbidden $E(\mathbf{k})$ appear in the $\mathbf{k}$ space: a band structure is obtained.

At that point two approaches are possible. In the nearly free approximation the results of the Sommerfeld theory are only modified by considering a weak potential perturbing the system. The tight binding approximation, on the other hand, describes a crystal as a collection of weakly interacting neutral atoms, the overlap of atomic wave functions leading only to corrections by comparison with isolated atoms. Since this approach primarily involves "localized states", the interpretation of the tight binding approximation is more easily transposed into chemical terms. Let's consider the simplest example, a band arising from the interaction of single atomic s-levels (2, 4). The Hund-Mulliken or molecular orbital approach consists in forming molecular orbitals from atomic orbitals. For H_2^+ it follows:

$$\Psi^* = \left(\frac{1}{2 + 2\alpha}\right)^{1/2} (1s_A + 1s_B) \tag{I.18}$$

$$\Psi = \left(\frac{1}{2 - 2\alpha}\right)^{1/2} (1s_A - 1s_B) \tag{I.19}$$

$1s_A$ is the ground-state hydrogen orbital centered on atom A. $1s_B$ is the orbital centered on B, and α is the overlap between the orbitals $1S_A$ and $1S_B$:

$$\alpha = \int 1s_A(1) \, 1s_B(1) \, d\tau \tag{I.20}$$

It may be demonstrated that the corresponding energies are given by:

$$E^* = \frac{\beta + \gamma}{1 + \alpha} \tag{I.21}$$

$$E = \frac{\beta - \gamma}{1 - \alpha} \tag{I.22}$$

with

$$\beta = \int 1s_A(1)\, h_e(1)\, 1s_A(1)\, d\tau \tag{I.23}$$

$$\gamma = \int 1s_A(1)\, h_e(1)\, 1s_B(1)\, d_\tau \tag{I.24}$$

h_e is the effective Hamiltonian of the system, β the atomic or coulombic integral, and γ the resonance integral. The major contribution to β is the energy of the electron on one of the separated atoms. It is a purely electrostatic term by nature. The resonance integral γ represents the energy of interaction between the orbitals $1s_A$ and $1s_B$. It is also called exchange integral. It can be shown that it is inherently negative. When the distance r between the two nucleons tends to infinity, the coulombic integral approaches the isolated atom value and the resonance integral tends to zero (Fig. I.5). When the two atoms come closer, the energy difference $E^* - E$ increases. If the overlap α may be neglected, the splitting between the bonding and the anti-bonding orbital is 2γ, as derived from Eq. I.23 and I.24.

This demonstrates the relationship existing between the orbital splitting and the resonance integral. The same type of result may be shown for an array of 1S single-level atoms (Fig. I.6). It may be demonstrated that for a face-centered cubic lattice of side a_c the energies of the electronic levels are given by (2):

$$E(\mathbf{k}) = E_S - \beta - 12\gamma + \gamma \cdot \mathbf{k}^2 \cdot a_c^2 \tag{I.25}$$

E_S: energy of the atomic s level

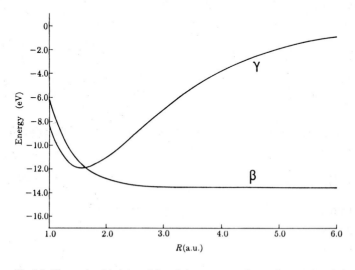

Fig. I.5. The coulombic integral β and the resonance integral γ as a function of the nucleus to nucleus distance r for H_2^+. [Reproduced with permission of (4)]

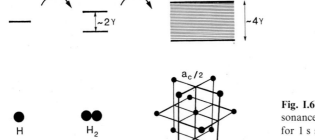

H H$_2$ a$_c$/2

Fig. I.6. Relationship between the resonance integral γ and the bandwidth for 1 s single level systems

A band is formed when **k** spans all the allowed values. The bandwidth, i.e. the spread between the maximum and minimum energies in the band, is thus proportional to the resonance integral γ.

More generally, the bandwith W is related to the integral γ by an expression of the form:

$$W = 2Z_c\gamma$$ (I.26)

Z_c: coordination number of the sites
γ: resonance integral

Mulliken proposed that γ should be roughly proportional to the corresponding overlap integral α.

In the case of H_2^+, the exchange integral γ and the orbital overlap α are equal within 30 % from 1 bohr radius (0.53 Å) to infinity. It is now possible to relate a molecular parameter, the orbital overlap, to a macroscopic electrical property, the mobility. For that, the concept of effective mass of the electron is very useful. When the electrons were considered as a gas, with no interaction with the lattice between two collisions, the energy of an electron was simply given by:

$$E(\mathbf{k}) = \frac{1}{2}mv^2 = \frac{\mathbf{p}^2}{2m}$$ (I.27)

with $\mathbf{p} = m\mathbf{v}$

and

$$E(\mathbf{k}) = \frac{\hbar^2 \cdot \mathbf{k}^2}{2m}$$ (I.28)

These relations are no longer valid when an interaction with a periodic potential is considered. An effective mass m* may be defined such that the movement of an electron in an applied field follows the classical kinetic equations but with an apparent mass m*:

$$m^* = \frac{\hbar^2}{\partial^2 E(\mathbf{k})/\partial \mathbf{k}^2}$$ (I.29)

It is then possible to use the form of the equations as found in the Sommerfeld theory:

$$E(\mathbf{k}) = \frac{\hbar^2 \cdot \mathbf{k}^2}{2m^*} \qquad (I.30)$$

$$\mu = \frac{e \cdot \tau_c}{m^*} \qquad (I.31)$$

From Eq. I.25, I.29, and I.31, it is possible to correlate the mobility of the charge carriers with the bandwidth and the molecular overlap between contiguous constituting molecules of a crystal:

$$m^* = \frac{\hbar^2}{2 \cdot \gamma \cdot a_c^2} \qquad (I.32)$$

$$\mu = \frac{2 \cdot \gamma \cdot a_c^2 \cdot e \cdot \tau_c}{\hbar^2} \qquad (I.33)$$

A macroscopic experimental parameter, the mobility of a charge carrier, is linked to a chemically defined structural feature, the molecular orbital overlap between adjacent molecules. Chemists have always been confronted with the problem of estimating orbital overlaps. An estimate suitable for the problem at hand can be obtained from simple considerations based on ionic or Van der Waals radii. These parameters, which may be determined from X-ray patterns, characterize all atoms, ions, or molecules [5]. At a distance corresponding to the Van der Waals radius, the attractive and repulsive forces counterbalance each other. For non polar molecules, the potential profile may be defined by the empirical Buckingham formula [6]:

$$\varrho(r) = -\frac{a}{r^6} + b \cdot \exp(-c \cdot r) \qquad (I.34)$$

a, b, c: constants
r: center-to-center distance between the non polar molecules.

The first term reflects the attractive Van der Waals interactions between neutral molecules and the exponential term expresses the short-range repulsion between the orbital cores. The molecular orbital overlap is also proportional to $\exp(-r)$ [7]. The Van der Waals term correctly describes the full potential up to the Van der Waals distance. A closer approach implies an orbital overlap and the exponential term becomes progressively predominant. It is easy to estimate roughly the extent of orbital overlap knowing the center-to-center packing distance. A particularly representative example is found by comparing neutral molecular crystals of tetracyanoquinodimethane (TCNQ) and their charge-transfer complexes with tetrathiofulvalene (TTF) [8]. In the first case, molecules are stacked at the Van der Waals distance and the mobility is close to 1 cm²/V · s, a value typical of pure molecular crystals. Indeed, for aromatic systems, at distances down to 3.4 Å, the overlap is weak and the corresponding mobility of the charge carriers is fairly low. The charge transfer complexes of TTF/TCNQ crystallize in segregated stacks of donors and acceptors. There is a partial transfer of electron from TTF to TCNQ, about $0.6\,e^-$ per TTF at room temperature. Previously empty π-antibonding orbitals are therefore populated, while previously fully occupied π-bonding orbitals are partly emptied. This favours π-π overlaps within the stacks. Interplanar distances smaller than the Van der Waals distances by 0.2—0.3 Å are indeed observed [8]. Correspondingly, the experimental mobilities are large, $\mu = 3$ cm²/V · s at room temperature and 300–450 cm²/V · s at 60 K [9].

In some circumstances a Lennard-Jones potential may be used:

$$\varrho(r) = -\frac{a}{r^6} + \frac{b}{r^{12}} \tag{I.35}$$

a, b: constants
the exponential term is replaced by $1/r^{12}$. This provides a certain convenience during numerical calculations on these systems.

For polar molecules the estimates of the size of orbital overlap must be modified by the inclusion of dipolar and quadripolar terms. For a pair of molecules AB, the dipole-dipole interaction energy is given by:

$$E_{d-d} = -\frac{2}{3kT}\frac{\mu_A^2 \cdot \mu_B^2}{r^6} \tag{I.36}$$

μ_A, μ_B: dipole moments of A and B
The dipole-quadrupole interaction energy follows:

$$E_{d-q} = -\frac{1}{kT}\cdot\frac{\mu_A^2 \cdot Q_B}{r^8} \tag{I.37}$$

μ_A: dipole moment of A
Q_B: quadrupole moment of B
A nonpolar molecule B may interact with a permanent dipole moment μ_A via the formation of an induced dipole:

$$E_{d-i} = -\frac{2\cdot\mu_A^2 \cdot \alpha_B}{r^6} \tag{I.38}$$

μ_A: dipole moment of A
α_B: polarizability of molecule B
Such simple considerations of Van der Waals and ionic radii, charge transfer interactions, and dipolar or multipolar interactions will, of course, not solve all problems of the properties of organic semiconductors. But to a satisfactory degree, as long as the underlying physical processes are kept clear, simple molecular models may furnish an useful guideline for designing new molecular semiconductors.

I.4 Limitations to Band Theory

The Drude and Sommerfeld theories are classical models. The conduction electrons are considered as a gas whose particles collide with the cations of the lattice. The trajectory of the electrons follows the classical equations of motion. The collisions are assumed to be the major scattering mechanism, i.e. the principal way of slowing the electrons motion.

In the semiclassical models, the periodic potential arising from the lattice ions is taken into account. The trajectory of the electrons no longer follows the classical equations of motion. The electron-ion collisions can no longer be the major scattering mechanism, since the effect of the lattice is introduced in solving the Schrödinger equation. In the semiclassical approach the following assumptions are made:

— The relaxation time approximation; the scattering mechanism is described by random, uncorrelated electronic collisions.

— The independent electron approximation; electron-electron interactions are ignored.

— The periodic potential in the crystal arises from a fixed immobile array of cations. Molecular and lattice vibrations are not considered.

More elaborate theories must relax these limitations. It is beyond the scope of this book to treat the various physical approaches which have been taken to achieve that goal. Electron-lattice distortion interactions, or electron-electron repulsions are, however, very important in determining many electrical properties of materials. No attempt to quantify these interactions will be made within the band model. It will be seen below that theories involving "localized states" may accomodate these parameters much more easily.

On the other hand, the Heisenberg uncertainty principle sets a limit to the applicability of band models. If ΔW is the effective bandwidth and τ the lifetime of the charge carrier, it follows that:

$$\tau \cdot \Delta W \geq \hbar \tag{I.39}$$

The relaxation time τ_c is limited by the lifetime of the charge carrier τ and thus Equ. I.39 still holds for τ_c. It has been estimated that for mobilities larger than $1 \text{ cm}^2/\text{V} \cdot \text{s}$ the band model is applicable [10, 11]. The mean free path of the free charge carriers are then larger than the lattice constant. The free electrons are not "localized" on one molecule. Their associated wave is spread over several constituting molecules. Molecular crystals or polymers, however, rarely show mobilities reaching $1 \text{ cm}^2/\text{V} \cdot \text{s}$. Some process localizes the charge carriers on a particular site of the lattice and the electrical conduction is more correctly described by a hopping or tunneling mechanism.

I.5 Hopping and Tunneling Mechanisms of Charge Migration

For most organic materials, the following phenomenological equation is found when the temperature dependence of the conductivity is studied:

$$\sigma = \sigma_o \exp\left(-\frac{\Delta E}{2kT}\right) \tag{I.40}$$

ΔE is some "activation energy" of an unspecified nature. In the band model, ΔE is temptingly assigned to be the energy difference between the occupied valence band and the empty conduction band (Fig. I.7). This band gap E_g is the energy necessary to create a pair of free charge carriers e^-/h^+. This assignment is in most cases unjustified and the apparent activation energy represents the sum of several elementary processes. The energy needed to generate charge carriers and the energy necessary to delocalize them are mixed into a single term.

Nuclear motion occurs on a time scale of 10^{-13} s, whereas electron motion occurs in about 10^{-15} s. During the transfer of electrons the nuclei appear to be "frozen" into their position. This furnishes the basis of all the localized-state models. Charges are trapped because their surrounding neighbors are oriented so as to maximize electrostatic interactions. Electrons or holes may also be trapped because of the molecular vibrations within a molecule. There is a competition between the delocalization process and the various relaxation mechanisms: lattice (phonon) or vibrational "scattering" among others. The relaxation times of the various processes indicate whether or not the mechanism considered may be efficient in trapping charge carriers. Since during the electron hopping no energy exchange may take place, a two steps mechanism must be involved:

$$A^- A \rightleftharpoons {}''A^- A'' \overset{k_{ET}}{\rightleftharpoons} {}''A A^{-''} \rightleftharpoons AA^-$$

The species $''A^- A''$ and $''AA^{-''}$ are thermally activated entities within which electron transfer may occur with no energy transfer. The rate-constant electron transfer k_{ET} within this "activated complex" is dependent upon the electronic overlaps of the pair of molecules. In physical terms, this conduction mechanism is described as a phonon-activated hopping mechanism. Electron transfer reactions have been the object of many studies in solution [12, 13], and the results are directly transposable to the solid state. Using the time-dependent perturbation theory it is possible to determine the hopping probability of the electron within the "activated complex" [14]:

$$P_h(t) = \sin^2 \pi \cdot v_h \cdot t \tag{I.41}$$

v_h: hopping frequency

P_h is the probability of the system to undergo the transition in a time t, and v_h is given by:

$$v_h = \frac{2}{h} \frac{|\gamma - \alpha\beta|}{1 - \alpha^2} \tag{I.42}$$

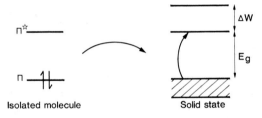

Fig. I.7. Example of formation of valence and conduction bands, VB and CB, respectively, arising from bonding and antibonding π orbitals

If an energy ΔE_{ac} is needed to generate the activated complex, the overall hopping probability becomes [15, 16]:

$$\mathscr{P}_h = v_h \exp\left(-\frac{\Delta E_{ac}}{kT}\right) \tag{I.43}$$

The diffusion constant $\mathscr{D}_h$ (in $cm^2\ s^{-1}$) is simply obtained through:

$$\mathscr{D}_h = \mathscr{P}_h a_c^2 \tag{I.44}$$

a_c: lattice constant

The mobility is then calculated from the Einstein relationship:

$$\mu_h = -\frac{e}{kT}\mathscr{D}_h \tag{I.45}$$

$$\boxed{\mu_h = \frac{e \cdot a_c^2}{kT} \cdot v_h \cdot \exp\left(-\frac{\Delta E_{ac}}{kT}\right)} \tag{I.46}$$

The mobility is temperature dependent. Equ. I.46 is often written as:

$$\mu_h = \mu_{h,o} \cdot \exp\left(-\frac{\Delta E_{ac}}{kT}\right) \tag{I.47}$$

with

$$\mu_{h,o} = \frac{e \cdot a_c^2}{kT} \cdot v_h \tag{I.48}$$

The conductance equation is finally obtained by adding the energy ΔE_{cc} needed to generate independent electrons or holes with no Coulomb interaction between the carrier and its charged counterpart [15]:

$$\sigma = e \cdot n \cdot \mu = e \cdot N \cdot \exp\left(-\frac{\Delta E_{cc}}{2kT}\right) \cdot \mu_{h,o} \cdot \exp\left(-\frac{\Delta E_{ac}}{kT}\right) \tag{I.49}$$

It is noteworthy that ΔE_{cc} may in fact represent a sum of several chemical processes including the ionization step and the charge separation process. Each additional step leads to a splitting of the exponential term into two components.

Typically, the vibrational relaxation times are of the order of 10^{-13} s. Therefore, the hopping frequency v_h will be a rate-determining factor only when it is less than $10^{13}\ s^{-1}$. This situation arises in weakly interacting molecules with small orbital overlap.

Tunneling may also be an efficient mechanism of charge transport. A model based on tunneling was first proposed by Eley [17, 18]. The charge carrier moves from site to site through a tunneling mechanism with a probability of occurence related to the shape, the height and the width of the potential barrier between the sites. The transition probability of tunneling P_{tu} for a triangular potential shape is given by [15]:

$$P_{tu} = a \cdot N_{tu} \cdot \exp\left(-\frac{2}{h} \cdot \sqrt{2m\,(E_{tu} - E)} \cdot l_{tu}\right) \tag{I.50}$$

N_{tu}: number of collisions of an electron with the barrier
l_{tu}: width of the barrier
E_{tu}: height of the potential barrier

P_{tu} is not explicitly temperature dependent. However, the barrier width l_{tu} and the barrier height E_{tu} may depend strongly on the initial energy state of the tunneling particle. A thermally activated tunneling mechanism is thus possible as a two step process: (i) excitation to a vibrationally upper level, (ii) tunneling.

Lattice vibrations may also limit the probability of transfer, since the mean inter-molecular distance, and thus, the resonance integrals, will vary with the vibrational state. Such an effect has recently been invoked to help rationalize the temperature dependence of conductivity in various molecular crystals [19–22].

The use of localized state or band models depends on the degree of molecular overlap. With a lesser extent of overlap a localized model should be appropriate, with greater overlap band theory becomes applicable. In localized state models, each electron hop is independent of the preceding one and there is complete randomization at every hopping. The band theory, on the contrary, involves cooperative motions (coherent transport).

An important general point must now be made. Reality generally provides systems where the conduction may be limited by more trivial factors. Impurities present in the material, or a large extent of disorder, may have devastating effects on conduction. It is already apparent that there are two distinct classes of organic semiconductors. Molecular crystals (naphthalene, anthracene, etc.) are materials of high purity and crystallinity but with poor π-π overlap. Polymers, in particular polyenes, have high covalent overlap between the subunits, but they are poorly characterized. This highlights a classical dilemma facing the designer of molecular semiconductors. Should the efforts be oriented towards the preparation of molecular monocrystals which have intrinsic but obtainable low mobilities or towards polymers with unobtainable good transport properties?

I.6 Charge Carrier Trapping Processes

a Molecular Crystals

Various mechanisms may oppose charge migration. If the residence time of the carrier on a specific site is large enough to allow time for a stabilizing mechanism to occur, the charge will be "localized" and some energy will be required to detrap it. The various "localization mechanisms" are defined by their relaxation times. In the case of anthracene-type crystals, the electronic polarization time (τ_e) is extremely small and electronic clouds always have enough time to relax. It is the predominant interaction between a charge and the surrounding lattice. The interaction energy is of the order of 1.5 to 2.0 eV (Fig. I.8) [6]. Intramolecular vibrations (τ_v) have time constants in the range of 10^{-13}–10^{-14} s and are therefore comparable with typical hopping times (τ_h) of charge carriers in aromatic-type single crystals. Vibrational relaxation may or may not be an effective mechanism of localization depending on the molecular structure of the organic compound and on the temperature. Vibration of the lattice occurs on the 10^{-11} and 10^{-12} s time scale and may also provide a trapping process.

Beside the previous localization mechanisms, which are intrinsic to the material and are unavoidable, charges may also be trapped by chemical impurities or structural disorder. The microscopic mobility μ_o, which depends on the bandwidth and on the orbital overlap, must be replaced by a drift mobility μ_d such that:

$$\mu_d = \Theta \cdot \mu_o \tag{I.51}$$

where Θ is the fraction of free carriers:

$$\Theta = \frac{n}{n + n_t} \tag{I.52}$$

n: density of free carriers
n_t: density of trapped carriers

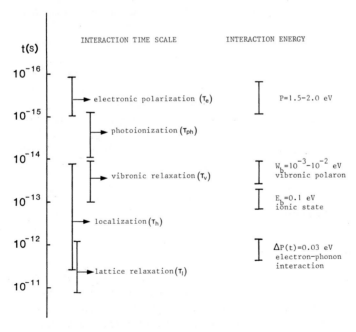

Fig. I.8. Comparison between various types of interaction affecting the charge carriers in an anthracene-type crystal: relaxation time scale and interaction energies. (After ref. 6)

In a band-scheme representation, the presence of traps introduces energy levels inside the band gap. They are generally characterized by their depth E_t, the energy needed to liberate the carrier, their concentration, and their distribution in the band gap. A distinction is made between shallow traps, where thermal detrapping may occur, and deep traps. The depth is defined by the energy difference between the impurity level and the edge of the conduction band for donors and the edge of the valence band for acceptors.

In neutral molecular crystals, it is possible to estimate the effect of local structural disorders. In the zero order approximation, the energy of interaction between a free charge carrier (e^- or h^+) and the surrounding lattice consists mainly of the polarization energy and is given by [6, 23]:

$$E_{pe} = E_{c-i} + E_{i-i} + E_{c-d} + E_{c-q} + E_s + E_{c-iq} + E_w \qquad (I.53)$$

E_{c-i}: charge-induced dipole interaction energy
E_{i-i}: induced dipole-induced dipole interaction energy
E_{c-d}: charge-dipole interaction energy
E_{c-q}: charge-quadrupole interaction energy
E_s: superpolarization effects
E_{c-iq}: charge-induced quadrupole interaction energy
E_w: energy contribution from higher-order multipoles

For typical molecular crystals, anthracene or naphthalene, the charge-induced dipole interaction is the major contribution to the polarization energy. The induced dipole-induced dipole interaction represents not more than 30–40% and the charge-quadrupole interaction approximately 6–7% of the total. The dependence on distance of these various interactions is known and the extent of the perturbation produced by a charge may be estimated. The polarization cloud due to a charge-induced dipole interaction spreads over approximately 12 unit cells, the induced dipole-induced dipole over no more than 2 unit cells [6]. Traps arising from structural defects perturb the electronic levels mainly through differences in the polarization energy. The lattice is expanded or compressed depending on the type of the strutural defects and these local shortenings or lengthenings are reflected in the polarization energy. If the lattice is compressed, the mean distance between the molecules is reduced and ΔE_{pe} will be positive, i.e. the charge is stabilized in the region of the structural defect. This type of distortion leads to a local trapping state. A region of extended lattice, on the other hand, produces an "antitrapping state", since ΔE_{pe} is now destabilizing (ΔE_{pe} negative). The trap depth may be estimated from simple electrostatic considerations [6]. Structural defects, which are not extended over a wide domain in the crystal (site vacancy, point defects), usually only lead to very shallow traps (ΔE_{pe} of the order of 0.01–0.05 eV) and therefore do not interfere drastically in the conduction process. Deep traps (0.5–1 eV) originating from polarization effects may only be formed from an aggregation of dislocations; several tens or even hundreds of molecules must be displaced to give such a large change in the polarization energy. The density of traps and their depth may be determined from thermally stimulated current studies (TSC) or space charge limited current measurements (SCLC). These methods have been applied in particular to the determination of the electrical properties of tetracene. Several forms of tetracene have been studied: single crystal, oriented polycrystalline layers, and amorphous or quasi-amorphous thin films. These latter forms are obtained by vacuum evaporation of tetracene onto a substrate either maintained at room temperature (oriented polycrystalline layers) or more or less cooled (amorphous or quasi-amorphous states) [6, 24] (Tables I.1 and I.2). In the single crystals only deep traps are observed. Their density is very small, of the order of 10^{13}–10^{14} cm^{-3}. The molecular density being approximately $2 \cdot 10^{21}$ cm^{-3}, such a concentration corresponds to 1 trap per 10^7 or 10^8 molecules. For polycrystalline and quasi-amorphous layers, shallow traps appear at concentrations larger than the previous deep traps. Although undetected by SCLC, deep trapping processes are still present in the quasi-amorphous state which is obtained when the substrate is maintained at very low temperature. The state is called amorphous because there is no long range order but only a short range order derived from the crystal structure [51]. The quasi-amorphous form is composed, in fact, of very small microcrystallites. There is a close relationship between the substrate temperature and the density of traps (Table I.2). A rise in substrate temperature decreases N_t and thus the structural disorder.

Significantly, there is not much change in electrical properties of molecular crystals in going from the polycrystalline to the amorphous state. This is certainly not the case for most inorganic materials. These compounds form crystalline states where atoms are covalently bound to each other. This type of bonding requires a rigorous geometry and the effect of disorder is to introduce "broken bonds". These defects act as deep traps. But the packing in molecular crystals is ensured by Van der Waals-type forces, so a change in geometry does not dramatically influence the local electronic levels and does not generate a deep trap. This advantage is gained at the price of weaker interactions between the subunits, narrower energy bands, and lower intrinsic mobilities.

The effect of chemical impurities on the electrical properties of molecular crystals may also be estimated [6, 23]. The difference in electron affinity or ionization potential

Table I.1   Typical values of trap depth E_t and concentration of traps N_t for tetracene under several crystalline states. Values have been determined from thermally stimulated current (TSC) or space-charge-limited currents (SCLC). (From Ref. [6]).

	from SCLC		from TSC	
Crystalline state	E_t^c (eV)	N_t (cm^{-3})	E_t^c (eV)	N_t (cm^{-3})
Single crystals	0.6	10^{13}–10^{14}		
Oriented	0.1	$4 \cdot 10^{14}$–$6 \cdot 10^{15}$		
polycrystalline[a]	0.3–0.4	$6 \cdot 10^{13}$–$2 \cdot 10^{14}$		
Quasi-amorphous[b]	0.1	$4 \cdot 10^{15}$	0.07	$5 \cdot 10^{16}$
	0.26	$2 \cdot 10^{14}$	0.4	$\sim 10^{15}$
	not detected		0.7	$\sim 10^{15}$

[a] evaporation in vacuo on substrate at T = 300 K; [b] substrate at T = 200 K (very small microcrystallites); [c] center of the trap distribution.

Table I.2 Trapping parameters determined by TSC technique for anthracene layers obtained in vacuo for various temperatures of the substrate (from Ref. [24]).

Substrate temperature (K)	E_t^a (eV)	N_t (cm^{-3})	Crystalline state
80	0	—	
130	0	$2 \cdot 10^{21}$	
140	0	$1 \cdot 10^{20}$	amorphous
160	0	$5 \cdot 10^{18}$	
180	0.07	$4 \cdot 10^{17}$	
200	0.07	$5 \cdot 10^{16}$	quasi-amorphous

[a] center of the trap distribution.

between the host molecular-crystal molecules and the guest impurities gives the depth of the trapping level introduced by the impurity [23, 25]:

$$E_t^e = (A_c)_{guest} - (A_c)_{host} \tag{I.54}$$

E_t^e: trap depth for electrons
$(A_c)_{guest}$: crystal electron affinity of the molecular crystal
$(A_c)_{host}$: crystal electron affinity of the impurity

The electron affinity in the crystalline state A_c is related to the gas-phase electron affinity through:

$$A_c = A_g + E_{pe}^e \tag{I.55}$$

A_g: gas-phase electron affinity
E_{pe}^e: electronic polarization energy

The difference in polarization energy between the guest and the host, ΔE_{pe}^e, is presumably caused by distortion of the lattice in the vicinity of the guest molecule. ΔE_{ep}^e is, in any case, expected to be small and to a first approximation:

$$E_t^e \sim (A_g)_{guest} - (A_g)_{host} \tag{I.56}$$

The same type of expression is found for the hole trapping process. In that case the trap depth for a hole is given by:

$$E_t^h \sim (I_g)_{guest} - (I_g)_{host} \tag{I.57}$$

I_g: ionization potential in the gas phase

Depending on the relative values of electron affinities of the guest or of the host, electron traps, $(A_g)_{guest} > (A_g)_{host}$, or antitraps, $(A_g)_{guest} < (A_g)_{host}$, are created in the host lattice.

It is possible to make further approximations to relate the electron affinity and ionization potentials in the solid state to the corresponding redox potentials in solution. One approximation that has been proposed [26] is:

$$A_c \sim E_{1/2}^{red} + 4.3 \tag{I.58}$$

$$I_c \sim E_{1/2}^{ox} + 4.3 \tag{I.59}$$

where the reduction potential $E_{1/2}^{red}$ and the oxidation potential $E_{1/2}^{ox}$ are expressed in volts versus a Saturated Calomel Electrode (SCE) taken as reference. The constant 4.3 includes the free energy change between an electron in the reference electrode and the same electron in vacuum at infinity. These expressions, while obtained at the price of drastic approximations, provide a very useful rough estimate of the solid-state properties of organic molecules.

The correlation between experimental trapping depths and theoretical values has been studied in the case of anthracene containing a known amount of impurities: tetracene, acridine, phenazine, anthraquinone, and phenothiazine [6, 27] (Fig. I.9). As a rule, higher homologues of anthracene form traps for both electrons and holes. Other impurities act only as electron traps (phenazine) or hole traps (phenothiazine). The agreement between the experimental trap depths and the values obtained from Equ. I.56 and I.57 is quite satisfactory.

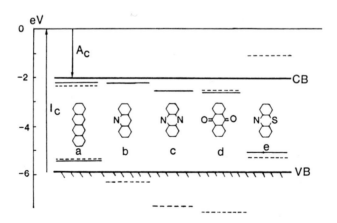

Fig. I.9. Trapping levels formed in anthracene by the guest impurities: **a** tetracene, **b** acridine, **c** phen-azine, **d** anthraquinone, **e** phenothiazine
Solid levels correspond to measured values and dashed levels are obtained theoretically from the electron affinity or the ionization potentials in the gas phase. (Eqs. I.56 and I.57) (After Ref. 6 and 27)

b Polymers

The behavior of polymeric semiconductors is far more difficult to rationalize. Many reviews have been devoted to the studies of polymeric organic conductors [28–35]. In most cases, no obvious relationship between the molecular structure of the polymer and its electrical properties has been found. Polymeric materials are insoluble, non volatile, and cannot be recrystallized nor sublimed. They contain, in consequence, a large concentration of chemical impurities which play a dominant role in the charge carrier transport processes. Even in a homogeneous series of polymers, results are very rarely reproducible, the mobility and the density of charge carriers may be found to vary over several orders of magnitude. Graphite, polydiacetylene, and polyacetylene are the only polymeric systems whose conducting and semiconducting properties have been studied in some detail.

The characteristic graphite structure is made of 2-dimensional layers of conjugated double bonds, each layer being separated by 3.35 Å (Fig. I.10). The carbon-to-carbon distance in a layer is 1.41 Å, a value intermediate between a single (1.54 Å) and a double bond (1.33 Å). A strong overlap is therefore operating and the delocalization of the electrons is complete. Graphite shows very large mobilities of the charge carriers. At room temperature the in-plane mobility is 13,000 cm^2/V · s, as compared to 35 cm^2/V · s for copper and 1600 cm^2/V · s for silicon [36]. Usual organic polymers have, on the other hand, mobilities of the order of 10^{-3}–10^{-5} cm^2/V · s. The difference in the amount of impurities and defects probably explains this difference of 10 orders of magnitude. Graphite has, however, a rather modest conductivity because of the very low carrier concentration of $2 \cdot 10^{-4}$ carriers per atom ($2 \cdot 10^{19}$ cm^{-3}). Doping is therefore a necessary process.

Polydiacetylene is obtained by irradiating (UV light or γ-rays) single crystals of the diacetylene monomer (see for example [39–41]) (Fig. I.11). Remarkably, a pre-organization of the reactants is necessary for the system to react. Polymerization does not occur when the diacetylene derivatives are dissolved in a solvent, but reaction does take place in amphiphilic multilayers [42]. Polymerization gives a polyenic compound of exceptional purity. The length of the conjugated chains reaches macroscopic dimensions. The average chain length has been determined by scanning electron microscopy to be about 5 μm [43]. Although early experiments severely underestimated the corresponding mobilities of polydiacetylenes [44, 45], recent measurements indicated a value of 6000 cm^2/V · s for the electron mobility in the chain direction [46]. On the other hand, the concentration of the free charge carriers is even lower than for graphite.

Polyene systems therefore furnish a very good way of obtaining good quality semiconductors. Covalent π-π overlaps are efficient in delocalizing charge carriers

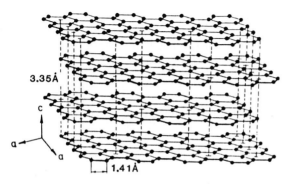

3.35Å

c

a

a

1.41Å

Fig. I.10. The hexagonal graphite lattice. (After Ref. 37 and 38)

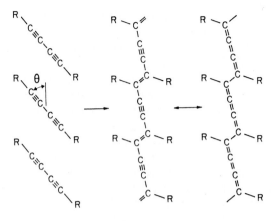

Fig. I.11. Solid-state polymerization of an array of diacetylene derivatives. [Reproduced with permission of (39)]

(see Sect. IV). However, in most cases, their intrinsic properties are masked by a significant concentration of chemical impurities and structural defects, which limit the mobility by trapping the free charge carriers.

Pendant group polymers have also been considered as a way of obtaining molecular semiconductors. Poly(2-vinyl pyridine) has been the most thoroughly studied from a theoretical point of view [47–49] (Fig. I.12). The electronic states of each site of the polymeric chain undergo fluctuations, because the local environment of the various molecular-pendant-group moieties is different. Dynamic fluctuations may also occur, and are associated with intramolecular polarization changes of the polymer matrix. In such a disordered system, the hopping integrals between the pendant groups γ_{ij} are not constant, and the extent of modulation may be expressed as:

$$(\Delta\gamma)^2 = \langle(\gamma_{ij} - \bar{\gamma})^2\rangle_{\text{average}} \tag{I.60}$$

with

$$\bar{\gamma} = \langle\gamma_{ij}\rangle_{\text{average}}$$

For the same reasons, the various electronic energies E_i differ from site to site:

$$\Delta^2 = \langle(E_i - \bar{E})^2\rangle_{\text{average}} \tag{I.61}$$

Charge carriers are trapped if the spread in energy due to the state of the polymeric material is larger than the corresponding hopping integrals [47–49]:

$$\Delta > c \cdot Z_c \cdot \gamma \tag{I.62}$$

c: dimensionless number depending on the dimensionality of system
Z_c: coordination number of a given site

A variety of molecular solids, including molecular crystals and pendant-group polymers, have been studied by photoemission spectroscopy [47–49]. From these

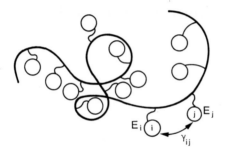

Fig. I.12. Schematic representation of an organic material of the pendant group polymer type:
γ_{ij}: hopping integral
E_i: electronic site energy

experiments, it has been possible to evaluate Δ, the spread in energy between the various sites. Both polymeric materials and glassy, condensed systems show the same spread in energy (0.5–1 eV). For pendant-group polymers, this undoubtedly leads to a trapping of the charge carriers, since the intersite overlap integral is approximately 10^{-2}–10^{-3} eV [47]. For molecular crystals, the hopping integral is of the order of 0.1 eV and then becomes comparable with Δ. The band model may thus be applicable for molecular crystals. This is never the case for pendant-group polymers. Localization of charges by a random potential is a quite general phenomenon and other disordered systems as glasses or liquids are subject to this effect [50].

In conclusion, it is no doubt possible to obtain a real molecular semiconductor. However, the usual definition related only to the overall conductivity is meaningless. An intrinsic organic semiconductor may be defined by the following conditions:
— The energy necessary to create free charge carriers must be less than 2 eV.
— The mobility must be limited by the intrinsic properties of the organic material and not by uncontrolled concentrations of chemical impurities or defects.
— Generation of free charge carriers must not involve impurity states.
— Overall conductivity must be in the range of 10^{-7}–10^2 Ω^{-1} cm^{-1}.

Molecular semiconductors having such properties altogether have been rarely or never described. Doped materials should also be considered. However, the obtention of the intrinsic properties prior to doping seems essential to really master the electrical properties of molecular materials in the semiconducting domain.

II Photoelectric Phenomena in Molecular Semiconductors

When light is shone on a semiconductor, a photoelectric effect may be observed. Photon energy is converted into electricity flowing in an external circuit. This phenomenon depends on the subtle interplay of a number of complex mechanisms. The aim of this chapter is to identify the individual process involved and to show how they can combine to produce the variety of photoelectric properties exhibited by molecular semiconductors.

The whole process begins with the production of an excited state. Excited states may be populated not only by light absorption, but also through a large number of ways: X-ray or γ-ray irradiation, electron bombardement, thermal or chemical excitations (Fig. II.1).

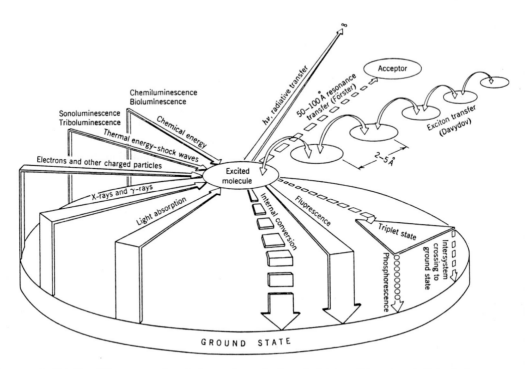

Fig. II.1. The different ways of producing an electronically excited state and its various destinies during its lifetime. [Reproduced with permission of (52)]

The next step is the production of charge carriers from the excited states. The excited states may collapse to a ground state by a number of unproductive pathways. The light energy may be degraded into heat by internal conversion, re-emitted from the singlet states by fluorescence or from the triplet levels by phosphorescence. Beside these processes, the excitation energy may migrate by various mechanisms to meet other fates. One of its destinies may be the production of free charge carriers. The basic properties of carriers have already been discussed, with the tacit assumption that they had been generated thermally. Extra considerations relevant to photo-generated charge carriers will be discussed.

In the photovoltaic effect, contacts of the molecular semiconductor with metals or other semiconductors are always used. A good understanding of the mechanisms involved in the junction formation is therefore necessary.

Finally all these considerations will be brought together with a discussion of how an organic solar cell might work.

II.1 Light Absorption

There are two ways of describing light absorption by a semiconductor. The first, proposed by Mott and Wannier, is a promotion of an electron from the valence band to the conduction band. An electron-hole pair is formed with a large e^-/h^+ distance and a correspondingly small coulombic interaction. The second, proposed by Frenkel, is an excitation of a molecule to a neutral, mobile localized excited state called an "exciton". The neutral state can be considered as a tight e^-/h^+ pair held together by a strong coulombic interaction. The Mott and Wannier model can be applied whenever the dielectric constant of the material is sufficiently large. In such a material the interaction between an electron in the conduction band and a hole in the valence band is given by the classical equation:

$$E_{ip} = -\frac{q \cdot q'}{\varkappa \cdot r}$$ (II.1)

E_{ip}: interaction energy of the ion pair
$\varkappa$:   dielectric constant

If the coulombic attraction is neglected, the absorption coefficient α_a of the material is related to the energy of the incident photon E_{hv} and to the band gap of the semiconductor through:

$$\alpha_a \sim (E_{hv} - E_g)^{1/2}$$ (II.2)

for a symmetrically allowed transition, and through:

$$\alpha_a \sim (E_{hv} - E_g)^{3/2}$$ (II.3)

for a symmetry forbidden transition [53, 54]. These relations are applicable only for "direct-gap" semiconductors in which the lattice vibrations (phonons) are not involved during the transition.

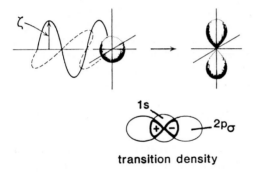

transition density

Fig. II.2. Orbital representation of a 1s hydrogen atom absorbing radiation causing 1s → 2p transition. The corresponding transition density is figured, this density is largely dipolar in character and interacts with light as a dipole. (After Ref. 55)

Organic materials usually have low dielectric constants; therefore, coulombic interactions are correspondingly large and the Mott-Wannier model is not generally applicable. However, heavily doped materials may show an increased dielectric constant; this model then may become relevant. In the general case, the exciton formalism is much more applicable to organic semiconductors. The classical laws used to rationalize the absorption properties of isolated molecules remain, on the whole, valid in the solid state. Excitation to electronically excited states occurs by interaction of light with molecular transition dipoles. An illustration of transition densities for an excitation from the 1s to $2p_\sigma$ orbitals is given in Fig. II.2. Transition densities may be visualized as the amount of overlap between the orbital of the ground state and the orbital of the excited state; their interaction with the electric component of the incident light leads to the formation of the excited state. The probability of light absorption by molecules is governed by the magnitude of the dipole strength $\mathscr{S}_d$ [55]:

$$\mathscr{S}_d = \mathbf{m}_t^2 = (\int \Psi_m \mathbf{M} \Psi_n \, d\tau)^2 \tag{II.4}$$

where $\mathbf{m}_t$ is the transition moment integral and Ψ_m and Ψ_n are the total electronic wave functions of the initial and final states. It is assumed in most cases that the wave functions may be factored into single electron functions and that only one of the electrons is excited. Under these conditions:

$$\mathbf{m}_t = \int \varphi_m^* \mathbf{M} \varphi_n \, d\tau \tag{II.5}$$

where φ_m and φ_n refer to the wave functions of the ground-state and excited-state orbitals of the single electron undergoing change during the excitation. The dipole moment operator $\mathbf{M}$ is defined by:

$$\mathbf{M} = \sum_i e \cdot \mathbf{r}_i \tag{II.6}$$

where r_i is the distance of the i-th electron from the center of the positive charge of the molecule. The transition moment m_t is therefore a measure of the charge displacement occurring during the passage from the ground state to the electronically excited state.

The probability for light absorption is given by [55]:

$$P_{mn} = \left(\frac{8\pi^3 \cdot e^2}{3h^2 \cdot c} \right) \cdot g_n \cdot \mathscr{S}_d \tag{II.7}$$

g_n: degeneracy of the n-th state, i.e. the number of suitable orbitals in the excited state with which the ground-state orbital may combine.

c: velocity of light

The oscillator strength $\hat{f}_{mn}$ is a parameter frequently employed in photochemistry and is defined by:

$$\hat{f}_{mn} = \left(\frac{8\pi^2 \cdot m_e \cdot c \cdot g_n}{3 \cdot h \cdot e^2} \right) \cdot \omega_{mn} \cdot \mathscr{S}_d \tag{II.8}$$

ω_{mn}: frequency in cm^{-1}

m_e: mass of the electron

$\hat{f}_{mn}$ is related to the experimentally determined molar extinction coefficient through:

$$\hat{f}_{mn} = \left[\frac{2302 \cdot m_e \cdot c^2}{\pi \cdot \mathscr{N} \cdot e^2} \right] F \int_{\omega_1}^{\omega_2} \alpha_a \cdot d\omega = 4.32 \cdot 10^{-9} F \int_{\omega_1}^{\omega_2} \alpha_a \cdot d\omega \tag{II.9}$$

$\mathscr{N}$: Avogadro's number

α_a: extinction coefficient

F: correction factor related to the refractive index of the medium [55]

ω_1 and ω_2 are the frequencies delimitating the band corresponding to the transition from states m to n.

Estimates of the intensity of the transitions must also take account of the various selection rules [56]. Radiative transitions involving a change of spin are strongly forbidden. The forbiddenness may also arise from symmetry or momentum changes.

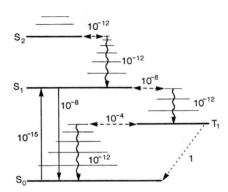

Fig. II.3. Schematic representation of the main energetic levels present in organic molecules. The typical time scales of the various transitions are indicated.

Arrowed lines: internal conversion (nonradiative transitions)

Consequently, the rate constants for electronic transitions span a 10-orders-of-magnitude time scale (Fig. II.3). The radiative transition from the excited triplet state to the ground state (phosphorescence) is spin forbidden and the rate constant is correspondingly very low. The transition probability may be increased by spin-orbit coupling if a heavy atom is present in the organic molecule. Nonradiative transitions occur on a very small time scale (picosecond range) for the vibrational de-excitations of the molecules in a given electronic state. Intersystem crossing between singlet excited states and triplet excited states may occur if a mixing of states of similar energies in both levels is possible. The kind of chemical reaction the excited molecule undergoes, depends on the characteristic times of the various processes involved, e.g. migration of energy, photoreaction with impurities, photodissociation into free carriers, internal conversion, intersystem crossing.

II.2 Energy Migration in Molecular Materials

a Mechanisms of Energy Migration

Transport of energy without transport of mass or charge mainly occurs in organic materials which possess "localized" energy states (reviews: [57–64]). There are two main types of energy migration in these materials. The first is the relatively trivial case of radiative transfer by emission-reabsorption processes. An excited molecule decays to its ground state by emitting a photon which is subsequently absorbed by another molecule. Such a process is governed by the fluorescence and absorption properties of the isolated molecules and needs to be discussed no further. The second type of energy transfer occurs with no intermediate radiative process. A number of different mechanisms operate for these radiationless transfers and will now be discussed in some detail. They all have in common the fact that, as for electron transfer, no exchange of energy with the external medium may occur during the energy transfer. Thus there must be resonance between the initial and final states. If a pair D, A of molecules is considered, the transition corresponding to the de-excitation of D* must have the same energy as the excitation of A. The reaction:

$$D^*, A \rightleftharpoons D, A^* \tag{II.10}$$

must be isoenergetic. Another way of stating this condition is to say that the probability of energy migration is related to the amount of spectral overlap, J_{ov}, between donor and acceptor (Fig. II.4). This is given by:

$$J_{ov} = \int_0^\infty F_D(v) \cdot G_A(v) \cdot dv \tag{II.11}$$

J_{ov}: spectral overlap
$F_D(v)$ and $G_A(v)$: spectral distributions of the donor emission and the acceptor absorption, respectively.

In most cases, *the approximation is made that only two electrons are concerned by the energy exchange.* The total energy of interaction between the initial state (D^*, A)

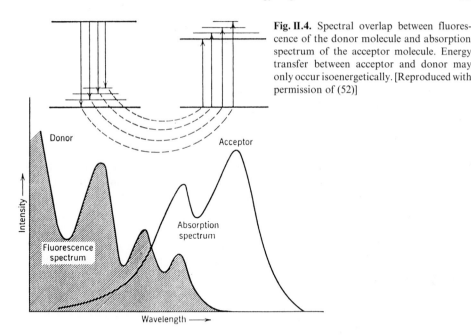

Fig. II.4. Spectral overlap between fluorescence of the donor molecule and absorption spectrum of the acceptor molecule. Energy transfer between acceptor and donor may only occur isoenergetically. [Reproduced with permission of (52)]

and the final state (D, A^*) is the sum of a coulombic term E_{coul} and an exchange term E_{exc}:

$$E_{tot} = E_{coul} - E_{exc} \qquad (II.12)$$

The coulombic term represents the interaction energy between the transition charge densities of the individual molecules and is given by:

$$\beta = \int \varphi_{D^*}(1) \cdot \varphi_A(2) \cdot h_e \cdot \varphi_D(1) \cdot \varphi_{A^*}(2) \, d\tau \qquad (II.13)$$

This purely electrostatic term can be expressed as the sum of interaction energies of multipoles of increasing orders:

$$E_{coul} = E_{d-d} + E_{d-o} + E_{o-o} + \dots \qquad (II.14)$$

E_{d-d}, E_{d-o}, and E_{o-o} are the energies of interaction dipole-dipole, dipole-octopole, octopole-octopole. The first term predominates and, in a first approximation, the others may be neglected, except in cases where the distance between interacting molecules is small. E_{d-d} is an interaction between transition dipole moments $\mathbf{m}_t^D$ and $\mathbf{m}_t^A$ and does not involve permanent dipoles:

$$E_{d-d} \sim a \, \frac{\mathbf{m}_t^A \cdot \mathbf{m}_t^D}{r^3} \qquad (II.15)$$

a: constant

Interactions between higher-order multipoles are given by:

$$E_{d-o} \sim b \frac{(m_t^A \cdot m_t^D)^{1/2}}{r^5} \tag{II.16}$$

$$E_{o-o} \sim c \frac{f(m_t^A, m_t^D)}{r^7}$$

b, c: constants

The exchange interaction E_{exc} arises from the indistinguishability of the electrons and from the symmetry properties of the electronic wave function on interchange of space and spin coordinates, it is expressed by:

$$\gamma = \int \varphi_{D*}(1) \cdot \varphi_A(2) \cdot h_e \cdot \varphi_D(2) \cdot \varphi_{A*}(1) \, d\tau \tag{II.17}$$

Such an interaction necessarily requires orbital overlap between molecules, with an exponential dependence usually assumed:

$$E_{exc} \sim \exp(-d \cdot r) \tag{II.18}$$

d: constant

The energy of interaction within the pair D*, A must still be related to the probability of energy transfer from D* to A. This is not possible in the general case since various relaxation mechanisms, including vibrational processes, are involved in the energy transfer. In view of their long relaxation times, lattice vibrations are of little importance for energy migration despite their important role in electron transfer processes. Intramolecular vibrations and vibrational relaxations, on the other hand, still remain as important relaxation mechanisms. Vibrations occur on a 10^{-13} s time scale and it sorresponds to atomic vibrations either in a vibrationally and electronically excited state or in the vibrationally relaxed state. Vibrational relaxation (10^{-12} s) corresponds to the rate at which a vibrationally excited state goes down to the lower vibrational level. Depending on the relative magnitude of the vibrational relaxation rate, the vibration rate and the energy transfer rate three cases are possible (Fig. II.5):

— Strong interaction case. The rate of energy transfer is larger than the vibration rate and the vibrational relaxation rate ($\tau_t \ll \tau_{vib} \ll \tau_{rel}$). The energy is transferred before relaxation to the vibrational ground state. The excited state experiences no intramolecular vibration.

— Weak interaction case ($\tau_{vib} \ll \tau_t \ll \tau_{rel}$). The rate of energy transfer is fast compared to vibrational relaxation, but molecular vibrations may take place before energy transfer occurs.

— Very weak interaction case ($\tau_{vib} \ll \tau_{rel} \ll \tau_t$). Full thermal equilibrium is attained before energy transfer.

α) Strong interaction case.

The term "exciton" is often reserved for the strong interaction case to describe the energy transport properties of an array of organic molecules [65–67]. In this case it

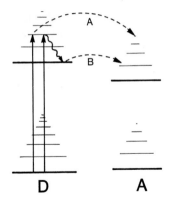

Fig. II.5. Energy transfer between D* and A. Three cases are possible depending on the relative magnitude of the vibrational relaxation rate, the vibration rate, and the energy transfer rate.
— Strong interaction case: path A, No vibration in D* occurs before energy transfer takes place.
— Weak interaction case: path A. Vibration occurs in D*.
— Very weak interaction case: path B. Both vibrational thermalization and vibration have time to take place in D*

is not possible to define a donor or acceptor molecule, nor can any simple transfer be considered since no component of an array can be excited on its own. For the couple D,A, the system is described by two stationary states:

$$\Psi_1 = c\varphi_{D*}\varphi_A + c'\varphi_D\varphi_{A*}$$

$$\Psi_2 = c\varphi_{D*}\varphi_A - c\varphi_D\varphi_{A*}$$

(II.19)

For the case of identical "donor" and "acceptor", the difference in energy between the two levels $2E_{el}$, called the *exciton splitting*, may be calculated at the condition that the total interaction energy may be decomposed into a pure electronic factor and vibrational factors:

$$(\beta - \gamma)_{tot} = (\beta - \gamma)_{el} \times \text{vibrational factors}$$

and

$$(\beta - \gamma)_{el} = E_{el}$$

(II.20)

The excitation energy is spread over the two molecules of the dimer case or over as many molecules as a crystalline array may contain. The excited-state geometry is therefore nearly identical to that of the ground state. The absorption spectrum is consequently very narrow since only the $0 - 0$ transition has a non-zero Franck-Condon factor [57]. If the excitation rests on molecule D at some initial time, the probability that the excitation is on A at some later time t is, from time dependent perturbation theory, given by:

$$P_{ex}(t) = \sin^2 \pi \cdot \nu_{ex} \cdot t$$

(II.21)

This expression is analogous to the equation found for electronic transfer (see Eq. I.41): v_{ex} may be related in the same way to the coulombic and exchange interaction energies between the initial and the final states. However, during the energy transfer both ground and excited states are involved. The reciprocal of the oscillation period between the states D^*,A and D,A^*, $T_{D^* \to A}$, may be thought of as a "transfer rate" although, as mentioned above, it is not possible to define an "acceptor" and a "donor" since both components are simultaneously excited.

$$n_{D^* \to A} = \frac{1}{T_{D^* \to A}} = \frac{4 \cdot E_{el}}{h} \tag{II.22}$$

β) Weak interaction case.

In the weak interaction case, the vibrations of the excited molecule must be taken into account. The probability of transition from the excited state in the vibrational level v' to the ground state in the vibrational state v'' is determined by the magnitude of the overlap integral:

$$\int \varphi_{v'} \varphi_{v''} = Q_{v'v''} \tag{II.23}$$

The transfer rate is then given by:

$$n_{D^* \to A} = \frac{4E_{el}}{h} \sum Q^A \cdot Q^D \tag{II.24}$$

where the sum is taken over all possible final resonant vibronic states and the appropriate distribution of initial states. In contrast to the strong interaction case, no profound changes in the absorption spectra are expected, and only a slight shift of the maximum absorption or a small splitting of certain vibronic bands (Davydov splitting) will occur.

γ) Very weak interaction case.

In the very weak interaction case, energy transfer takes place from a fully relaxed excited state. The interaction energy is smaller than the vibronic bandwidth ΔE_{rel}. The transfer rate is expressed as:

$$n_{D^* \to A} = \frac{2\pi \cdot E_{el}^2}{\hbar \cdot \Delta E_{rel}} \sum Q^{A^2} \cdot Q^{D^2} \tag{II.25}$$

where the sum is over all possible final resonant states and a Boltzmann distribution of initial states. In the very weak interaction case, simple relationships between the rate of energy transfer and the parameters of the isolated organic molecules may be found.

If the molecules are relatively far apart, E_{el} is dominated by Coulomb interactions as discussed by Förster [68, 69]. If the molecules are close, then exchange interaction predominates as discussed by Dexter [60, 70].

The Förster mechanism is effective up to 50–100 Å between the energy donor and the acceptor. The interaction between D* and A is generally approximated as a dipole-dipole term. The rate constant for energy transfer is related to the transition moments of the isolated molecules and to the spectral overlap between the emission and absorption spectra:

$$k_{D^* \to A}(\text{dipole–dipole}) = J_{ov} \frac{2\pi}{\hbar} c \frac{\mathscr{S}_d^D \cdot \mathscr{S}_d^A}{r^6 \cdot \bar{v}^2} \theta_a^2(\theta)$$ (II.26)

c: constant
$\bar{v}$:   average wavenumber for the transitions
J_{ov}:   spectral overlap
$\mathscr{S}_d^D, \mathscr{S}_d^A$: oscillator strengths for the donor and acceptor transitions.

$\theta_a(\theta)$ is the angular dependence of the dipole-dipole interaction. The rate of energy transfer varies with the distance between the molecules as r^{-6}. The selection rules are similar to those corresponding to the excitation of a single molecule. Spin multiplicity must be conserved during the transfer, the spin allowed transitions are thus:

$$D^* \text{ (singlet)} + A \text{ (singlet)} \to D \text{ (singlet)} + A^* \text{ (singlet)}$$ (II.27)

$$D^* \text{ (singlet)} + A \text{ (triplet)} \to D \text{ (singlet)} + A^* \text{ (triplet)}$$ (II.28)

The Dexter mechanism relies on exchange interactions which necessarily require orbital overlap. Thus it operates only at very short distances (5–10 Å) with typical exponential distance dependence:

$$k_{D^* \to A}(\text{exchange}) = J_{ov} \frac{2\pi}{\hbar} \cdot K_e \cdot \exp\left(-\frac{2r}{\mathscr{L}}\right)$$ (II.29)

K_e: quantity with the dimension of an energy [57]
$\mathscr{L}$: effective average orbital radius for the initial and final electronic states.
The previous selection rules are relaxed and conservation of spin within a molecule is no longer required. The weaker constraint of conservation of overall multiplicity in the D*, A and D, A* couples now applies. Thus, triplet energy transfers are allowed:

$$D^* \text{ (triplet)} + A \text{ (singlet)} \to D \text{ (singlet)} + A^* \text{ (triplet)}$$ (II.30)

b Effect of Traps

Chemical impurities or structural disorders may induce a barrier to the energy migration step just as they did in the transport of charge carriers. The mechanisms involved are, however, very different, since both the ground and the excited states participate in the energy transfer. Moreover, for energy transfer the diffusing particle is neutral. The scattering mechanisms are therefore not the same as for when a charge moves through the lattice. The influence of disorder may be estimated, for molecular crystals, by considering the effect of a locally compressed or extended lattice on the relative distances between the constituent molecules [6, 71, 72]. The trapping of electrons or holes was due to an increase or a decrease of the electronic polarization energy ΔE_{pe} in the vicinity of the defect. The mechanisms of exciton trapping are quite different. The local deformation of the lattice causes the energy levels of the excited and ground states to be slightly different from those in the bulk (site shift term). If energy levels are stabilized, the exciton is trapped at the defect. The excited molecule interacts with the surroundings through Van der Waals interactions (if no permanent dipoles are present); the distance dependence of such interactions is thus of the form:

$$E \sim \frac{a}{r^6} \qquad\qquad (II.31)$$

a: constant

The site shift term shows the same distance dependence. Local disorders in crystals also have an influence on the probability of energy transfer from molecule to molecule (resonance term). This factor is dependent upon the mechanism of the energy transfer: Dexter- or Förster-type processes. Structural defects may be experimentally detected by fluorescence spectroscopy studies [6]. Even after careful purifications, molecular crystals, such as anthracene, show residual long-wave bands in their luminescence spectra. These bands are strongly dependent upon the crystal growing conditions; they are also altered by a mechanical deformation or a thermal treatment of the crystal. These bands arise from the luminescence properties of excitons trapped in the vicinity of defect levels. It is thus possible to determine the depth of the exciton trap and, by varying the external conditions on the crystal, their nature. In the case of anthracene, thorough studies have been carried out (Table II.1). Local deformations of the crystal only lead to shallow traps for triplet exciton (of the order of 10–60 meV). The presence of deep traps is associated with chemical impurities or the formation of excimers. It is interesting to compare the trap depths for excitons or charge carriers induced by the same structural defect (Table II.2). In all cases, the same defect forms traps three to five times deeper for charge carriers than for singlet excitons. *Energy transfer inside an organic material is far less sensitive than charge transport to structural*

Table II.1 Mean depth of triplet exciton traps in an anthracene single crystal as determined from long-wave shifts in phosphorescence spectra (From Ref. [6] and [73]).

depth E_t^{ex}		Nature of the traps
cm^{-1}	eV	
129	0.015	Traps of structural origin, due to local deformations of the crystal
305	0.038	
346	0.043	
516	0.064	
761	0.094	Most likely, trap in the environment of an oxidized anthracene molecule (chemical impurity)
2500	0.301	Most likely, a center of excimer origin

Table II.2 Singlet exciton trap depth, E_t^{ex} as determined by the method of defect fluorescence and the corresponding estimated trap depth of charge carriers E_t for various structural defects in an anthracene crystal. (From Ref. [6]).

cm^{-1}	exciton trap depth E_t^{ex} (eV)	charge-carrier trap depth E_t (eV)
130	0.016	~ 0.05
240	0.03	~ 0.1
275	0.034	0.13
320	0.04	0.15
616	0.08	0.30
1565	0.19	1.05
1660	0.20	1.1
1600–2000	0.2–0.25	1.2–1.4

disorders as the neutral exciton does not interact with the surrounding as much as the charge carriers. Consequently, structural disorder may generate hardly any deep traps for excitons. It is even reasonable to envisage the possibility of exciton states in a liquid phase [74, 75]. A percolation mechanism in weakly disordered molecular crystals or organized mesophases has been described recently [76–79].

II.3 Photogeneration of Charge Carriers

There are numerous mechanisms for the photogeneration of carriers. They can be classified depending whether one quantum leads to one charge carrier or if several quanta are necessary [23]:

One-quantum processes.
— Excitation across an energy gap to form an electron-hole pair.
— Spontaneous ionization of an exciton:

$$A^*, A \rightarrow A^+, A^- \rightarrow \text{free charges} \tag{II.32}$$

— Thermal ionization of an exciton
— Field ionization of an exciton
— Collison of an exciton with a dissociation center
— Liberation of a trapped carrier by a photon or an exciton
Two-quantum processes.
— Photon/exciton interaction:

$$A^*, A \xrightarrow{h\nu} A^+, A^- \rightarrow \text{free charges} \tag{II.33}$$

— Collisons of two excitons:

$$A^*, A^* \rightarrow A^+, A^- \rightarrow \text{free charges} \tag{II.34}$$

The primary quantum yield of charge carrier formation, η, is defined as the number of free electrons and holes which are formed per absorbed light quantum. In most organic solids, this quantum yield is very small because many other ways of de-activation of the excited state are possible. Generally, the expression of η explicitly takes into account the intersystem crossing rate constant to the triplet state (k_T) and the zero-order non-radiative transitions (k_n). The other quenching processes are assumed to follow pseudo-first-order kinetics (k_Q), with quenchers of concentration [Q]. If k_c is the rate constant for free carrier production, then:

$$\eta = \frac{k_c}{k_c + k_n + k_T + k_Q[Q]} \qquad (II.35)$$

The gain factor $\mathscr{G}$ is defined as the number of carriers passing through the outer circuit divided by the number of photons absorbed by the photoconductor during the same period of time. $\mathscr{G}$ may be orders of magnitude lower than the primary quantum yield because of the trapping processes occurring in the bulk material.

Because of the low dieletric constant effective in organic semiconductors, the energy needed to separate the initial ion pair may be greater than the energy needed to electronically excite one molecule:

$$A^*, A \xrightarrow[\substack{\text{ionization}\\\text{step}}]{} A^+, A^- \xrightarrow[\text{ion pair}\\\text{separation}]{} A^+, ..., A^- \qquad (II.36)$$

The Onsager approach is generally used to estimate the probability of charge carrier formation as a function of the energy available [80, 81]. Different stages are postulated. A photon is first absorbed by a molecule, and, before any relaxation may take place, a carrier is ejected from the excited state with some kinetic energy:

$$A \xrightarrow{h\nu} \underset{\substack{\text{not}\\\text{relaxed}}}{"A^*"} \to A^+ + e^- \qquad (II.37)$$

The kinetic energy of the generated carrier strongly depends on the wavelength of the incident photon, since the excess energy available is directly related to the energy of the incident photon. The charge carrier dissipates its excess energy by collisions with the medium. During this time the carrier has travelled over a distance r_o from its parent countercharge. The carrier still experiences a coulombic interaction given by the classical expression:

$$E_{ip} = -\frac{q \cdot q'}{\varkappa \cdot r_o} \qquad (II.38)$$

$\varkappa$: dielectric constant
r_o: thermalization length

The carrier may either recombine to the parent countercharge or escape from the ion pair by diffusion. The condition for free carrier formation is that the coulombic attraction is less than the thermal energy kT. The thermalization length must therefore be such that:

$$r_o \geq -\frac{q \cdot q}{\varkappa \cdot kT} \qquad (II.39)$$

Geminate recombination, the recombination of a charge carrier with its parent countercharge, is thought to govern the quantum yield of charge carrier formation in many organic solids. Charge-charge interactions are effective at long range and their importance is magnified by the low dielectric constants of the organic materials. The amount of geminate recombination is diminished if an external electric field is applied to the organic material. This furnishes a way of studying the quantum yield of carrier formation [82–84]. Figure II.6 illustrates such studies for thin films of amorphous selenium. The thermalization length depends strongly on the excitation wavelength, from 70 Å at 400 nm to only 8.4 Å at 620 nm. Correspondingly, the quantum efficiency, or the number of carriers generated per photon absorbed, varies with the magnitude of the external field applied, from about 0.5 at high fields (10^9 V/cm) to 10^{-4}–10^{-5} at low fields and at 620 nm. Fairly high electric fields are therefore necessary to produce a significant effect on the probability of geminate recombination.

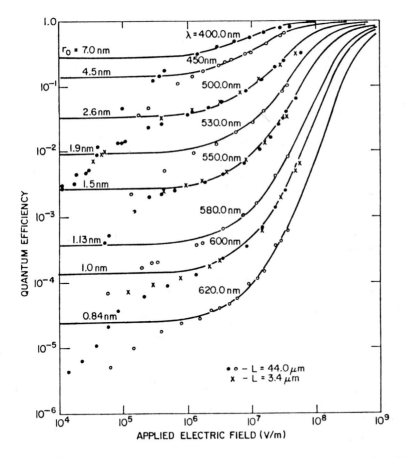

Fig. II.6. Quantum efficiency of photogeneration of charge carriers as a function of the wavelength of the incident photons and of the applied field in amorphous selenium films. Data concern films of two different thicknesses. Solid lines are the theoretical Onsager dissociation efficiencies for the initial separation r_o indicated. [Reproduced with permission of (82)]

Onsager's approach has been modified to fit particular cases. For example, the model has been applied to the one-dimensional case for studying the formation of charge carriers at interfaces [85, 86].

Some organic materials, such as phthalocyanines, however, show a remarkable insensitivity of the quantum efficiency to excitation wavelength [87]. In this case, part (or all) of the excess excitation energy is dissipated by internal conversion, only the remainder is used to eject the charge carrier. Vibrational relaxation of upper levels of excited states occurs on a picosecond time scale (10^{-12} s), and relaxation is therefore usually in competition with carrier ejection. Geminate recombination fluorescence or exciplex formation may further complicate the processes involved [88, 89].

II.4 Semiconductor Junctions

Light absorption, energy migration, and carrier production make up one aspect of photovoltaic behavior. But equally necessary for the production of a useful photovoltaic device is the presence of a semiconductor junction. In this section the essential properties of such junctions will be outlined. Section II.5 will then bring the elements of the chapter together with a discussion of how organic solar cells may work in practice.

a p-n Junctions: Formation and Electrical Properties

In the case of intrinsic semiconductors no ionizable impurities are present in the material (Fig. II.7a). It may be demonstrated that the number of free electrons, n, in the conduction band is given, within the Boltzmann approximation, by:

$$n = N_c \cdot \exp \left(- \frac{E_c - E_F}{kT} \right) \tag{II.40}$$

E_F is the Fermi level, E_c the energy at the bottom of the conduction band, and N_c the effective density of states near the bottom of the conduction band. The number of free holes, p, in the valence band is similarly given by:

$$p = N_v \cdot \exp \left(- \frac{E_F - E_v}{kT} \right) \tag{II.41}$$

E_v: energy at the top of the valence band
N_v: effective density of states in the valence band

Free carriers are formed by thermal or photochemical excitation of electrons from the valence band to the conduction band, and the electroneutrality condition requires that:

$$n = p = n_i \tag{II.42}$$

The position of the Fermi level is thus obtained by:

$$E_F = E_i = \frac{E_c + E_v}{2} + \frac{kT}{2} \cdot \ln \frac{N_v}{N_c} \tag{II.43}$$

Consequently, the Fermi level lies very close to the middle of the band gap for intrinsic semiconductors since the effective densities of states in the valence and conduction bands are in most cases approximately equal (Fig. II.7a).

Now consider doped semiconductors. They are made by adding either donor (n doping) or acceptor (p doping) impurities to the semiconductor (Fig. II.7b and c). New energetic levels are introduced in the band gap, their positions correspond to the energy which is necessary to remove an electron from the valence band to the accepting impurity (p type) or to ionize the donor, the electron being released into the conduction band (n type). The number of free electrons and holes is no longer equal, but since the charge carriers are still in thermal equilibrium, then:

$$n \cdot p = n_i^2 \tag{II.44}$$

The Fermi level is no longer near the middle of the gap but is displaced towards the conduction band (n doping) or the valence band (p doping) (Fig. II.7).

Now form a p-n junction by joining two parts of a semiconductor, one p-doped, the other n-doped (Fig. II.8) [96–99]. Donor impurities on the n-side tend to give electrons to the acceptor centers of the p side. Electrons flow from the n-type to the p-type semiconductor creating an electric field ξ_{bi} (built-in electric field) at the interface. In a region extending to both parts of the interface most of the impurities are ionized, positively in the n part and negatively in the p part. This area is called the *space-charge or depletion region*. The space-charge region is depleted in majority carriers because of the recombination which must occur between electrons and holes to

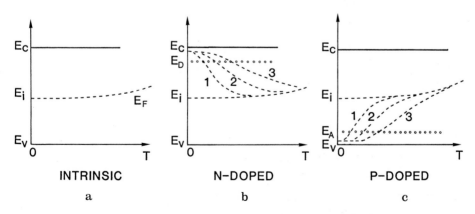

Fig. II.7. The variation of the Fermi levels E_F as a function of temperature for **a** intrinsic, **b** n-doped, and **c** p-doped semiconductors. The curves 1, 2, and 3 correspond to increasing impurity concentrations.
E_i: intrinsic Fermi level at 0 K. (After Ref. [506])

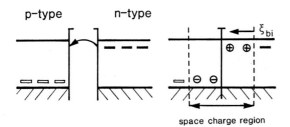

p-type   n-type

Fig. II.8. Scheme representing the formation of a p-n junction

space charge region

obey Eq. II.44. This lowering of the concentration of free charge carriers induces a higher resistivity of this region of the semiconductor. The built-in electric field ξ_{bi} is directed from the n- to the p-type region; it thus produces a force opposing further diffusion of electrons. ξ_{bi} corresponds to a gradient of the electrostatic potential $\left(-\dfrac{\partial V}{\partial x} \text{ in one dimension}\right)$, a schematic representation of the corresponding potential variation in an ideal abrupt p-n junction is shown in Fig. II.9. It is generally assumed that in the regions outside the space charge area, the potential is constant. The built-in potential is therefore given by:

$$V_{bi} = V_n - V_p \tag{II.45}$$

where V_n and V_p are the potentials of the neutral regions in the n- or p-part of the semiconductor. Band diagrams are more often expressed in terms of electron energy $(-eV)$, as shown in Fig. II.10. The Fermi levels on the two sides of the junction must be equal at equilibrium (the chemical potentials of the two sides of the interface must be equal). It is therefore possible to quantitatively estimate the value of the built-in potential knowing the energies corresponding to the Fermi levels in the two sides of the junction (Fig. II.11). The energy shift needed to equalize the two Fermi levels is such that:

$$e \cdot V_{bi} = E_F(n) - E_F(p) \tag{II.46}$$

$E_F(n)$ and $E_F(p)$ are the Fermi levels of the isolated n- and p-type counterparts. The width of the space-charge region may be easily calculated, since the overall

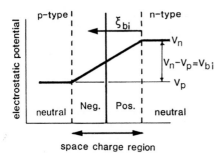

space charge region

Fig. II.9. Electrostatic potential $V(x)$ in an ideal abrupt p-n junction. (After Ref. [96])

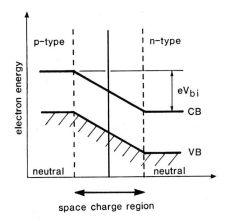

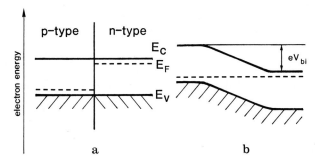

Fig. II.10. Band-energy diagram in an ideal abrupt p-n junction. (After Ref. [96])

Fig. II.11. Scheme representing the formation of a p-n junction. The value of the built-in potential is obtained through equalization of the Fermi levels.
(After Ref. [96]).
a: Band diagram of a p-n junction before the attainment of equilibrium.
b: Band diagram at equilibrium showing equalized Fermi levels and the space-charge region

electrical neutrality requires that the number of positive charges in the n-region must be equal to the number of negative charges in the p part:

$$N_a \cdot x_p = N_d \cdot x_n \tag{II.47}$$

N_a, N_d: densities of ionized acceptors or donors in the space-charge region
x_p, x_n: distance of the space-charge region extending into the p side or n side, respectively.

A simpler way of obtaining the same result may be found by following a chemical way of reasoning. Consider an organic material composed of molecules A and containing either donor I_D or acceptor I_A impurities. A junction is formed by joining the two counterparts (Fig. II.12). A transfer of electrons from the donors to the acceptors occurs:

$$I_A + I_D \rightleftarrows I_A^- + I_D^+ \tag{II.48}$$

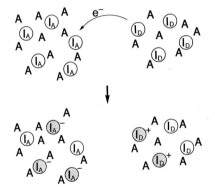

Fig. II.12. Schematic representation of the formation of a p-n junction in an organic material composed of molecules of A type, doped with acceptor, I_A, or donor, I_D, impurities

The amount of charge transfer may be determined if the solid-state redox potentials of the impurities, E_A and E_D, can be estimated:

$$I_A + e^- \rightleftarrows I_A^- \qquad E_A \tag{II.49}$$

$$I_D^+ + e^- \rightleftarrows I_D \qquad E_D \tag{II.50}$$

The densities of ionized acceptors and donors are obtained from Nernst's law, in the case of a monoelectronic transfer:

$$\log K_{eq} = \frac{E_A - E_D}{0.059} \tag{II.51}$$

with

$$K_{eq} = \frac{[I_A^-][I_D^+]}{[I_A][I_D]} \tag{II.52}$$

$[I_A]$, $[I_D]$: densities of impurities

It has been seen previously that the reducing or oxidizing properties of a molecule in the solid state may be approximated by the redox potentials in solution. Consequently, E_A and E_D may be estimated from standard electrochemical measurement (cyclic voltammetry or polarography). The classical image of junction formation by equalization of the Fermi levels is formally equivalent to writing the mass action law of Eq. II.52. If the absolute difference between the oxidation and reduction potentials is less than 0.25 V, partial charge transfer between the two kinds of impurities occurs ($K_{eq} \leq 10^4$). This approach, however, rapidly reaches its limitations. For example, the probability of electron transfer depends on the distance of the impurity center from the interface. This is not taken into account by Eq. II.52. Such a model must therefore be used with care; it must only be taken as a way of visualizing the chemical mechanism of the formation of the junction.

At equilibrium, two opposite fluxes of charges flow through the junction. The diffusion current J_{n1}, flowing from the n- to the p-side, is due to the difference in concentration of charge carriers between the n and the p part of the junction. $\mathbf{J}_{n1}$ is related to the oxidizing and reducing abilities of the two doping agents. Classically the flux of the particles is related to the concentration gradient ∇C by:

$$\mathbf{F} = -D \cdot \nabla C \qquad \text{(II.53)}$$

$\mathbf{F}$: flux of particles
D: diffusion coefficient

The corresponding current density is therefore:

$$\mathbf{J}_1 = q \cdot \mathbf{F} \qquad \text{(II.54)}$$

q: elementary charge

The diffusion current for the electrons is then:

$$\mathbf{J}_{n1} = e \cdot D_n \cdot \nabla n \qquad \text{(II.55)}$$

D_n: diffusion coefficient for electrons
n: density of electrons

The same relationship may be found for hole diffusion:

$$\mathbf{J}_{p1} = -e \cdot D_p \cdot \nabla p \qquad \text{(II.56)}$$

D_p: diffusion coefficient for holes
p: density of holes

The opposite flux, the generation or drift current, is caused by the built-in electric field ξ_{bi}, generated by the ionized impurities of different sign between the two sides of the junction. The corresponding current density $\mathbf{J}_{n2}$ is proportional to the electrical field ξ_{bi}, to the number of free electrons or free holes, and to their mobilities through the usual Eq. I.10:

$$\mathbf{J}_{n2} = e \cdot n \cdot \mu_n \cdot \xi_{bi} \qquad \text{(II.57)}$$

μ_n: electron mobility

$$\mathbf{J}_{p2} = e \cdot p \cdot \mu_p \cdot \xi_{bi} \qquad \text{(II.58)}$$

μ_p: hole mobility

At equilibrium, the overall hole and electron currents must cancel:

$$e \cdot D_n \cdot \nabla n + e \cdot n \cdot \mu_n \cdot \xi_{bi} = 0 \qquad \text{(II.59)}$$

$$-e \cdot D_p \cdot \nabla p + e \cdot p \cdot \mu_p \cdot \xi_{bi} = 0 \qquad \text{(II.60)}$$

In the case of intrinsic semiconductors, the diffusion coefficients may be related to the mobility of the charge carriers through the Einstein relation:

$$q \cdot D = \mu \cdot k \cdot T \qquad \text{(II.61)}$$

$k \cdot T/q$ is equal to 0.0259 V at room temperature. In the simplified one dimensional case, Eq. II.59 and II.60 may be integrated within the following assumptions:
— the system is an ideal abrupt junction; the passage from the n to the p side is not graded.
— the concentrations of charge carriers in the neutral regions are at equilibrium.
— the concentration of charge carriers varies from p_p to p_n and from n_p to n_n across the space-charge region. p_p and p_n are the concentration of holes on the p or n side of the junction, respectively. Under these conditions:

$$V_{bi} = \frac{kT}{e} \cdot \ln \frac{p_{p_o}}{p_{n_o}} = \frac{kT}{e} \cdot \ln \frac{n_{p_o}}{n_{n_o}} \tag{II.62}$$

p_{p_o}, p_{n_o}: thermal equilibrium hole concentrations in the neutral p and n regions.
n_{p_o}, n_{n_o}: thermal equilibrium electron concentrations in the neutral p and n regions.

If N_a and N_d are the concentrations of ionized donors or acceptors, respectively, on the two sides of the junction, then:

$$p_{n_o} = \frac{n_i^2}{N_d} \tag{II.63}$$

$$p_{p_o} = N_a \tag{II.64}$$

n_i: density of charge carrier in absence of dopants

and so:

$$\boxed{V_{bi} = \frac{kT}{e} \ln \frac{N_a \cdot N_d}{n_i^2}} \tag{II.65}$$

The built-in potential is entirely determined by the concentration of ionized dopants on the n and p side of the junction. From the Poisson equation V_{bi} may be expressed in a different way:

$$V_{bi} = \frac{2\pi e}{\varkappa_s} \left[N_d \cdot x_n^2 + N_a \cdot x_p^2 \right] \tag{II.66}$$

$\varkappa_s$:   dielectric constant of the semiconductor
x_n, x_p: extent of the space charge region on the n and p sides, respectively.

It may be demonstrated that the total width of the space-charge region $l_{sc} = (x_n + x_p)$ is [96]:

$$l_{sc} = \frac{\varkappa_s \cdot kT}{2\pi e^2} \left(\frac{N_a + N_d}{N_a \cdot N_d} \right) \ln \left(\frac{N_a \cdot N_d}{n_i^2} \right)^{1/2} \tag{II.67}$$

l_{sc}:   width of the space-charge region
N_a, N_p: densities of ionized impurities

It is now possible to discuss the mechanism of current flow in a p-n junction. For reasons of simplicity, only electron currents are considered. Two electron currents flow in opposite directions. The diffusion current, J_{n1}, is due to the flow of electrons from the donor to the acceptor dopants. It is therefore composed of those majority carriers, electrons on the n-side of the semiconductor, which have enough energy to surmount the energy barrier (equal to $e \cdot V_{bi}$ at equilibrium) (Fig. II.13a). The second current, the generation or drift current J_{n2}, is composed of minority carriers, electrons on the p-side, which diffuse to the space-charge region and are swept down the energy barrier into the n-type region (Fig. II.13b). At equilibrium, J_{n1} and J_{n2} are equal. The diffusion current J_{n1} is made up of the small fraction of the majority electrons having sufficient energy to overcome the barrier eV_{bi}, while the generation current is composed of a large proportion of the minority electrons diffusing "down hill" up to the neutral n region. Consequently, the generation current J_{n2} is almost independent of the magnitude of the energy barrier, while the diffusion current J_{n1} is strongly correlated with this barrier.

If an external potential is applied to the junction, it may support or oppose the built-in potential. One then talks of a reverse bias or a forward bias, respectively (Fig. II.14a and b). With increasing reverse bias, the diffusion current J_{n1} decrease while the generation current J_{n2} is more or less unchanged. Forward bias decreases the height of the energy barrier and thus the diffusion current J_{n1} increases while the generation current J_{n2} is approximately constant. The overall net current, composed of both electrons and holes, therefore increases with increasing forward bias and decreases with increasing reverse bias (Fig. II.15). At high reverse bias, the total electron current through the junction is constant and is equal to the generation current J_{n2}, the diffusion current J_{n1} becoming negligible. The same conclusion is true for holes. The overall current at high reverse bias is given by:

$$J_{sat} = J_{n2} + J_{p2} \tag{II.68}$$

J_{sat}: is the saturation current

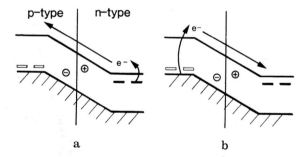

p-type   n-type

a   b

Fig. II.13. Schematic representation of the diffusion current J_{n1} and of the drift or generation current J_{n2}.

a Diffusion current J_{n1}: the fraction of the majority carriers having enough energy (electrons on the n side) surmounts the energy barrier at the junction, going from the n side to the p side.

b Drift or generation current J_{n2}: minority carriers (electrons on the p side) are swept down the energy barrier by the built-in electric field, going from the p to the n side. Electrons are thermally generated from the valence band

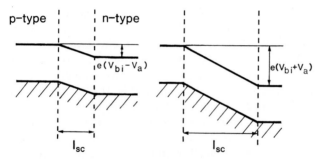

Fig. II.14. Effect of an external applied field V_a on the energy-band diagram of an abrupt p-n junction:
Left: Forward bias: the magnitude of the energy barrier has decreased to $e(V_{bi} - V_a)$; the width of the space-charge region has decreased correspondingly.
Right: Reverse bias: the energy barrier has increased to $e(V_{bi} + V_a)$; the width of the space-charge region has increased.
(After Ref. [96])

It is worth pointing out that the current transport through p-n junctions is mainly due to the minority carriers. Under forward bias, the electron current from the n to the p side represents an "injection" of minority carriers at the p-side of the junction. With reverse bias it corresponds, conversely, to an extraction of minority carriers.

It is possible to calculate the current flowing across the junction within the previous model (see for example [95, 96]). The classical Schockley diode equation relating the current density J to the applied voltage gives in one-dimension:

$$J = e \cdot \left(\frac{D_n}{L_n} n_{p_0} + \frac{D_p}{L_p} p_{n_0} \right) \cdot \left[\exp\left(\frac{e \cdot V_a}{kT} \right) - 1 \right] \qquad (II.69)$$

D_n, D_p: diffusion coefficients of the electrons and holes
L_n, L_p: diffusion lengths of the charge carriers defined by $L = \sqrt{D\tau}$ τ: carrier lifetime
V_a: applied potential
n_{p_0}, p_{n_0}:   equilibrium electron and hole concentrations in the neutral p and n regions, respectively

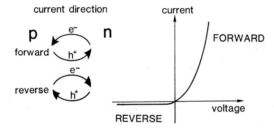

Fig. II.15. Current-voltage plot for a p-n junction: effect of a reverse or a forward bias

The saturation current, which is equal to the generation current at high reverse bias, is given by:

$$J_{sat} = e \cdot \left(\frac{D_n}{L_n} \cdot n_{p_0} + \frac{D_p}{L_p} \cdot p_{n_0} \right)$$

(II.70)

The Schockley equation is obtained within the following assumptions:

(i) the junction is abrupt,

(ii) the densities of minority carriers injected in the neutral regions are small compared to the majority carrier densities,

(iii) there is no generation or recombination of charge carriers in the space-charge layer. Most of these hypotheses must be considered with suspicion when applied to molecular semiconductors where the concentration of free charge carriers is generally very low and where majority and minority carriers are often present in approximately equal concentrations. These problems are more thoroughly brought out in a subsequent section.

b Schottky Junctions: Semiconductor-Metal Contacts

Space-charge regions may also be formed at interfaces between a semiconductor and a metallic electrode. Band diagrams before and after contacts are represented in Fig. II.16 for an n-type semiconductor. When the metal and the semiconductor are brought together, electrons flow from the semiconductor to the metal until the Fermi levels are equalized. The space-charge layer in the semiconductor is made up of the remaining positively charged ionized impurities. The characteristics of the Schottky junctions are generally defined by the work function Φ_M of the metal, the energy necessary to remove an electron from the top of the fermi distribution in the metal, and by Φ_S the work function of the semiconductor. Φ_S is defined by:

$$\Phi_S = E_{vac} - E_F(S)$$

(II.71)

E_{vac}:   vacuum energy level
$E_F(S)$: Fermi level of the semiconductor

The electron affinity of the semiconductor, A_s, is given by:

$$A_s = E_{vac} - E_c$$

(II.72)

E_c: energy of the conduction band edge

From Fig. II.16b, it is possible to estimate the energy barrier ΔE_{SM} for an electron going from the semiconductor into the metal:

$$\Delta E_{SM} = \Phi_M - \Phi_S$$

(II.73)

and also the energy barrier ΔE_{MS} for an electron going from the metal to the semiconductor:

$$\Delta E_{MS} = \Phi_M - A_s$$

(II.74)

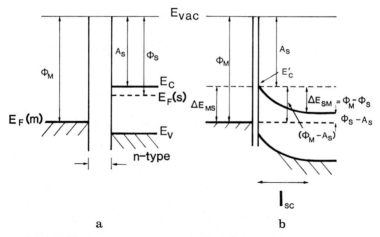

Fig. II.16. Band diagrams corresponding to system metal/n-type semiconductor.
a before contact
b after contact
E_{vac}:   vacuum energy level
Φ_M: work function of the metal
A_s: electron affinity of the semiconductor
Φ_s: work function of the semiconductor
l_{sc}:   width of the space-charge region
ΔE_{MS}: energy barrier between metal and semiconductor
ΔE_{SM}: energy barrier between semiconductor and metal

In this example the semiconductor work function Φ_S is larger than the metal work function Φ_M. The contact is said to be "rectifying". The current-voltage plot is asymmetric because of the formation of a space-charge region.

For the case where Φ_M is larger than Φ_S, the contact between the n-type semiconductor and the metal is "ohmic", no space-charge layer is formed. In fact, electrons must be transferred from the metal to the semiconductor in order to equalize the Fermi levels (Fig. II.17). However, as the semiconductor is n type, the dopant centers

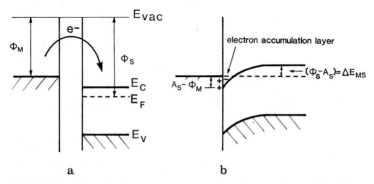

Fig. II.17. Ohmic contact between an n-type semiconductor and a metal. In this case, the work function of the metal Φ_M, is larger than the work function of the semiconductor

may donate and not accept electrons; consequently, the space-charge region may not develop. An electron accumulation layer is formed in the semiconductor close to the interface. The spatial extent of the accumulation layer is far less than the space-charge depth. Whatever the external applied voltage, no energy barrier to the current arises and the current-voltage plot is symmetric.

Although the mechanisms occurring in p-n junctions and Schottky contacts are very similar, an important difference exists. *Current transport through p-n junction has been seen to be mainly governed by the minority carriers. In Schottky contacts, on the other hand, the majority carrier current is predominant.* For the n-type semiconductor/metal junction represented in Fig. II.16, the majority carriers (electrons) flow from the semiconductor to the metal at forward voltages.

The current-voltage plots for Schottky junctions still obey a Schockley-type equation:

$$J = J_{sat} \cdot \left[\exp \left(\frac{e \cdot V_a}{kT} \right) - 1 \right] \tag{II.75}$$

The calculation of the saturation current J_{sat} is, however, far more complex than for p-n junctions [95, 96]. From thermionic emission theory it may be demonstrated that [95]:

$$J_{sat} = A^* \cdot T^2 \cdot \exp \left(- \frac{\Delta E_{MS}}{kT} \right) \tag{II.76}$$

A*: Richardson constant

c Insulator-Metal Contacts

Molecular crystals or polymeric systems very often have low intrinsic conductivities. The overall conductivity may be brought into the semiconductivity range by "heavy" doping. However, the dopants generally act also as recombination centers, and their concentration thus limits the mobilities of the charge carriers. The charge transport may be regarded as a hopping of carriers from site to site. These materials are more appropriately called "heavily doped insulators" than "semiconductors". Most of the previous equations established for describing junctions are not applicable in this case.

The equilibrium concentration of charge carriers in insulators is very small because of the exponential dependence of the density of carriers with the band gap E_g. With an energy gap of 4 eV, the fraction of electrons excited across the gap at room temperature is of the order of 10^{-35} (for an intrinsic semiconductor); if E_g is 0.25 eV, this value is of the order of 10^{-2}. The gap in most semiconductors is less than 2 eV; beyond that the materials are considered as insulators. Since charge carriers cannot be generated thermally, the current transport through insulators is governed by the concentration of charge carriers injected from the contacts. Consider the I—V plot of an insulator sandwiched between two metallic electrodes (Fig. II.18) [100–104]. At low voltages there is negligible injection of carriers from the electrodes; the thermal equilibrium density of charge carriers in the absence of voltage is greater than the density injected through the contacts. The current therefore obeys Ohm's (regions A — B′ or A — B in Fig. II.18). When this condition is not fulfilled, the current becomes space-charge limited: the organic material between the electrodes is no longer neutral, excess charges have been injected. These charges develop a potential which influences the I–V characteristics. The mechanism of formation of this space charge is different from the one previously found in p-n junctions or Schottky contacts since no ionized impurities are involved. The space charge generated by injection of charges has two components: the mobile charge carriers and eventually the filled traps. The filling of traps by injected charge carriers gives electrically charged centers, which may contribute to the formation of the space charge. First consider the trap-free case: all the injected electrons (or holes) remain free in the conduction band. If a charge Q is present in the organic insulator, the corresponding voltage may be expressed as [100–104]:

$$Q = C \cdot V \qquad (II.77)$$

C: capacitance of the device

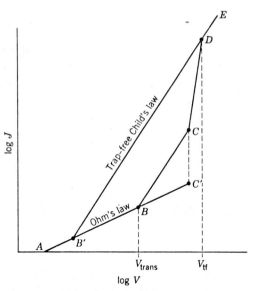

Fig. II.18. Typical current-voltage plot for insulators containing traps (log J — log V curves) [Reproduced with permission of (23)].

A—B′: Ohm's law domain, no traps in the material. The density of injected carriers is lower than the density of carriers at thermal equilibrium in absence of voltage.

A—B: Ohm's law domain, traps in the material. The filling of traps also contributes to the space-charge formation.

B′—D—E: Space-charge limited current (SCLC) region in the trap-free case.

B—C: SCLC, shallow traps.

C—D: SCLC, deep traps.

V_{trans}: Transition voltage where trapping processes become important.

V_{tf}: Voltage at which traps are all filled

At a voltage corresponding to B' (Fig. II.18) the concentration of injected carriers becomes greater than the majority carriers thermally generated, Ohm's law is no longer obeyed, and the current becomes space-charged limited. The current may be estimated from the time τ required for the charge to go from one electrode to the other:

$$I_{SCL} = \frac{Q}{t_t}$$ (II.78)

I_{SCL}: space-charge-limited current
t_t: transit time for the charge carriers

The transit time t_t may be expressed in terms of the carrier mobility μ and the field $\xi = \frac{V}{d}$ in the material (d: electrodes spacing):

$$t_t = \frac{d^2}{\mu \cdot V}$$ (II.79)

d: thickness of the organic material
V: external potential applied
μ: mobility of the charge carriers

The two electrodes form a capacitor whose capacitance may be approximated by:

$$C = \frac{\varkappa_s}{4\pi \cdot d}$$ (II.80)

$\varkappa_s$: dielectric constant for the organic material

The space-charge-limited current becomes:

$$I_{SCL} \sim \varkappa_s \cdot \mu \cdot \frac{V^2}{d^3}$$ (II.81)

In the trap-free case, the current is thus proportional to the square of the applied voltage (Trap-free Child's law, part B' — D — E of Fig. II.18).

Now consider a material containing shallow traps. The space charge is composed of the mobile free carriers but also of trapped charges. The residence time inside the traps is lower than the transit time, because trapped charges are readily freed thermally (shallow trap condition). Under these conditions Eq. II.81 is still applicable, provided the mobility is replaced by an effective mobility:

$$\mu_{eff} = \theta \cdot \mu$$ (II.82)

ϑ: proportion of free charge carriers

The current is still proportional to the square of the applied voltage.

When deep traps are considered, the previous approximation cannot be made. The residence time of the charges in the traps becomes longer than the transit time. I_{SCL} therefore depends on the density of traps and on their distribution inside the band gap [100–104].

The discussion so far have adopted a phenomenological approach with the implicit assumptions that:

(i) both electrodes were made of the same metal,

(ii) the electric field ξ generated in the organic layer was constant, irrespective of position and equal to $\dfrac{V}{d}$,

(iii) the injected charge density was determined by an overall capacitance. To remove these restrictions requires to consider a system of nonlinear differential equations based on the Poisson and the conductance equations [110]. These may be solved either analytically [105, 106] or numerically [107–109]. The establishment of the system of differential equations has been made in most cases for sandwiches M/organic layer/M, where M is a metallic electrode leading to an ohmic contact. There is no barrier to carrier injection from the electrodes into the organic layer, but there is an infinite charge density in the insulator at the interface insulator/metal. Different cases have been considered depending on the presence or absence of traps within the organic material.

The trap-free case has been widely studied [105–110]. Depending on the applied voltage, two regimes may be distinguished. When the current arising from the injected charges is predominant, the phenomenological model and the resolution of the differential equations give very similar results. When the conditions previously exposed are fulfilled, the space-charge-limited current follows the relation given by Eq. II.81. At low voltages, an empirical formula of the form:

$$I_{SCL} \sim \varkappa_s \cdot \mu \cdot \frac{V}{d^3} \tag{II.83}$$

is obeyed. The current is directly proportional to the applied voltage, as in the ohmic law:

$$I_\Omega \sim e \cdot \mu \cdot n_o \cdot \frac{V}{d} \tag{II.84}$$

n_o: thermal equilibrium density of charge carriers

However, in the first case, a $1/d^3$ dependence on the thickness of the organic layer is found [108], whereas a $1/d$ dependence should be obtained with the ohmic law. Current arising from injected charges is stronger than the intrinsic current for wide gap materials. Although the number of injected charges is greater than the number of intrinsic charge carriers which are thermally generated, a same I–V dependence is found. Both regimes may be distinguished by studying the thickness dependence of the current.

The effect of traps may also be taken into account. First consider, as previously, two identical ohmic electrodes. The concentration and the nature of the traps do not vary with their distance from the metallic electrodes. The distribution in energy of the traps may be expressed as [110]:

$$\varrho_f = a \cdot \varrho_{tot}^b \tag{II.85}$$

ϱ_f: free charge density
ϱ_{tot}: total charge density

In the trap-free case, $a = b = 1$. For shallow traps of density N_t at an energy E_t below the valence band, a and b are given by:

$$a = \left(\frac{N_v}{N_t}\right) \cdot \exp\left(-\frac{E_t}{kT}\right) \quad \text{and} \quad b = 1 \tag{II.86}$$

For an exponential distribution of traps inside the band gap characterized by the distribution function:

$$h(E_t) = \left(\frac{c}{kT \cdot b}\right) \cdot \exp\left(-\frac{E_t}{kT \cdot b}\right) \tag{II.87}$$

$b > 1$ and a is given by:

$$a = e \cdot N_v \cdot \left[\frac{\sin\left(\dfrac{\pi}{b}\right)}{e \cdot c\left(\dfrac{\pi}{b}\right)}\right]^{b} \tag{II.88}$$

c: constant

These equations apply only to hole injection from the electrodes; the equations corresponding to electron injection can be derived similarly. Within the constraints of the model, it is found that at low voltages:

$$I \sim \frac{V}{d^{2b+1}} \tag{II.89}$$

and at high voltages [110]:

$$I \sim \frac{V^{b+1}}{d^{2b+1}} \tag{II.90}$$

b is a constant which is characteristic of the trap energy distribution within the band gap (Eq. II.87); d is the electrode spacing.

However, the spatial distribution of traps within the organic material is expected to be nonuniform. The concentration of traps next to the electrodes would normally be higher than the bulk average. The effects of the spatial variation of trap density on the space-charge-limited current has been much studied using the phenomenological model [111–114]. More complete solutions of the problem, taking into account the possibility of spatial variation of the concentration of carriers (diffusion current), have since been described [115, 116]. Calculations based on such models give a potential profile across the organic layer as shown in Fig. II.19. Contacts at the electrodes are assumed to be ohmic and the injected carriers are holes. Two cases are possible depending on whether the traps are symmetrically or unsymmetrically distributed between the two electrodes. If the density of traps is higher near one electrode than the other, a rectification at low voltage may be observed: the I–V characteristics shows

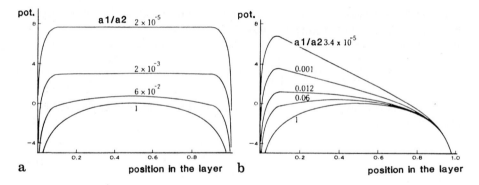

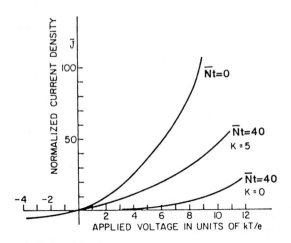

Fig. II.19. Potential profiles in an organic material sandwiched between two ohmic electrodes for two different spatial distributions of traps.
The parameters a and b are defined in the text. In both cases b = 1 (shallow traps) and:
a Symmetrical case: a = a_1 near electrode 1
              a = a_2 bulk
              a = a_1 near electrode 2
b Unsymmetrical case: a = a_1 near electrode 1
              a = a_2 rest of the organic layer. (After Ref. [115])

Fig. II.20. Current-voltage plots for a system M_1/insulator/M_2, where M_1 and M_2 are two different metallic electrodes.
K corresponds to the equilibrium constant:

$$K = \frac{\overline{n}_f \cdot (\overline{N}_t - \overline{n}_t)}{\overline{n}_t}$$

$\overline{N}_t$:   normalized overall density of traps
$\overline{n}_f$:   normalized density of free carriers
$\overline{n}_t$:   normalized density of trapped carriers
$\overline{N}_t = 0$: corresponds to the trap-free case
K = 0: traps permanently filled
(The plots are only valid for unipolar injection). (After Ref. [117])

the behavior of a diode although both electrodes are ohmic (no barrier to injection of carriers). An asymmetrical distribution of traps may be due to different surface states if the two electrodes are identical; when the two metallic electrodes are different, the less noble metal is expected to give rise to a higher concentration of nearby traps than the less easily oxidized metal.

Rectification may be obtained even in the trap-free case when two different metallic electrodes are used [117] (Fig. II.20). When uniformly spatially distributed trapping states are considered, the rectification ratio decreases with increasing number of trapping states.

II.5 Photovoltaic Effect

Becquerel in 1839 first discovered that the potential of an electrode immersed in an electrolyte is varied when this electrode is exposed to light. Thus it was shown that the conversion of light quanta into electricity was possible. However, it took more than a century to produce a solar cell converting efficiently sunlight into electrical power ($\sim 6\%$) [90]. Single-crystal silicon cells can now exhibit conversion efficiencies as high as 19% [91], with one photon out of 5 being converted into an electron flowing in the external circuit. The state-of-the-art in the domain of molecular semiconductors is not at that level. Power conversion efficiencies are more in the order of $10^{-1}–10^{-5}\%$. All too little is known about the processes occurring within an organic solar cell, and it will surely require much basic research before these processes are well understood. Nevertheless, there is a renewed interest in organic solar cells since the work of Ghosh and Morel on merocyanines, who have obtained conversion efficiencies around 1% under standard white light conditions ($100 \ mW/cm^2$) [92–94]. This was several orders of magnitude better than had commonly been obtained previously. More generally, molecular conductors and semiconductors have progressively lost their bad reputation during the last ten years. They had been known as poorly characterized materials with no reproducible electrical behavior. The immense amount of work done on charge-transfer complexes and polyacetylene allowed a clearer understanding of the basic electrical properties of the organic materials, and solid-state physicists became more willing to accept their pecularities. It has been natural to apply the formalism developed for inorganic solar cells to their organic counterparts, and this approach is discussed in the next section. However, the limitations of such an approach are readily apparent and other models will be discussed in the subsequent sections.

a Molecular Solar Cells: Classical Formulation

Figure II.21 shows a p-n junction illuminated by photons with an energy greater than the band gap. The photons absorbed on both the p and n sides generate an e^-/h^+ pair; because of the field ξ_{bi}, the minority carriers are swept down the energy barrier. The separated photogenerated carriers set up an electric field which is opposite to the built-in electric field ξ_{bi}; the difference in potential between the two sides of the junction is reduced from V_{bi} to a smaller value $V_{bi} - V_f$. The effect would be the same if an external forward-bias voltage of value V_f were applied across the

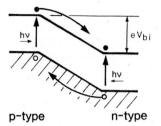

p-type   n-type

Fig. II.21. p-n junction irradiated with photons whose energy hν is greater than the band gap E_g. The minority carriers generated by illumination are swept down the energy barrier. (After Ref. [96])

junction. The maximum open-circuit voltage under illumination V_{oc} is therefore less than or equal to the built-in potential V_{bi}.

A photovoltaic effect may also arise in Schottky junctions if the contact is irradiated through a metallic electrode thin enough to be semitransparent (Fig. II.22). Photo-effects may arise from three main processes. Firstly, light may be absorbed in the metal and excite electrons from the metal to the semiconductor over the energy barrier ΔE_{MS}. This contribution is generally fairly small and is usually negligible compared to the other mechanisms. Secondly, light may be absorbed in the space-charge layer of the semiconductor: the e^-/h^+ generated is separated by the built-in electric field of the junction before recombination may occur. Thirdly, light may be absorbed in the bulk semiconductor; in this case, the minority carriers (the holes in Fig. II.22) must diffuse up to the junction to be collected [97]. Irradiation is, as for p-n junctions, equivalent to the application of a forward voltage.

The equivalent electrical circuit of a solar cell is shown in Fig. II.23 [97]. The photo-current is represented by a current generator I_{ph} which is opposite in direction to the forward-bias current of the diode I_{dark}. The shunt resistance (R_{sh}) may arise from surface leakages, diffusion spikes at the interface, or metallic bridges along micro-cracks [97]. The series resistance (R_s) has various origins: contact resistance at the interfaces; bulk resistance of the semiconductor (neutral region); sheet resistance

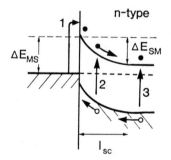

Fig. II.22. Energy-band diagram of a Schottky contact with an n-type semiconductor. Under illumination, three mechanisms giving a photocurrent are possible.
(1): The photons excite electrons from the metal to the semiconductor over the barrier ΔE_{MS}.
(2): Light is absorbed within the space-charge layer of the semiconductor, the e^-/h^+ pair generated is separated by the built-in electric field before recombination can occur.
(3): Light is absorbed in the bulk semiconductor; the minority carriers (holes) diffuse to the junction where they are collected. (After Ref. [97])

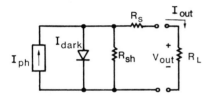

Fig. II.23. Equivalent circuit of a solar cell.
I_{ph}: photocurrent; R_s: series resistance; R_{sh}: shunt resistance; R_L: load resistance. (After Ref. [97])

of the thin semitransparent electrodes. The relationship between the output voltage V_{out} and the output current I_{out} is given by [97]:

$$I_{out}\left(1 + \frac{R_s}{R_{sh}}\right) = I_{ph} - \left(\frac{V_{out}}{R_{sh}}\right) - I_{dark} \qquad (II.91)$$

If R_s and R_{sh} are neglected and if the dark current is expressed as the semi-empirical equation:

$$I_{dark} = I_{oo} \cdot \left[\exp\left(\frac{q \cdot V_a}{A_o \cdot k \cdot T}\right) - 1\right] \qquad (II.92)$$

V_a: applied voltage across the junction
A_o: junction perfection factor (adjustable parameter, $A_o = 1$ for a "perfect diode")
I_{oo}: preexponential factor

a simple expression for I_{out} may be found:

$$I_{out} = I_{ph} - I_{oo} \cdot \left[\exp\left(\frac{q \cdot V_{out}}{A_o \cdot kT}\right) - 1\right] \qquad (II.93)$$

The short circuit current is such that:

$$I_{sc} = I_{ph} \qquad (II.94)$$

and the open-circuit voltage under illumination V_{oc} is given by:

$$V_{oc} = A_o \cdot \left(\frac{kT}{q}\right) \cdot \ln\left(\frac{I_{sc}}{I_{oo}} + 1\right) \qquad (II.95)$$

The maximum power extractable from the cell is a major parameter characterizing a solar cell:

$$P_{out}(max) = I_{max} \cdot V_{max} \qquad (II.96)$$

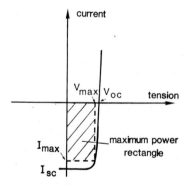

Fig. II.24. Typical I–V curve for a solar cell under illumination.

I_{sc}: short-circuit current

V_{oc}: open-circuit voltage

V_{max}, I_{max}: voltage and intensity corresponding to the maximum power output.

$$FF = \frac{V_{max} \cdot I_{max}}{V_{oc} \cdot I_{sc}}$$

I_{max} and V_{max} are the current and voltage corresponding to the maximum power output. The fill factor (FF) is often used to characterize the maximum power output:

$$FF = \frac{V_{max} \cdot I_{max}}{I_{sc} \cdot V_{oc}} \tag{II.97}$$

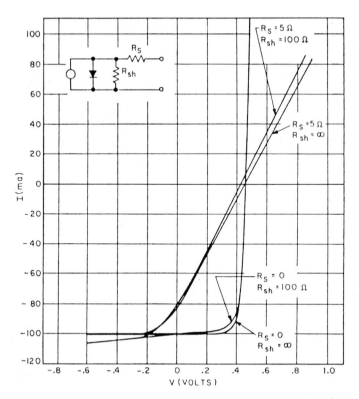

Fig. II.25. Theoretical I–V characteristics for different values of the shunt resistance (R_{sh}) or the series resistance (R_s). The equivalent circuit is shown in the insert. [Reproduced with permission of (95)]

The fill factor may be represented on the I–V curve obtained under illumination (Fig. II.24); it is a measure of the "squareness" of the plot.

When R_{sh} and R_s are not negligible, the $V_{out} - I_{out}$ relationship becomes almost impossible to be determined analytically. The effects of R_s and R_{sh} on the I–V characteristics of a silicon solar cell are shown in Fig. II.25. In this example, $I_{ph} = 100$ mA and T = 300 K [95]. A shunt resistance as low as 100 Ω does not appreciably affect the fill factor. On the other hand, a series resistance of 5 Ω considerably reduces the maximum power output [95].

b Molecular Solar Cells: Localized States Formulation

In this section only metal/organic semiconductor junctions will be considered since only a very few cases of molecular p-n devices have been described.

For most molecular materials it is more fruitful to divide the photovoltaic effect into a succession of elementary steps. First, a photon excites a given molecule to an upper electronically excited state. The probability of excitation is related to the absorption coefficient of the material which, in turn, determines the penetration length of the photon l_a. The excitation migrates from molecule to molecule over a distance l_{ex} without dissociation to free charge carriers. It is assumed, for the moment, that there is no space-charge region near the metal-semiconductor interface (Fig. II.26) or, equivalently, that its extent within the semiconductor (l_{sc}) is negligible compared to the other characteristic distances. A fraction $\dfrac{l_{ex}}{l_a + l_{ex}}$ of the "excitons" can diffuse up to the front surface where they undergo chemical changes. Near the electrode, the excited state may dissociate by ejecting one electron (or hole) into the metallic electrode. This charge transfer process may be characterized by the rate constant k_{CT}. The excited state may alternatively exchange the excess energy with the metal via a non-radiative energy transfer with rate constant k_{ET}. The rates of these two quenching processes do not have the same dependence on the distance between the excited state and the electrode. Energy transfer arises through a long-range dipolar

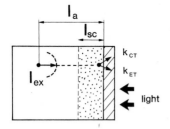

Fig. II.26. Schematic representation of the photovoltaic effect for a Schottky contact (localized states picture).

l_{sc}: space-charge-region depth
l_a: light penetration depth
l_{ex}: exciton diffusion length
k_{CT}: charge transfer to the electrode (rate constant)
k_{ET}: energy transfer to the electrode (rate constant)

mechanism with a $1/r^3$ distance dependence. Fluorescence quenching studies [118, 119] show that for an interface anthracene-aluminum, the exciton reaction probability at 30 Å from the junction is still 90% that of the probability at the junction itself. The charge-transfer process, on the other hand, will mostly occur through a tunneling mechanism. If d is the distance of the exciton from the interface then [118]:

$$k_{CT} = k_{CT,0} \cdot \exp\left(-\frac{d}{d_0}\right) \tag{II.98}$$

The critical distance d_0 is given by:

$$d_0 = \frac{\hbar}{2(2m^* \cdot \Delta E)^{1/2}} \tag{II.99}$$

where m^* is the effective electron mass and ΔE is the energy gap between the tunneling electron and the conduction level within the barrier. The exponential dependence implies a very rapid fall-off with the distance of the probability of charge transfer.

Various other mechanisms of decay of the excited state are possible in the bulk. These processes may be brought together as a single term:

$$k_Q = \sum_i k_i \tag{II.100}$$

k_i are the characteristic rate constants for the various decay processes and k_Q is the corresponding overall quenching rate constant.

The fraction $\dfrac{l_{ex}}{l_a + l_{ex}}$ of the "excitons" reaches the interface where a given amount of them may undergo a surface reaction. The fraction of excitons undergoing any surface reaction is given by:

$$\left[1 + \frac{l_{ex}}{a_c \cdot k_Q \cdot \tau_\infty}\right]^{-1} \tag{II.101}$$

a_c: lattice parameter
τ_∞: intrinsic exciton lifetime

The photocurrent may then be written as [118–121]:

$$I_{ph} = e \cdot \Phi_0 \cdot \left[1 + \frac{l_{ex}}{a_c \cdot k_Q \cdot \tau_\infty}\right]^{-1} \left[\frac{l_{ex}}{l_a + l_{ex}}\right] \cdot \eta_{CT} \cdot \varphi_{CT} \tag{II.102}$$

η_{CT}: probability that a charge carrier is formed when an exciton reacts
φ_{CT}: probability that the charge carrier generated contributes to the photocurrent
Φ_0: incident photon flux

η_{CT} is strongly dependent on the distance between the exciton and the metallic electrode; the expression for the tunneling mechanism case has been given previously.

Equation II.102 does not take into account the presence of a space-charge region. It has been seen that this region may arise from the ionization of impurities (usual Schottky barrier), from the carriers injected from the electrode (insulator-metal contact) or from a high trap density in the vicinity of the interface. In all these cases, the mechanisms of exciton dissociation near the interface are far more complicated and various other processes intervene. Consider an organic molecular crystal A "doped" with an accepting impurity I_A. The classical space-charge region may be written schematically as:

$$A, I_A^-/\text{metal}$$

The photovoltaic effect under irradiation may then be written as:

$$A, I_A^-/\text{metal} \xrightarrow{h\nu} A, I_A/\text{metal} (+e^-) \tag{II.103}$$

There is a question as to whether the photon directly excites the electron from I_A to the metallic electrode or whether some intermediates are formed:

$$A, I_A^- \xrightarrow{h\nu} A^*, I_A^- \tag{II.104}$$

$$A^*, I_A^- \to A^+, I_A^{2-}/\text{metal} \to A^+, I_A^-/\text{metal} (+e^-) \tag{II.105}$$

This latter mechanism has been postulated in the case of metallophthalocyanines doped with chloranil. Absorption may also occur in the charge-transfer band of the pair A, I_A such that:

$$A, I_A \xrightarrow{h\nu_{CT}} A^+, I_A^- \tag{II.106}$$

or

$$A^+, I_A^- \xrightarrow{h\nu_{CT}} A, I_A \tag{II.107}$$

Electron transfer from or to the metallic electrode via these photoexcited charge-transfer complexes is then possible:

$$A^+, I_A^-/\text{metal} \to A^+, I_A/\text{metal} (+e^-) \tag{II.108}$$

The ability of the photons to detrap the charge carriers is strongly dependent upon the electric field effective in the space-charge region. Thus, the probability of photo-stimulated electron transfer should also show a strong dependence on the distance from the interface.

A few chemical mechanisms arising in doped materials have been discussed above. In the case of pure insulator/metal contacts more quantitative predictions may be given. No donors or acceptors are present in the molecular material and space charges near the interfaces arise entirely from carriers injected by the electrode. Intrinsic generation of charge carriers in the dark can be ignored. It has been seen that photo-excited molecules need further energy to dissociate into free charge carriers because of the strong coulombic interactions within the primary ion pair. The model consequently assumes that "intrinsic photo-generation" of charge carriers across the band gap can be ignored [122]. The electrodes are assumed to inject a monopolar carrier (hole or electron) into the insulator. In the dark, the injection rate of carriers from the electrodes is taken to be constant. The photoinjection is assumed to occur at a rate proportional to the incident light intensity. The detailed physico-chemical mechanism of photoinjection need not be specified in this model. Some care must, however, be taken in the formulation. Direct photoinjection, which may be represented as:

$$A^*/\text{metal} \longrightarrow A^+/\text{metal} \, (+e^-) \tag{II.109}$$

is indeed a fairly exceptional mechanism [123, 124], and most of the processes attributed to direct photoinjection have lately been shown to be, in fact, photodetrapping processes:

$$e^- \text{ trapped} \overset{h\nu}{\frown} /\text{metal} \tag{II.110}$$

However, in this model, the only hypothesis concerning the photoinjection is that its rate is proportional to light, and no assumptions are made about the chemical mechanism. The photovoltage developed under irradiation is proportional to the ratio of the dark to the photoinduced injection rates at the interface. If the incident photon flux is Φ_o, it may be demonstrated that the photovoltage V_{ph} is given by [122]:

$$V_{ph} = \left(\frac{kT}{e}\right) \cdot \ln \left[\frac{1 + a\Phi_o}{1 + b\Phi_o}\right] \tag{II.111}$$

This formula takes into account the absorption of light at the front electrode (term a) and at the back electrode (term b), where a and b are the ratios of photo-injection quantum efficiency (per incident photon) to dark injection rate. An insulator containing no ionizable acceptors or donors may therefore show a photovoltage under irradiation. It is generally fairly difficult to settle whether the photovoltaic effect arises from this present mechanism or whether it arises from the presence of a space-charge region produced by ionized impurities. It will be seen later that capacitance measurements can prove helpful in deciding on the mechanism.

c Effect of Surface States

In the previous models, the properties of the surface have not been distinguished from the properties of the bulk. However, it is well known that molecules or atoms at the surface are often drastically different from their fully coordinated homologue in the bulk. In the band model, the crystal surface leads to additional levels within the band gap called "surface states". The wave functions corresponding to these levels are localized in the vicinity of the surface and do not extend into the semi-conductor. Molecular crystals are less sensitive to surface states than covalently bonded semiconductors. In the case of molecular crystals, the position of the surface states may be estimated by considering the energy of polarization of a charge generated at the surface or in the bulk. The difference is an indication of the depth of the surface level inside the band gap.

Surface states may induce a space-charge region at an interface with air. Consider an n-type semiconductor having accepting surface states, neutral when empty and negatively charged when filled (Fig. II.27). At equilibrium, the surface becomes negatively charged whilst a space-charge region constituted of positively ionized donors extends into the semiconductor. The position of the Fermi level is determined by the number of surface states. If, at equilibrium, all the acceptor states are filled, the Fermi level lies above the surface-state level. On the other hand, if the acceptor states are not all ionized at equilibrium, the Fermi level rises until it coincides with the surface-state energy. The Fermi level is "pinned". The electron transfer from the donors to the surface states induces a band bending, $e\Phi_{surf}$, Φ_{surf} being the surface potential. The width of the space-charge region l_{sc} is related to the bulk donor density N_D and to the density of surface acceptors N_{surf} through [96]:

$$l_{sc} = \frac{N_{surf}}{N_D} \tag{II.112}$$

$$\Phi_{surf} = \frac{2\pi \cdot e \cdot N_{surf}^2}{\varkappa_s \cdot N_D} \tag{II.113}$$

$\varkappa_s$: dielectric constant of the semiconductor

The effect of surface states is also important for classical Schottky contacts. Consider a junction between an n-type semiconductor and a metal whose work function Φ_M

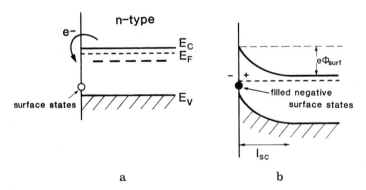

a   b

Fig. II.27. Band diagrams of an n-type semiconductor with acceptor surface states.
a before equilibrium
b after equilibrium.
The energy bands are "bent" upward by an amount $e\Phi_{surf}$ where Φ_{surf} is the surface potential. (After Ref. [96])

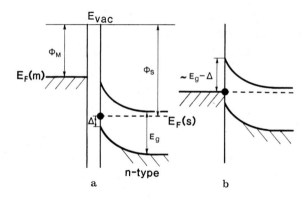

Fig. II.28. Band diagram of a Schottky contact between an n-type semiconductor and a metal. The work function of the metal Φ_M is smaller than the work function of the semiconductor Φ_S. The presence of surface states "pinned" the Fermi level at an energy Δ above the valence-band edge.
a before equilibrium with the metal
b after equilibrium

is smaller than the semiconductor work function Φ_S. An ohmic contact should be obtained. However, surface states may induce a pinning of the Fermi level. Assume that E_F is "pinned" at an energy Δ above the valence-band edge (Fig. II.28). After equalization of the Fermi levels in the metal and in the semiconductor, an energy barrier to the flow of electrons from the metal to the semiconductor still exists. The energy barrier is approximately equal to $E_g - \Delta$ where E_g is the band gap of the semiconductor. What should have been an ohmic contact now becomes rectifying [96].

d Characterization of Junctions by the Capacitance Method

Electrical determinations using alternating current (ac) are irreplaceable for studying the transport properties of molecular semiconductors, ac determinations permit the simultaneous measurement of both the resistive and the capacitive components of an electrical circuit. Figure II.29 shows a system composed of two resistances R_1 and R_2 and two capacitors C_1 and C_2. Such a network may be transformed mathematically into either of two equivalent circuits: the parallel circuit characterized by R_p and C_p and the series circuit characterized by C_{ser} and R_{ser}. The basic parameters R_1, R_2, C_1, and C_2 do not depend on the frequency of the applied ac current but R_p, C_p, C_{ser}, and R_{ser} are all frequency dependent. The equivalent resistances and capacitances are directly correlated with the measured in-phase and out-of-phase currents. The complex admittance of the parallel equivalent circuit is given by:

$$G_p + i\omega C_p \tag{II.114}$$

with

$$G_p = \frac{1}{R_p} \tag{II.115}$$

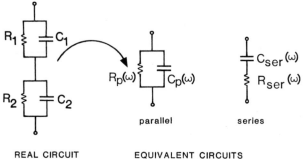

parallel   series

REAL CIRCUIT   EQUIVALENT CIRCUITS

Fig. II.29. Equivalent network analysis of a system composed of two resistances and two capacitances

ac measurements allow the determination of the "intrinsic" properties of materials even when monocrystals are not available [125]. Powdered molecular semiconductors may be characterized by four parameters: R_{cont} and C_{cont} represent the resistance and the capacitance of the contacts between the microparticles and R_{bulk} and C_{bulk} are the intrinsic parameters of the material, within the microcrystallites. At high frequencies, the higher resistances are "short-circuited" by the capacitances, R_p and C_p tend towards R_1 and C_1 when $R_2 > R_1$ and $C_2 \gg C_1$. At low frequencies, R_p and C_p tend towards $R_1 + R_2$ and C_2 (Fig. II.30). By varying the frequency it is possible to measure all the parameters.

The capacitance method is widely used for studying Schottky junctions [126–128].

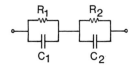

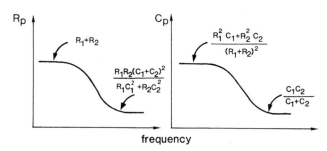

frequency

Fig. II.30. Frequency dependence of the equivalent parallel resistance R_p and of the equivalent parallel capacitance C_p for the illustrated electrical network.
$R_2 > R_1$
$C_2 \gg C_1$

In the simplest case, when surface states and trapping processes are ignored and for wide-band semiconductors, basic junction parameters may be readily calculated from capacitance determinations. The number of charges in the depletion region, Q_{sc}, is proportional to the number of ionized donors N_d and to the depth of the space-charge layer l_{sc}, for an n-type semiconductor:

$$Q_{sc} = q \cdot N_d \cdot l_{sc} \tag{II.116}$$

By definition, the capacitance is given by:

$$C = \frac{\partial Q_{sc}}{\partial V_a} \tag{II.117}$$

V_a: applied voltage

For an abrupt junction it may be demonstrated that [95]:

$$l_{sc} = \sqrt{\frac{2\varkappa_s}{q \cdot N_d} (V_{bi} \pm V_a)} \tag{II.118}$$

$$C = \sqrt{\frac{q \cdot \varkappa_s \cdot N_d}{2(V_{bi} \pm V_a)}} \tag{II.119}$$

$\varkappa_s$: dielectric constant of the semiconductor
N_d: density of ionized donors
V_{bi}: built-in potential

The ($\pm$) signs come from the reverse and forward-bias conditions. By plotting $1/C^2$ vs V, it is possible to calculate the density of the ionized impurities and the built-in potential:

$$\boxed{\frac{d(l/C^2)}{(-dV)} = \frac{2}{q \cdot \varkappa_s \cdot N_d}} \tag{II.120}$$

It is worth pointing out that the larger the dielectric constant of the material, the more the space-charge region extends into the semiconductor. These equations become rather more complex if surface states or traps are taken into account [126–128].

The previous equations are not valid for narrow-band semiconductors or for insulators. Molecular materials are generally best described as doped insulators. In metal/doped insulator/metal devices, Schottky barriers may occur at the metal-insulator interfaces (Fig. II.31) [128]. The Schottky barriers are formed through the electron flow taking place from the donor impurities of the insulator to the electrode. The concentration of impurities is directly related to the depth of the space-charge layer (Table II.3). At a concentration of impurities of 10^{17} cm^{-3}, the depth of the space-charge region is 1800 Å, assuming a dielectric constant of 15 and a built-in potential of 2 V [128]. Such a density represents 1 impurity for 10^4 molecules in the

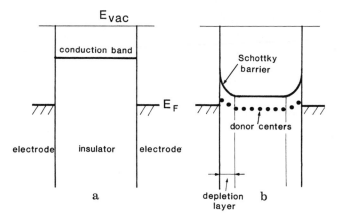

Fig. II.31. Energy-band diagrams of metal-insulator-metal system.
a: undoped insulator
b: doped insulator with Schottky barriers at the interfaces (donor impurities). (After Ref. [128])

case of a typical molecular crystal. This state of purity is rarely reached in organic chemistry.

The equivalent electrical circuit of a metal/doped insulator/metal device is shown in Fig. II.32. The space-charge region is represented only by a capacitance C_j, the corresponding resistive term is neglected. The bulk doped insulator is characterized by R_b and C_b. In the model used, the magnitude of R_b is determined by the number of charge carriers thermally excited from donor centers. Such a mechanism implies that most of the carriers are trapped, free charge carriers are generated by thermal detrapping. Within these constraints, R_b is given by [128]:

$$R_b = R_o \cdot \exp\left(\frac{\Delta E_{act}}{kT}\right) \qquad (II.121)$$

R_o is a resistive parameter depending on the mobility of the charge carriers, the film thickness, the temperature, and the densities of donors and acceptors, ΔE_{act} is the activation energy needed to liberate the charge carriers from the traps. The determinations of C_p and G_p as a function of the temperature and the frequency allows the measurement of all the basic parameters of the system. At high temperature, the impurities are ionized in the bulk doped insulator (between the two space-charge regions) and the bulk resistance R_b becomes sufficiently low to shunt the capacitance

Table II.3 Depth of the space-charge region (l_{sc}) for several values of the density of impurities (N_d), $\varkappa_s = 15$, $V_{bi} = 2$ V. (From Ref. [128]).

N_d (cm^{-3})	10^{15}	10^{17}	10^{19}	10^{21}
l_{sc} (Å)	18,000	1800	180	18

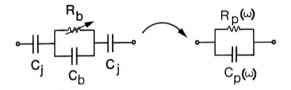

Fig. II.32. Electrical circuit corresponding to a metal/doped insulator/metal system.
R_b: bulk resistance;
C_b: bulk capacitance;
C_j: capacitance of the junction;

$$R_b = R_o \cdot \exp\left(\frac{E_{act}}{kT}\right)$$

at low frequency. At high temperature and low frequency, the measured capacitance depends only on the capacitance of the junction giving:

$$C_p \sim \frac{C_j}{2} \tag{II.122}$$

The capacitance is independent of the overall insulator thickness. The determination of C_j allows the calculation of the depth of the space-charge regions and the density of the ionized donors. At very low temperatures, very few impurities are ionized, R_b therefore becomes very large. The bulk capacitance C_b does contribute to the overall capacitance of the system and, at low frequency, C_p is given by the geometrical capacitance:

$$C_p \sim \left[\frac{2}{C_j} + \frac{1}{C_b}\right]^{-1} \tag{II.123}$$

The equivalent capacitance, this time, depends on the thickness of the insulator. At high frequency, the same expression for C_p is obtained whatever the temperature. At low frequency, the temperature at which the capacitance begins to vary with the temperature is directly related to the activation energy ΔE_{act}, providing a way of determining this value [128].

The various characteristics of Schottky contacts are fairly easily obtained from ac studies. However, the equivalent electrical networks which should represent the devices are very rarely clearly defined. The presence of oxide layers on the electrodes may, for example, entirely negate most of the previous interpretations. The reality of the equivalent circuit used must therefore be checked from the chemical point of view before elaborating any model of junction formation.

III Metallophthalocyanines

Phthalocyanines (from the Greek naphtha and *cyanide*, rockoil and dark blue) are dyes that have been known for many years. The first phthalocyanine was produced accidentally in 1907 during a study of the properties of 1,2-cyanobenzamide. On heating an alcoholic solution of the benzamide a highly insoluble blue product precipitated [129]. Twenty years later, a copper phthalocyanine was obtained during an attempted preparation of 1,2-dicyanobenzene from dibromobenzene and CuCN [130]. But it was Linstead and coworkers in the 1930's who fitted these earlier observations into a systematic scheme showing that a vast range of phthalocyanines were all based on the one structure shown in Fig. III.1 [131]. A huge number of different metallophthalocyanines (abbreviated as PcM) has been produced and studied, with a concomitantly large literature of several thousand publications. Apart from their intrinsic use as dyes the PcM's show a number of special properties which account for the great interest they have always aroused.

 i) They are easily crystallized and sublimed, resulting in materials of a purity (10^{14}–10^{16} traps per cm^3) exceptional in organic chemistry. Only a few anthracene-type crystals may be purified to a similar extent.

 ii) They show an exceptional thermal and chemical stability. In air PcM's undergo no noticeable degradation up to 400–500 °C and in vacuum most complexes do not decompose below 900 °C [132]. Strong acids (conc. H_2SO_4) or strong bases do not affect them. Only with very strong oxidizing agents (dichromate or ceric salts) can the molecules be broken down to phthalimide or phthalic acid [131, 133].

 iii) They show remarkable optical properties. The conjugated π system, containing 18 electrons in the macrocyclic ring, leads to very intense absorption bands at 400 nm and 700 nm with extinction coefficients of the order of $2 \cdot 10^5$ in solution.

 iv) They provide an astonishingly versatile chemical system. As shown in Fig. III.2, elements from groups I_A to V_B can all combine with the phthalocyanine ring

Fig. III.1. Molecular structure of metallophthalocyanines. Abbreviation used: PcM

I_a	II_a	III_a	IV_a	V_a	VI_a	VII_a	VIII			I_b	II_b	III_b	IV_b	V_b	VI_b	VII_b	0
H																	He
Li	Be											B	C	N	O	F	Ne
Na	Mg											Al	Si	P	S	Cl	Ar
K	Ca	Sc	Ti	V	Cr	Mn	Fe	Co	Ni	Cu	Zn	Ga	Ge	As	Se	Br	Kr
Rb	Sr	Y	Zr	Nb	Mo	Tc	Ru	Rh	Pd	Ag	Cd	In	Sn	Sb	Te	I	Xe
Cs	Ba	La	Hf	Ta	W	Re	Os	Ir	Pt	Au	Hg	Tl	Pb	Bi	Po	At	Rn
Fr	Ra	Ac															

Lanthanides 1f	Ce	Pr	Nd	Pm	Sm	Eu	Gd	Tb	Dy	Ho	Er	Tm	Yb	Lu
Actinides 5f	Th	Pa	U	Np	Pu	Am	Cm	Bk	Cf	Es	Fm	Md	No	Lw

Fig. III.2. The various metallophthalocyanines known. Approximately 70 derivatives have been synthesized.
Outlined italic: PcM synthesized before 1965; Boldface: after 1965

and more than 70 different PcM's are known. The nature of the sequestered metal ion has a profound influence on the physico-chemical properties of the PcM. For example, the oxido-reduction behavior of the macrocyclic ring or the nature of the photochemical excited state may be altered drastically by changing the metal ion in the complex.
v) By varying substituents on the ring the range of properties of the PcM's may be expanded even further. An infinite number of variations is possible with PcM's as the theme.

Monographs on the general properties of metallophthalocyanines are available in the literature [136–139].

III.1 Syntheses and Physico-Chemical Properties

a Syntheses

The methods of synthesis originally described by Linstead are still the most widely used for preparing metallophthalocyanines [131, 133–135]. Many PcM's are obtained in one step by reacting 1,2-dicyanobenzene, or a closely related derivative, with a finely dispersed metal (Fig. III.3). The reaction carried out at 250–300 °C, is exothermic. The reactants are usually heated in high-boiling-point solvents (trichlorobenzene, chloronaphthalene, etc.). The metal may be replaced by the

corresponding salt, but the additional anion, at high temperature, commonly decomposes to give side reactions. The use of chloride derivatives frequently leads to
halogenation of the aromatic nuclei. Isoindole and benzamide derivatives may be
used instead of dicyanobenzene as starting materials [136–139]. Most of the PcM's
may be purified by repeated sublimations in a stream of nitrogen (7 torr, 400–500 °C).
Large needle-shaped crystals are thus obtained. The overall yields for the synthesis
of PcM's vary from 20% to 90% depending on the metal used. PcLi$_2$ may be used as
an intermediate for the formation of other metal derivatives by simply reacting it
at room temperature with the corresponding metal salt dissolved in absolute alcohols:

$$PcLi_2 + MX_2 \rightarrow PcM + 2\,LiX \qquad\qquad (III.1)$$

PcLi$_2$ itself is formed in high yield by reacting 1,2-dicyanobenzene in alcohol with
two equivalents of lithium amylate or n-propanolate [140, 141]. PcLi$_2$ is one of the
only metallophthalocyanines which is soluble in cold organic solvents such as absolute
ethanol, amyl alcohol, or acetone; the PcM's formed during the displacement reactions
precipitate from the reaction medium and may be easily separated. The first non-
metallic phthalocyanine complex has recently been synthesized [142, 507] from PcH$_2$
by the displacement reaction:

$$PcH_2 + PBr_3 \xrightarrow{\text{Pyridine}} PcP^{III}Br \xrightarrow{\text{air}} PcP^V \qquad\qquad (III.2)$$

Phthalocyanine complexes of lanthanide (4f elements) and actinide (5f elements)
ions may be prepared [143–148]. Both types are prepared by reacting 1,2-dicyano-
benzene with the tetraiododerivative of the metal. These complexes are most often
dimeric with the metal "sandwiched" between the two phthalocyanine rings. Lanthanide ions are normally in the +III state, the dimer is therefore a neutral molecular
radical. Rare-earth-element derivatives have also been prepared starting from PcLi$_2$
[508]. The first dimeric complexes of metallophthalocyanines were described in 1936
by Linstead et al. [135]. The stannic dimer was prepared by a displacement reaction:

$$PcSnCl_2 + PcNa_2 \rightarrow Pc_2Sn \qquad\qquad (III.3)$$

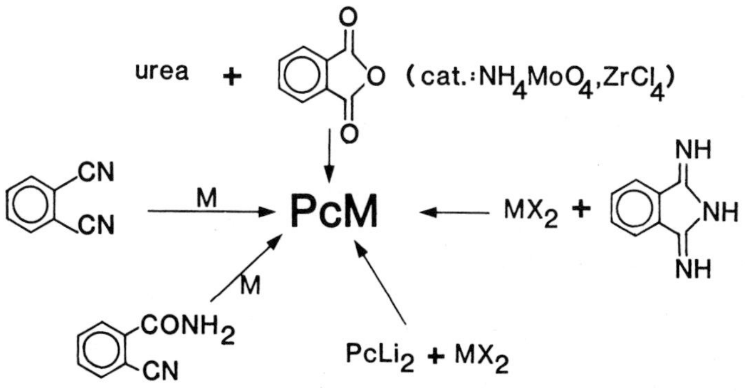

Fig. III.3. Main chemical pathways used to synthesize metallophthalocyanines

PcM's may be synthesized from a mixture of phthalic anhydride and urea in the presence of a catalyst (NH_4MoO_4 or $ZrCl_4$). PcCu, for example, has been prepared in excellent yield (>70%) from a mixture of phthalic anhydride, urea, boric acid, and copper chloride using ammonium molybdate as the catalyst [139, 149].

Metallophthalocyanines are generally very difficult to demetallate; only a few metal complexes (Li_2, Mg, Ag, Pb, Sb, Mn, Sn, Hg) are decomposed when dissolved in concentrated sulfuric acid. The complexes of Cu, Zn, Co, Ni, Pt, Pd, Al, Ga, In resist such a treatment. PcH_2 may be prepared directly in high yields (75–90%) by moderate heating (130 °C) of a solution of 1,2-dicyanobenzene in alkylalkanolamine [150].

An electrosynthesis of PcM's has recently been performed [151]. 1,2-dicyanobenzene is electrolyzed at −1.6 V (vs SCE) in a two-compartment cell, the phthalocyanine is formed in the cathodic region. The PcM's are probably formed by a mechanism involving the anion radical of 1,2-dicyanobenzene. Another original method of synthesis of PcM's consists in bombarding the unmetallated derivative PcH_2 with positive ions (In^+, $InCl^+$); the corresponding metal complex is thus formed [152, 153]. Extraction of radiometal nuclides from reactor irradiated PcM's has also been described [509].

The solubility of PcM's is increased by introducing peripheral substituents ($-CF_3$, $-CH_2OEt$, $-CH_2O\Phi$) on the aromatic rings [154–158]. The metallophthalocyanines are synthesized from the corresponding substituted dicyano-derivative. Tetranitro-, -chloro, -amino, and -hydroxy phthalocyanines have been prepared from the substituted phthalic acid derivatives in the presence of urea and ammonium molybdate in nitrobenzene at 180–190 °C [139]. Water-soluble PcM's may be obtained in the

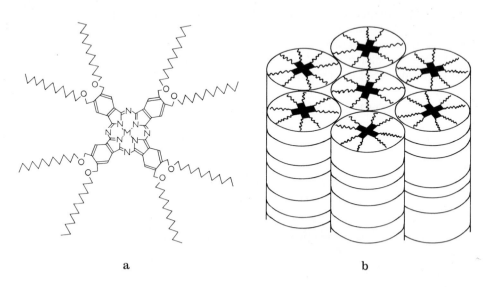

a   b

Fig. III.4. Liquid crystalline mesophases obtained from octasubstituted metallophthalocyanines.
a: Derivative used.
b: Type of liquid crystals obtained: the phthalocyanines are stacked in columns isolated from each other by the molten paraffinic chains (Discotic structure). (After Ref. [160])

same way by using 4-sulfophthalic acid [139, 159]. Phthalocyanines substituted by long paraffinic chains have recently been described. These derivatives form liquid crystals in which macrocyclic rings are stacked in columns, the different columns form a hexagonal array (Fig. III.4) [160]. The phthalocyanine subunits are super-posed, surrounded by the molten hydrocarbon chains. These compounds are highly soluble, often as aggregates, in conventional solvents.

The transport properties of organic materials are highly dependent upon their state of purity. Surprisingly, only a few studies have been devoted to the characteriza-tion of the impurities contained in PcM's. Trace metal content has been determined by plasma atomic emission [161]. The amount of Na, Ca, Fe, Cu . . . has been shown to vary with the purification procedures. Extraction of the PcM's with hot dimethyl-formamide leads to a significant decrease in the concentration of metals such as Na, Mg, or Ca. Reprecipitation from concentrated sulfuric acid reduces the content of non-transition elements, but increases the concentration of transition metals in PcH_2 [161].

μ-Bridged dimeric phthalocyanines are also known. The first dimeric species of the type PcM-L-MPc was synthesized early on by Linstead et al. [135]; it was a complex of formula PcAl-O-AlPc with a μ-oxo-bridge linked to $PcAl^{III}$ subunits.

Numerous dimeric, trimeric, and tetrameric systems have since been described: $[PcMn]_2O$, 2Py [162–164], $[PcGeO]_2X$ [165], [PcAlOSiPcOAlPc]X [166], $[PcSiO]_3X$ [165], $[PcSiO]_4X$ [167], $[PcSiO]_5X$ [165], where X stands for various oxysilane deriva-tives. The μ-oxo-manganese (III) derivative is obtained by heating PcMn in pyridine in the presence of air [163]. The silicon oligomers are obtained in a two-step synthesis according to:

$$PcSiCl_2 \xrightarrow{OH^-} PcSi(OH)_2 + 2Cl^- \qquad (III.4)$$

$$PcSi(OH)_2 + PcSi(OH)\left[OSi(Me)(OSiMe_3)_2\right] \xrightarrow{300\,°C} [PcSiO]_n \quad (III.5)$$

(n = 2–5) oligomers

The oligomers are soluble in most organic solvents. They are purified by chromato-graphy over alumina; petroleum ether-benzene mixtures are used as eluents [165].

The molecular structure of two of the μ-oxo dimers has been determined. In $[PcMn]_2O$, the two phthalocyanine rings are parallel and staggered by 49° with respect to each other [163]. The manganese is in its +III state. The Mn-O-Mn angle is 178°; the Mn-O distance is slightly shorter (1.71 Å) than would be predicted for a single covalent bond. $[PcSiO]_3X$ also shows a linear PcSi-O-SiPc arrangement with an interplanar spacing of 3.324 Å [168, 169], slightly shorter than the previous one. Polymers of similar structure may also be formed: $[PcSiO]_x$ [170–173], $[PcGeO]_x$ [174–176], $[PcSnO]_x$ [175]. Polymerization occurs by heating the corresponding dihydroxy monomer $PcM(OH)_2$ under vacuum at 300°–400 °C. $PcM(OH)_2$ may also be heated in nitrobenzene at 175 °C, in which case the degree of polymerization is approximately 11, as compared with around 100 for the polymerization under vacuum [176]. The ring-to-ring distance in the polymers has been estimated from X-ray powder measurements [175, 177]. The interplanar spacing is closely related to the ionic radii of the complexed ions: $[PcSiO]_x$ 3.32 Å; $[PcGeO]_x$ 3.50 Å; $[PcSnO]_x$ 3.83 Å. Silicon polymers are exceptionally stable; they are unaffected, for example, by treatments in aqueous HF at 100 °C, aqueous NaOH (2M) at reflux or concentrated sulfuric acid at room temperature [175, 177].

Other bridging ligands may be used to link the phthalocyanine subunits. Examples are provided by [PcAlF]$_x$ [178], [PcGaF]$_x$ [178], and $\left[\text{PcFe N} \diagup \diagdown \text{N} \right]_x$ [179, 180].

The fluoro derivatives are obtained by treating the corresponding hydroxo compounds with concentrated HF. High-purity polymers are obtained by vacuum sublimation (10^{-3} torr; 540 °C, M = Al; 430 °C, M = Ga) [177]. The X-ray crystal structure of PcGaF has been described recently [181]; the gallium ion achieves an octahedral coordination through bonding to the four nitrogen atoms of the Pc and two equidistant fluorine atoms. The Ga-F distance is 1.936 Å. The phthalocyanine rings are eclipsed while all other collinearly stacked structures show a staggered conformation. The dianion of acetylene $^-C \equiv C^-$ has been proposed as bridging ligand. On the basis of MO calculations, it has been shown that this group should provide an efficient conjugation path for the electrons [182, 183].

Two-dimensional polymers may also be prepared. Tetracyanobenzene is the privileged starting material for that purpose (Fig. III.5). Polyphthalocyanines have been known for many years [184–189]. The materials, however, were very poorly characterized: subsequent analysis has demonstrated that the synthetic procedures reported give rise to impure compounds. In most cases only monomeric phthalocyanine derivatives are produced, and when a polymeric material is obtained, it is quite impure [190]. Improved methods of synthesis have been reported more recently [191, 192, 510]. A gas-phase reaction between tetracyanobenzene and metal sheets (Cu, Ni, Co, Fe, Ti) has been described [193]. It has been claimed that such a procedure directly gives a polymeric phthalocyanine coating on the metal.

Complexes containing more than four subunits of phthalonitrile per atom of metal have been synthesized. "Superphthalocyanines" containing six phthalonitrile

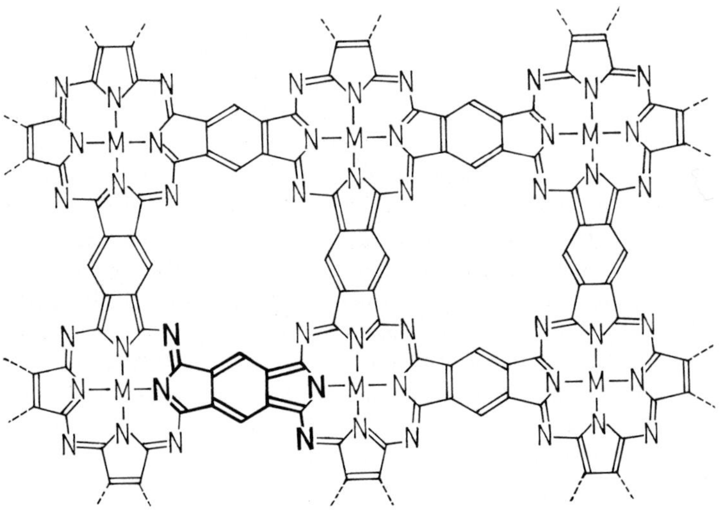

Fig. III.5. Polyphthalocyanines formed from tetracyanobenzene

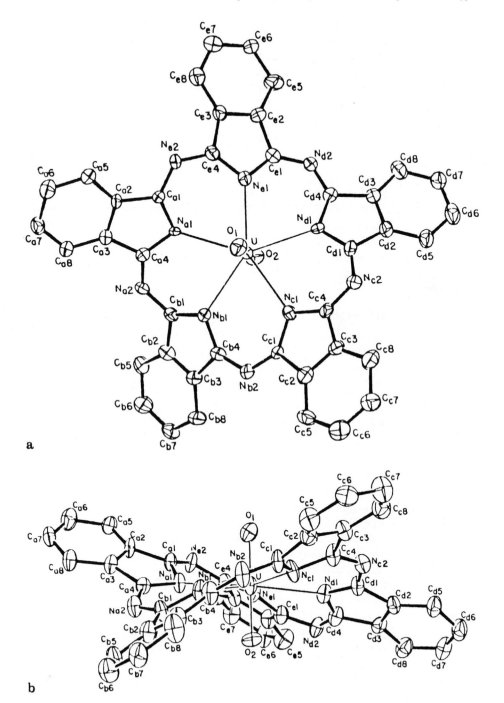

Fig. III.6. Crystal structure of an uranyl superphthalocyanine (super PcUO$_2$).
a: View perpendicular to the mean plane.
b: View approximately normal to the axis of the uranyl group. [Reproduced with permission of (145)]

moieties have been prepared by heating phthalonitrile at 160–170 °C with the corresponding metallic salts (Cu^{2+}, Co^{2+}, Ni^{2+}) [194, 195]. A pentameric uranyl superphthalocyanine complex has also been described [145, 196] (Fig. III.6). The equatorial coordination of UO_2^{2+} is achieved by five nitrogen atoms; the macrocyclic ligand is severely distorted from planarity. The uranyl superphthalocyanine reacts with transition metal salts according to:

$$\text{super } PcUO_2 + MX_2 \rightarrow PcM + UO_2X_2 + \quad\quad\quad \quad\quad\quad (III.6)$$

The superphthalocyanine is decomposed into metallophthalocyanine, and the excess phthalonitrile molecule is released. Super-$PcUO_2$ shows a strong absorption band in the near IR (log ε = 4.82 at 914 nm), considerably shifted away from the usual metallophthalocyanine bands. For comparison, Pc_2U has a maximum absorption wavelength at 643 nm [139].

Other dicyano-aromatic derivatives may be used instead of phthalonitrile. Naphthalocyanines have been synthesized from dicyanonaphthalene and iron dipivaloyl-methane [197].

b Structure and Morphology

Since 1935, the structures of most of the PcM's have been determined from X-ray diffraction measurements [198–201]. Three polymorphic forms are known, they are designated by the letters α, β, and x (Fig. III.7). Large-size monocrystals are, in most cases, of the β-type. They are generally grown by sublimation under a stream of nitrogen (7 torr) at a temperature of 400–500 °C. The crystals are needle-shaped, typically 1 cm long, 0.1 cm wide, and 0.01 cm thick. PcM's crystallize in a base-centered monoclinic lattice. The large area surfaces are (001) faces, the needle direction is the b axis (Fig. III.8). The phthalocyanines of Mn, Be, Fe, Co, Ni, Cu, and H_2 are isomorphous, they differ only by slight variations in the angle between the b axis and the perpendicular of the phthalocyanine ring (Table III.1) [205]. PcPt and PcCr, however, exhibit monocrystals belonging to the α-crystalline form [201, 206, 207]; the angle with the b axis is then considerably reduced ($\sim 20°$). PcM's form polycrystalline films of the α type [208–212] when evaporated under vacuum (10^{-5}–10^{-6} torr) onto a substrate maintained at room temperature. The x modification is obtained from the α form by neat milling [204]. The domains of stability of each crystalline form have been determined [202, 213–218]. PcH_2 sublimed in high vacuum onto a substrate cooled to liquid nitrogen temperature gives amorphous films [215]. No long-range order exists in the material, though this does not preclude short-range ordering. The crystallization to the α-form occurs between 50 °C and 140 °C, and this phase is stable up to 207 °C where the β form is formed [214, 215]. The α-crystalline form is directly obtained if the PcM is sublimed at a pressure less than 50 torr onto a substrate held at room temperature [211]. High-resolution electron microscopy studies have shown that vacuum-deposited polycrystalline films of PcZn present three different polymorphs derived from the α form [511].

The infrared spectra of the α and β forms are significantly different allowing the determination of the kinetic of the α to β transition. In an alcohol atmosphere the corresponding rate constants are changed [202]. The nature of the complexed metal

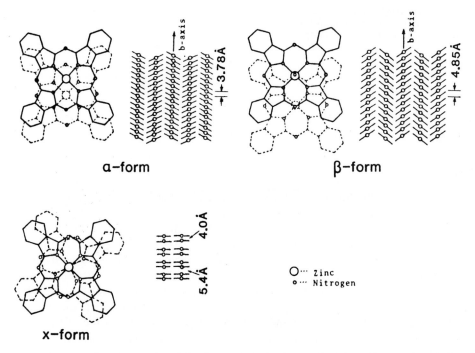

Fig. III.7. Schematic representation of the three main molecular stackings found for metallophthalo-cyanines. (After Ref. [202, 203, 204])

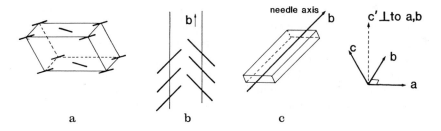

Fig. III.8. Basic parameters of the lattice of β-PcH$_2$.
a: Unit cell
b: Stacking in the ab plane
c: Needle axis

as well as the state of purity also have an effect on the α → β transition. The different polymorphs contain different amounts of impurities [217, 218]. It will be shown later on that this is of the utmost importance in determining the electrical properties of the various crystalline forms.

Epitaxial orientation of the PcM's has been produced. It is achieved by vacuum sublimation onto cleavage faces of muscovite [219] or onto (100) and (111) faces of copper [220].

Table III.1 Angles between the b axes and the normals to the phthalocyanine planes. (From Ref. [205]).

Compounds		
PcMn	116 K	47.9
	295 K	47.7
PcFe		47.3
PcCo		47.3
PcCu		46.5
PcZn		48.4
PcH_2		45.7

In most metallophthalocyanines the macrocyclic ring is planar to within 0.3 Å. The symmetry of the molecule is approximately D_{4h}. The metal-nitrogen bond is between 1.8 and 2.0 Å for most of the square planar metal complexes; this is shorter than for the corresponding porphyrin homologues by 0.07 ± 0.03 Å [221].6-coordinate tetragonal and 5-coordinate square pyramidal complexes are possible by axial ligation of various molecules [512] (Fig. III.9). Several molecular structures of such derivatives have been determined: $PcSnCl_2$ [222], $PcCo(picoline)_2$ [223], PcFe-(picoline)$_2$ [224], PcOs(CO)(pyridine) [225] (Fig. III.10). In this latter, the osmium ion is octahedrally coordinated and is slightly displaced by 0.15 Å out of the plane of the phthalocyanine towards the carbonyl group. The distances between the osmium ion and the nitrogen atom of the phthalocyanine ring are not equivalent (1.978 Å, 2.034 Å, 1.983 Å, 2.027 Å); the Os^{II}-N(pyridine) distance (2.202 Å) is significantly larger.

Out-of-plane metal complexes may be formed when the size of the metallic ion greatly exceeds the size of the cavity available. Pb^{2+}, which has an ionic radius of 1.20 Å, forms out-of-plane complexes with the phthalocyanine ligand [226, 227]. Most of the divalent transition metals have ionic radii of the order of 0.7–0.8 Å and can form in-plane complexes.

Two crystalline forms of PcPb are known (Fig. III.11). In the monoclinic form, PcPb molecules are packed in columns, the Pb atoms forming a one-dimensional chain. In the triclinic form, two independent molecular columns exist in the crystal.

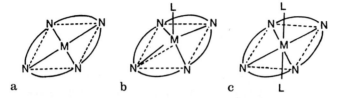

Fig. III.9. Schematized representation of the various coordination geometries which are possible for metallophthalocyanines.

a: 4-coordinate square planar
b: 5-coordinate square pyramidal
c: 6-coordinate tetragonal. (After Ref. [221])

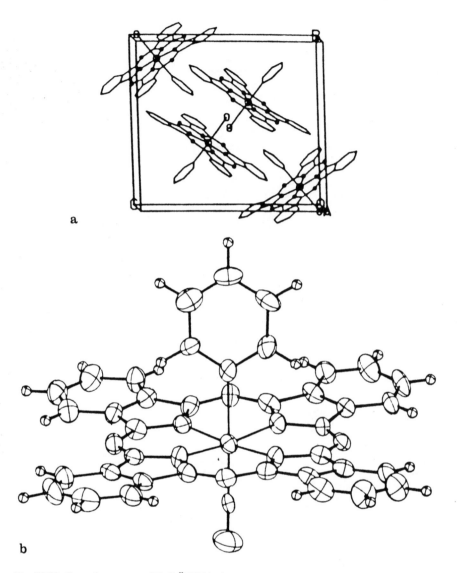

Fig. III.10. Crystal structure of PcOs^{II}(CO)(py).
a: Packing in the unit cell
b: Individual component. [Reproduced with permission of (225)]

The triclinic form has an electrical conductivity which is about 10^8 lower than the monoclinic one [227]. The Pb-N distance is different in the two-crystalline forms, 2.36 Å for the triclinic and 2.21 Å for the monoclinic form. Another monoclinic modification has been described [513, 514].

Precise determinations of the bond lengths in PcM's have been achieved by low-temperature X-ray diffraction [205] and neutron diffraction [228, 515] allowing the estimation of the extent of the π delocalization. Neutron diffraction studies at 4.3 K

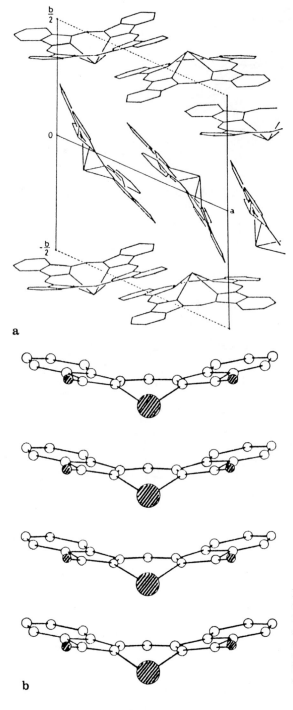

a

b

Fig. III.11. Crystal structures of the triclinic and monoclinic forms of PcPb.

a: Triclinic form; projection onto the (001) plane. [Reproduced with permission of (227)].

b: Monoclinic form. [Reproduced with permission of (226)]

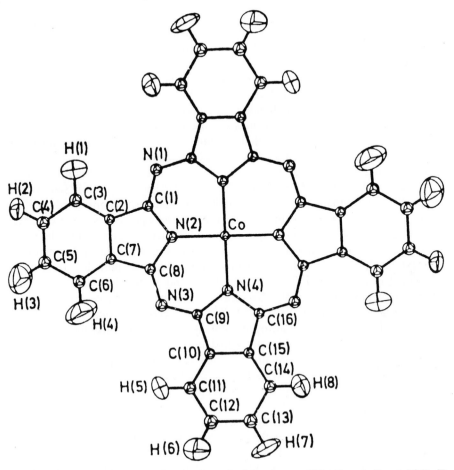

Fig. III.12. Molecular structure of PcCo determined by neutron-diffraction studies at 4.3 K. [Reproduced with permission of (228)].

Co-N(2)	1.908(2)	N(4)-C(9)	1.377(5)
Co-N(4)	1.915(2)	N(4)-C(16)	1.379(1)
N(1)-C(1)	1.312(4)	C(1)-C(2)	1.459(5)
N(1)-C(16')	1.321(4)	C(2)-C(3)	1.394(2)
N(2)-C(1)	1.382(1)	C(2)-C(7)	1.400(4)
N(2)-C(8)	1.380(5)	C(3)-C(4)	1.401(5)
N(3)-C(8)	1.321(2)	C(4)-C(5)	1.405(4)
N(3)-C(9)	1.320(3)	C(5)-C(6)	1.391(2)
C(6)-C(7)	1.392(5)	H(2)-C(4)	1.089(4)
C(7)-C(8)	1.448(3)	H(3)-C(5)	1.056(9)
C(9)-C(10)	1.446(3)	H(4)-C(6)	1.081(7)
C(10)-C(11)	1.391(5)	H(5)-C(11)	1.081(7)
C(10)-C(15)	1.394(4)	H(6)-C(12)	1.072(9)
C(11)-C(12)	1.394(2)	H(7)-C(13)	1.094(3)
C(12)-C(13)	1.412(4)	H(8)-C(14)	1.080(7)
C(13)-C(14)	1.389(5)	Co ... N(1)	3.384(2)
C(14)-C(15)	1.395(2)	Co ... N(3)	3.340(3)
C(15)-C(16)	1.443(4)	N(2) ... N(4)	2.733(2)
H(1)-C(3)	1.075(7)	N(2) ... N(4')	2.672(4)

avoid the effects of asphericity of the valence electrons that are significant in X-ray diffraction data. In the case of PcCo [228], the cobalt ion coordinates to a rigorously planar array of nitrogen atoms, with Co-N bond lengths of 1.908 Å and 1.915 Å (Fig. III.12). The C-N bond lengths of the isoindole moieties [C(1)-N(2), 1.382 Å and C(8)-N(2), 1.380 Å] differ only by 0.002 Å. The C-N bonds of the bridging nitrogens [C(1)-N(1), 1.312 Å, C(8)-N(3), 1.321 Å] are 0.07 Å shorter than those of the isoindole moieties. The C-C bonds of the aromatic rings vary from 1.391 Å to 1.405 Å. By considering all the previous distances, it may be seen that the extent of π delocalization in the whole phthalocyanine ring is very high. By comparison, the expected difference between a single and a double bond is approximately 0.2 Å.

The molecular structures of various dimeric phthalocyanines obtained with 4f and 5f elements have been determined [139, 229–231]. The metal ion is sandwiched between the two phthalocyanine rings; it is simultaneously coordinated to both central nitrogens of the two subunits. In Pc_2Nd^{III}, the eight Nd-N bond distances vary from 2.39 to 2.49 Å, one of the phthalocyanine macrocycles is slightly deformed towards the neodymium ion while the other is planar (Fig. III.13) [231]. The existence of an extraacidic hydrogen has been postulated. The two phthalocyanine rings are oriented in a staggered configuration. Pc_2U and Pc_2Sn show the same sandwich-type structure. In Pc_2U, the distance between the two planes defined by the four pyrrole nitrogen atoms is 2.80 Å [229, 230]. The macrocyclic rings are not planar. If the plane of the four pyrrole nitrogen atoms is taken as reference, the carbon and nitrogen atoms belonging to the 16-membered aromatic core are at distances varying between 0.09 Å and 0.28 Å from the reference plane, the peripheral aza-bridging atoms vary between 0.1–0.35 Å (Fig. III.14). The uranium ion is equidistant from all the pyrrole nitrogen atoms, the U-N distance is 2.43 Å. The phthalocyanine rings are staggered by 37°.

Beside the solid phases previously described, substituted PcM's may also form liquid crystals by heating [160] or they may from aggregates in solution [232–238]. The tetrasulfonic derivatives of PcM's are soluble in water and form aggregates if the concentration exceeds 10^{-5} M [232–238]. Dimerization constants are in the range 10^5–10^7 M^{-1}. The size of the aggregates is not known with certainty, but the largest particles contain approximately 20 phthalocyanine molecules [236–239]. ESR studies on aggregates of paramagnetic metal complexes allowed the determination of the metal to metal distance [232]. The values found (4.3–4.5 Å) are significantly larger than the Van der Waals distances for aromatic derivatives (3.4 Å) and a face-to-face configuration of the phthalocyanine moieties is improbable. Tetraoctadecyl-sulfonamido-phthalocyanine complexes also form aggregates in nonpolar solvents (240, 241). These solvents may be classified according to their ability of giving rise to aggregates, the following order is found: CCl_4 > benzene > toluene > chloroform > dioxane > tetrahydrofuran.

c Spectroscopic Properties

The symmetry of the orbitals involved in the complexation of a transition-metal ion by a phthalocyanine macrocycle is shown in Fig. III.15. σ coordination of nitrogen lone pairs directed toward the center of the ring occurs with the central metal atom to form in-plane σ orbitals. Interaction of the metal orbitals with nitrogen $p_π$ orbitals

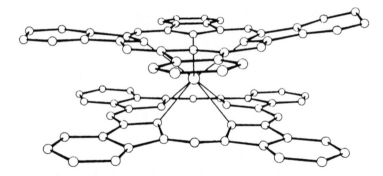

Fig. III.13. Molecular structure of Pc_2Nd^{III}. [Reproduced with permission of (231)]

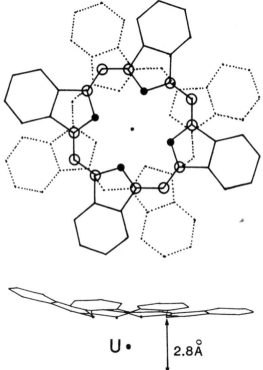

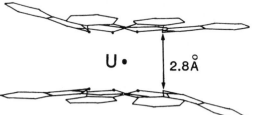

Fig. III.14. Molecular structure of Pc_2U. Full circles: reference plane; the four pyrrole nitrogen atoms.
Open circles: carbon and nitrogen atoms belonging to the 16-membered aromatic ring. (After Ref. [230])

gives rise to perpendicular-to-the-plane overlaps. The macrocyclic ligand, through the σ orbitals, is clearly a donor of electrons to the metal. The π orbitals of the ligand may act both as π donor or π acceptor.

Beside the pyrrolic nitrogen orbitals, which ensure the coordination of the central metal, the π orbitals of the ligand form a conjugated system. In a D_{4h} symmetry, the molecular orbitals have a_{2u}, a_{1u}, e_g, b_{2u}, or b_{1u} symmetries (Fig. III.16).

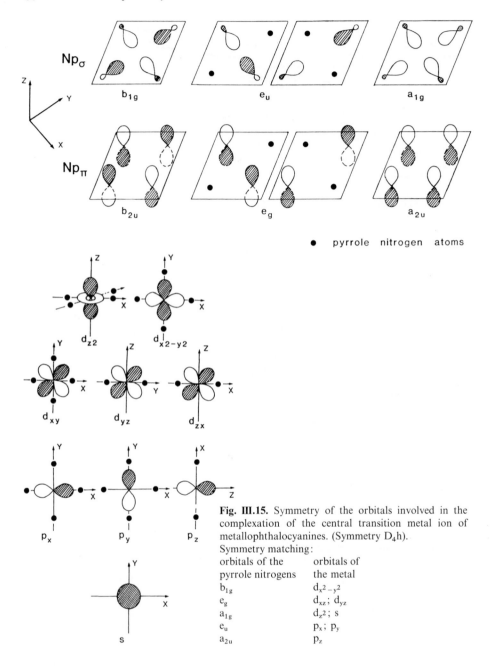

Fig. III.15. Symmetry of the orbitals involved in the complexation of the central transition metal ion of metallophthalocyanines. (Symmetry D_4h).
Symmetry matching:

orbitals of the pyrrole nitrogens	orbitals of the metal
b_{1g}	$d_{x^2-y^2}$
e_g	d_{xz}; d_{yz}
a_{1g}	d_{z^2}; s
e_u	p_x; p_y
a_{2u}	p_z

Experimentally, the absorption spectra of the usual metallophthalocyanines are composed of two very intense bands (log $\alpha_a \sim 5$) in the 300–400 nm region (Soret band) and in the 650–700 nm region (Q band) (Fig. III.17). Because of the high extinction coefficients (α_a), films as thin as 30 Å are visible to the naked eye [136, 242]. The peak around 700 nm is split into a doublet in the case of PcH_2. This suggests

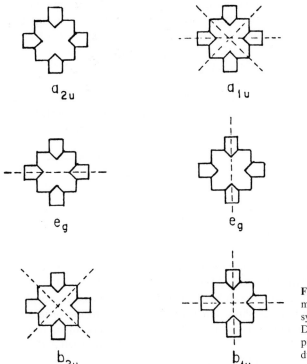

a_{2u}　　　　　a_{1u}

e_g　　　　　e_g

b_{2u}　　　　　b_{1u}

Fig. III.16. Molecular orbital symmetries for the conjugated π system of phthalocyanines in a D_{4h} symmetry. The nodes are pictured by dashed lines. [Reproduced with permission of (136)]

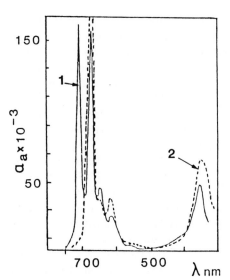

Fig. III.17. Electronic absorption spectra of PcH_2 (1) and PcZn (2) in solution. [Reproduced with permission of (136)]

that PcH_2 has a lower symmetry than metallophthalocyanines and that symmetrically degenerate bands are involved in the transition centered at 700 nm. In addition, most PcM's show three other bands in the UV region. They are labeled N (36,400 cm^{-1}, 275 nm), L (40,800 cm^{-1}, 245 nm), and C (47,600 cm^{-1}, 210 nm). The N band is the most sensitive to substitutions of the central metal atom [516].

Various models have been used to rationalize the optical properties of PcM's. The simplest one relies on the free-electron gas model, the π electrons are assumed to freely move in a closed ring-shaped path [136, 243, 244]. The energy difference between the highest occupied state and the lowest empty state is given by:

$$\Delta E = \frac{8mc}{h} \cdot \left(\frac{l_{cp}^2}{N_\pi} + 1 \right) \tag{III.7}$$

c: velocity of light
l_{cp}: circumference path
N_π: number of π electrons involved

More satisfactory are the various molecular orbital treatments [245–247]. Ab initio calculations [248, 249], extended Hückel (EH) calculations, and Pariser-Parr-Pople (PPP)-type calculations [250, 251] have all been carried out in the case of porphyrin complexes. The various approaches show some discrepancies as to the ordering of the orbitals and as to the amount of d character of the uppermost occupied orbitals. Only extended Hückel [246, 247] and PPP calculations [252, 253] have been performed for PcM's. A closely related ligand, the tetraazaporphyrin, has recently been studied by a non-empirical method [254]. The results obtained from the extended Hückel method are shown in Fig. III.18 for PcNi and PcCo [246]. Low-lying levels Np_σ of symmetry b_{2g}, e_u, a_{1g}, and b_{1g} corresponding to the bridge-nitrogen atoms appear beside the π orbitals of the macrocycle and the σ orbitals arising from the interaction of the inner nitrogen atoms with the central metal ion. In PcNi and PcCo complexes, the d_{xy} orbitals of the metal combine with the $b_{2g}(Np_\sigma)$ bridge-nitrogen orbitals to give the states $b_{2g}^*(d_{xy})$ and $b_{2g}(d_{xy})$. The two main absorption bands of PcM's are both ascribed as $\pi - \pi^*$ transitions from a_{1u} to e_g for the Q band and from a_{2u} to e_g for the Soret band. Both PPP and EH calculations give the same results for the nature of the transitions; however, only the PPP-CI calculations give good quantitative agreement with the experimental data. The EH model does nonetheless predict the low-lying transitions in the correct order, and it should also describe satisfactorily the transitions involving n and d orbitals [247]. For PcMg and PcZn [246], the Q band at 660 nm is a relatively pure $a_{1u}(\pi) \rightarrow e_g(\pi^*)$ transition. However, the top filled bridge-nitrogen Np_σ orbitals are expected to give rise to $e_u(Np_\sigma) \rightarrow e_g(\pi^*)$ transitions in about the same region. Such excitation bands have been observed in Shpol'skii matrices at approximately 1600 cm^{-1} above the Q band [255]. In the near UV region, PcNi shows unusual absorption bands. They are thought to arise from $d \rightarrow \pi^*$ transitions [246, 247]. This assignment is in agreement with a model relating the redox potentials of the complexes with the observed charge-transfer bands [256].

EH calculations usually overestimate the extent of the delocalization of the d orbitals on the macrocyclic ring. For PcCu, it is found that the metal centered orbital $b_{1g}(d_{x^2-y^2})$ is only 28 % localized on Cu; ESR experiments, on the contrary, indicate

LIGAND   METAL

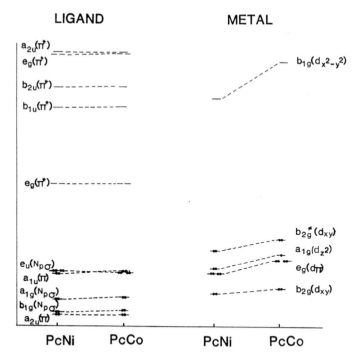

Fig. III.18. Highest filled and lowest empty molecular orbitals of PcNi and PcCo obtained from extended Hückel calculations. (After Ref. [246])

70% of localization on the metal [257, 258]. PPP calculations are in agreement with this last value [259]. Nonempirical Xα calculations on a related derivative, the copper complex of tetraazaporphyrin, demonstrate that the b_{1g} state is really the highest occupied level and that this level includes a considerable d-electron contribution (52%) [254].

Magnetic circular dichroism (MCD) measurements largely confirm the previous assignment of the absorption bands [260, 261]. MCD spectra of the α form of PcH_2, PcCo, PcNi, and PcCu have been carried out between 12,000 and 40,000 cm^{-1}. The Q band is assigned as a $^1A_{1g} \rightarrow {}^1E_u(\pi \rightarrow \pi^*)$ transition. The excited state ^1E_u can be factored into the molecular orbital configuration $(a'_{1u}e'^*_g)^1(e^*_g)^1$. The broad near UV band (B band or Soret band) is ascribed to a transition of the type $^1A_{1g} \rightarrow {}^1E_u$ ($\pi \rightarrow \pi^*$). There is some evidence that two ($\pi \rightarrow \pi^*$) transitions of similar intensities lie under the B band. The broadening might, however, be due also to unidentified (n $\rightarrow \pi^*$) transitions. MCD is particularly useful for studying charge-transfer (CT) bands of PcM's [262]. The band around 450 nm in $PcCo^{III}(CN)_2$ is commonly assigned as a CT band, but it has been shown to be, in fact, a transition arising from the phthalocyanine ring. Similarly, doubt is shed upon the CT nature of the band at 450 nm in $PcFe^{II}(CN)_2$. On the other hand, it has been confirmed that the transition centered around 470 nm in $PcCo^I$ is really a metal-to-ligand charge-transfer band.

Near infrared absorption peaks have been observed for PcCu [263, 264]. These bands are absent in PcH_2. These peaks are thought to arise from spin-forbidden singlet-triplet transitions. The strong spin-orbit coupling due to the heavy central metal ion relieves the spin forbiddenness of the transition. The observation of a phosphorescence band at 1120 nm for PcCu further supports this assignment [265, 266]. Inelastic electron tunneling spectroscopy has been used to detect such singlet-triplet transitions [267].

Important spectroscopic changes are induced on going from solution into the solid state. The relatively narrow Q band is transformed into a broad peak showing more or less clearly a splitting [268, 270]. According to Davydov theory, the amount of splitting is a measure of the interaction energy between molecules having different site symmetries [67]. The bands split into as many components as there are non-translationally equivalent molecules in the unit cell. The Davydov splitting varies from $1350 \ cm^{-1}$ for PcH_2 to $2230 \ cm^{-1}$ for PcZn [268]. There exists a correlation between the tendency of the central metal ion to form out-of-plane bonding and the magnitude of the Davydov splitting [268]. Davydov splitting is observed both for the Q and B (Soret) bands but not for the N band corresponding to higher energies [260]. The transformation of the α form to the β form also induces large differences in the transmission spectra [269].

Luminescence properties of PcM's have been extensively studied by Russian workers [271–275]. The spin multiplicity of the excited state is determined by the nature of the central metal ion and its effect on the magnitude of the spin-orbit coupling [276–278]. The luminescence properties of the most representative PcM's are shown in Table III.2. Light closed shell metallophthalocyanines exhibit high fluorescence yields (0.03 to 0.7) and are only weakly phosphorescent. Fluorescence lifetimes τ_f of PcAlCl, PcGaCl, and PcInCl follow the Stückler-Berg relationship relating the intensity of the absorption band to the fluorescence lifetime [279, 281]. For most of the PcM's, fluorescence lifetimes are of the order of a few nanoseconds [228, 279, 282, 283], for PcH_2 $\tau_f = 6.5$ ns, for PcMg $\tau_f = 7.2$ ns, and for PcZn $\tau_f = 3.8$ ns [284]. Phosphorescence lifetimes τ_p are strongly dependent upon the cation complexed. For lighter metals (Mg, Zn, Cd) τ_p is as long as 0.3 to 1.1 ms. For heavier metals, the phosphorescence quantum yield increases to the detriment of the fluorescence quantum yield, and there is a concomitant decrease in τ_p. The sum $\Phi_F + \Phi_T$, where Φ_F is the fluorescence quantum yield and Φ_T the triplet yield, is generally very close to unity. This indicates that the singlet excited state either reemits a photon to return to the ground state or is interconverted to the triplet state; but, no thermal degradation of the light energy occurs. It is known that for symmetry reasons, $S_1 \rightarrow S_0$ radiationless transitions are inefficient for rigid planar π systems [286, 287]. Porphyrin derivatives show the same behavior [284]. Metallophthalocyanines are, however, considerably more fluorescent than their porphyrin counterparts [285].

Phosphorescence quantum yields are in most cases extremely low, radiationless decay of the lowest excited state back to the ground state by thermal degradation seems to be the major de-excitation pathway for the triplet states. Figure III.19 shows the different relaxation mechanisms for the excited states of PcPd together with the corresponding time scales over which they are effective [277]. In this example, spin-orbit coupling splits the triplet state T_1 into three sublevels and intersystem crossing populates solely the highest triplet component. This latter does not radiate to the ground state, but in turn populates the lower-lying triplet sublevels. For sym-

Table III.2   Luminescence properties of metallophthalocyanines.

PcM M =	Fluo.[a] max. (nm)	Phospho.[b] max. (nm)	Fluo. lifetime (ns)	Phospho. lifetime (μs)	Φ_F	Φ_P	Φ_T	Ref.
H_2[c]	699				0.7			[278]
				140			0.14[d]	[280]
Mg[c]	683	1100		1000	0.6	$5 \cdot 10^{-6}$		[278]
Zn[c]	683	1092		1100	0.3	$1 \cdot 10^{-4}$		[278]
Cd[c]	692	1096		350	$3\text{–}8 \cdot 10^{-2}$	$2\text{–}4 \cdot 10^{-4}$		[278]
Cu[c]	?	1065		3	10^{-4}	$1 \cdot 10^{-3}$		[278]
				0.035[d]	0		0.7[d]	[280]
VO[c]	?	1140–1180		?	?	$5 \cdot 10^{-6}$		[278]
Pd[c]	663	990		25	$5 \cdot 10^{-4}$	$3 \cdot 10^{-3}$		[278]
Pt[c]	?	944		7	?	$1 \cdot 10^{-2}$		[278]
RhCl[c]	662	987		16	$1 \cdot 10^{-3}$	$2 \cdot 10^{-3}$		[277]
IrCl[c]		957		3.5		$8 \cdot 10^{-3}$		[277]
AlCl[e]			6.8		0.58		0.4	[279]
GaCl[e]			3.8		0.31		0.7	[279]
InCl[e]			(0.37)[f]		0.031		0.9	[279]

[a]: Position of the fluorescence maximum; [b]: position of the phosphorescence maximum; Φ_F: fluorescence quantum yield; Φ_P: phosphorescence quantum yield; Φ_T: triplet yield; [c]: in 1-chloronaphthalene at 77 K; [d]: in 1-chloronaphthalene; [e]: in 1-chloronaphthalene at room temperature; [f]: calculated value, see Ref. [279].

metry reasons, only two of the three triplet components may radiate to the ground state, and, consequently, at liquid helium temperatures, two different phosphorescence lifetimes are observed. The vibrational relaxation of the excited states is very rapid. The vibrational relaxation rate constants are $3.3 \cdot 10^{11} \text{ s}^{-1}$ for the singlet state and $10^{11}\text{–}10^{12} \text{ s}^{-1}$ for the triplet level. Most of the photoreactions of the excited state therefore occur from the lowest vibrational level. The rate constants of the nonradiative processes span a fairly large timescale from $1.7 \cdot 10^{11} \text{ s}^{-1}$ for the radiationless transitions from the singlet state to 10^4 s^{-1} for the nonradiative decays from the triplet levels.

The total luminescence yields increase with increasing nuclear charge within a group (PcPd; PcPt; PcRhCl; PcIrCl) (Table III.2). As spin-orbit coupling becomes stronger, both radiative and nonradiative rates increase, but the radiative rates increase faster. This results in higher phosphorescence and total luminescence yields [277]. PcNi and PcCo, on the other hand, do not show any luminescence. This has been attributed to the presence of low-lying energy states arising from metal d electrons. In PcNi and PcCo, the metal states lie below the triplet state of the macrocyclic ligand providing radiationless paths to the ground state [278].

The Stokes shifts are a measure of the geometrical distortions of the excited state as compared to the ground state. The Stokes shifts as measured by the difference in wavelength between the 0-0 fluorescence band and the 0-0 absorption band, are usually fairly small for PcM's [247, 278]; PcMg and PcZn show a shift of 5 nm, PcCd of 8 nm. Because of its lower symmetry, PcH_2 gives rise to two Q bands, Q_x and Q_y. Q_x, the lower energy peak, has an unusually small Stokes shift [247, 283,

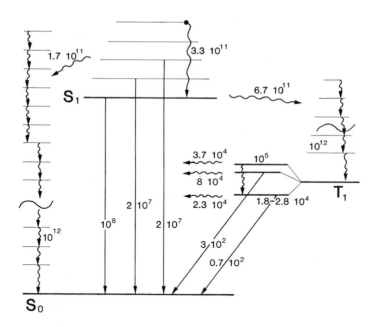

Fig. III.19. Decay pathways of PcPd in 1-chloronaphthalene at liquid helium temperature. The numbers are the corresponding rate constants in (second)$^{-1}$. Straight arrows represent radiative processes and wavy arrows radiationless decays. (After Ref. [277])

288]. It has been assumed that, on excitation, the phthalocyanine ring expands. In PcH_2 the protons can readjust themselves to the right geometry. Metal complexes, which are more rigidly bound, cannot undergo a similar readjustment.

Laser flash photolysis experiments have been carried out on very dilute solutions (10^{-6} M) of PcM's [280, 289]. At higher concentrations (10^{-5}–10^{-4} M), solutions of PcH_2 and PcCu give short-lived (100 ns) and long-lived (2.5 µs) transients. This has been interpreted as evidence for the formation of microcrystalline particles dispersed in the solution. These microcrystals would give rise to "excitons" characterized by the long-lived transient [280]. Studies of PcH_2 and PcZn in homogeneous solutions and in microemulsions also demonstrate the influence of aggregation on the decay modes of excited states of PcM's [289].

d Photoelectron Spectroscopy

X-ray Photoelectron Spectroscopy (XPS), also known under the name Electron Spectroscopy for Chemical Analysis (ESCA), is a powerful method for surface analysis. The sample is irradiated with monoenergetic soft X-rays, this irradiation produces a photoemission of electrons which are then analyzed in energy. The most common X-ray sources are the MgKα (1253.6 eV) and AlKα (1486.6 eV). The kinetic energy E_{KE} of the electrons photoemitted is given by:

$$E_{KE} = E_{hv} - E_{BE} - \Phi_{sp} \qquad (III.8)$$

E_{hv}: energy of the incident X-ray
E_{BE}: binding energy of the atomic orbital from which the electron originates
Φ_{sp}: spectrometer work function

XPS measurements allow the determination of a set of ionization energies, each corresponding to a particular orbital of the molecule under study. The Fermi level corresponds, by definition, to the zero binding energy. XPS therefore allows the location of E_F at interfaces of Schottky contacts, for example [290].

Thin films of PcM's have been extensively studied by XPS [291–297]. Two peaks in a ratio of 3:1 arise from the C(1s) orbitals [293, 295]. These two peaks have been attributed to the 24 aromatic carbons and to the 8 nitrogen-bonded carbons, respectively. The two C(1s) peaks of PcCu and PcH_2 coincide: this indicates that the complexation negligibly affects the chemical environment of the carbon atoms (Table III.3) [293, 295]. In contrast, the N(1s) peaks are shifted in going from PcH_2 to PcCu. PcH_2 shows two different N(1s) peaks, while PcCu demonstrates only one peak. In PcH_2, there are three chemically different types of nitrogen atoms: the two pyrrole nitrogen atoms, the two aza-nitrogen atoms coordinated to the two central hydrogens, and the four meso-bridged nitrogen atoms. The splitting of the N(1s) peak in the case of PcH_2 is a strong evidence in favour of the tight bonding of the hydrogens with the corresponding nitrogen atoms, the bridged-structure where every hydrogen is coordinated simultaneously to two nitrogen atoms is therefore highly improbable (Fig. III.20).

The N(1s) peaks are more sensitive to the nature of the metal ion complexed than those of the C(1s) (Table III.3) [293, 295]. The maximum shift in N(1s) peaks is, however, only 0.4 ± 0.1 eV. On the other hand, changes in the oxidation state of the central metal induce significant shifts in the XPS metal peak itself. In the tetraphenylporphyrin (TPP) series for example, the shift of the $Fe(2p_{3/2})$ peak between TPP $Fe^{III}Cl$ and TPP $Fe^{II}(piperidine)_2$ indicates that the net charge on the iron in the oxidized and reduced forms differ by about 0.44 electron [293].

XPS measurements at low binding energies (0–30 eV) allow the study of valence band electrons. In the 6–30 eV region all PcM's have approximately the same spectra

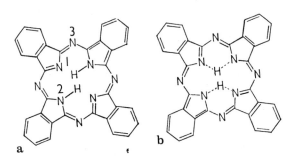

a   b

Fig. III.20. Possible molecular structures of PcH_2.
a: Bonded structure
b: Bridged structure
1: Pyrrole nitrogen atom
2: Aza-nitrogen atom
3: Meso-bridged nitrogen atom

Table III.3   Binding energies of core levels for metallophthalocyanines. (From Ref. [296]).

	Binding energies				
	$C(1s)^{a,c}$	$N(1s)^a$		Other core lines^b	
	(eV)	(eV)		(eV)	
PcH_2	284.8	398.9	400.3		
PcFe	284.7	399.1		$2p_{3/2}:709.2$	$2p_{1/2}:722.7$
PcCo	284.7	399.3		$2p_{3/2}:780.6$	$2p_{1/2}:796.5$
PcNi	284.8	399.5		$2p_{3/2}:856.3$	$2p_{1/2}:873.8$
PcCu	284.8	399.2		$2p_{3/2}:935.8$	$2p_{1/2}:955.8$
PcPt	284.9	399.4		$4f_{7/2}:77.3$	$4f_{5/2}:84.0$

^{a}: Data errors $\pm\,0.1$ eV; ^{b}: data errors $\pm\,0.4$ eV; ^{c}: signal arising from the 24 aromatic core C-atoms (lowest binding energy).

[296] (Fig. III.21 a). They are closely similar to the theoretical spectrum expected from a mixture of benzene and pyrrole (curve 3). The 2p-electron spectra of PcM's are therefore approximately a superposition of the spectra of the constituents of phthalocyanines, benzene, and pyrrole with almost no modifications. But XPS spectra differ considerably in the 0–6 eV range. There is apparently a close relationship between the intensity of the photoemission peaks and the number of d-electron states (Fig. III.22). Using a crystal field type approximation these peaks have been assigned to quasilocalized d orbitals [296].

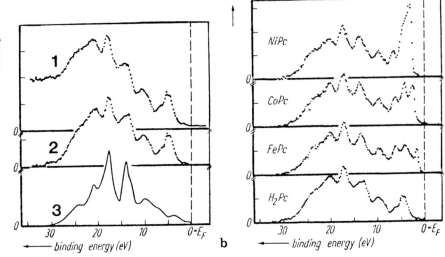

Fig. III.21. Photoelectron spectra of metallophthalocyanines in the valence-band region.
a 1: Raw data of PcH_2
   2: Same spectrum corrected for inelastic scattering
   3: Theoretical spectrum obtained by superposition of XPS spectra of pyrrole and benzene.
b: XPS spectra after correction for inelastic scattering for different transition metal complexes.
Irradiation: monochromatized AlKα rays (1487 eV)
Resolution: 0.55 eV. (After Ref. [296])

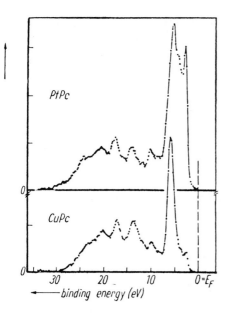

Fig. III.22. Photoelectron spectra of PcCu and PcPt after correction for inelastic scattering in the valence-band region. (Irradiation: monochromatized AlKα rays). (After Ref. [296])

Far UV radiation is energetic enough to ionize valence-band electrons and may be used instead of X-ray for studying the photoemission properties of materials. In the UV photoelectron spectroscopy (UPS), the most commonly used sources are He I (21.2 eV), He II (40.8 eV), and Ne I (16.8 eV). Results from UPS studies seem to contradict those obtained by XPS [298–300]. UPS spectra of PcH$_2$, PcFe, PcCu, and PcPt are surprisingly similar at low binding energies. The spectra are essentially those of the π electrons, although a weak contribution from the d electrons may be identified. PcCu and PcH$_2$ have been studied using a synchrotron source at 75 eV [301]. Although both spectra are very similar, the difference spectrum does show up effects arising from the Cu orbitals. The discrepancy between the XPS and UPS spectra may well be caused by the different cross sections of the orbitals towards radiations of different energies. Thus, the localized d orbitals interact more strongly with the high energy X-ray and the delocalized π electrons interact more strongly with the low energy UV radiation; hence, the emphasis is on d-orbital effects in XPS and on π-orbital effects in UPS.

It has not been possible to obtain a satisfactory agreement between the UPS data and the theoretical binding energies calculated by the extended Hückel method [299]. However, ab initio calculations and particularly the Xα local density method have been shown to be promising for describing the electronic structure of metallophthalocyanines and related compounds [300].

An energy reference is fairly difficult to find for the photoelectron energy distribution curves obtained by XPS or UPS. In the absence of charging effects, the zero binding energy of Fermi level of insulators and semiconductors may be at any position within the band gap depending on the sample purity [296]. An "external" reference may be provided by evaporating a gold-metal layer onto the PcM's thin films [298]. The Fermi energy of PcM's then varies strongly with the nature of M. It is found that

E_F (with respect to the vacuum level) is 3.8 eV for PcH_2, 4.1 eV for PcMg, and 4.3 eV for PcPb [298]. The Fermi level seems to move closer to the top of the valence band in a metal substituted phthalocyanine. It may be asked whether these changes in E_F arise from some intrinsic property of the materials or whether they only reflect a difference in purity between the various phthalocyanines. The central metal has a strong effect on the ability of PcM to bind O_2 or other impurities. O_2 acting as a dopant, the Fermi level will be correspondingly shifted. The variations in E_F previously observed could be due to different concentrations of oxygen in the various samples.

e Oxidation-Reduction Properties

Gaseous ionization potentials corresponding to the equilibrium:

$$(PcM)_g \rightleftarrows (PcM^+)_g + e^- \tag{III.9}$$

have been determined by electron impact measurements [302]. The nature of M has little or no effect on the gaseous ionization potentials of PcM's. Surface ionization potentials, defined as the energy required to move an electron from the crystal surface into the vacuum, have also been determined (Table III.4). The gaseous ionization potentials are remarkably independent of the nature of the central metal. The surface ionization potentials show a more pronounced variation. Finally, redox potentials determined in solution by cyclic voltammetry or polarography are strongly dependent upon the nature of the central metal (Table III.4). The oxidizing or reducing abilities of PcM's are therefore correlated to the solvation properties of the PcM's. The solvent dependence of phthalocyanine redox equilibria has been described [517]. It has been shown previously that the crystal ionization potential may be estimated from the oxidation potential in solution:

$$I_c \sim E_{1/2}^{ox} + 4.3 \tag{III.10}$$

Table III.4 Gaseous ionization potentials (I_g), crystal ionization potentials (I_c), surface ionization potentials (I_s), and oxidation potentials in solution ($E_{1/2}^{ox}$) of various metallophthalocyanines.

	I_g[a]	I_c[b]	I_s[c]	$E_{1/2}^{ox}$[d] vs SCE
	(V)	(V)	(V)	(V)
PcH_2	7.36	5.2	5.2	0.9
PcZn	7.37	4.98	5.00	0.68
PcCu	7.37	5.17	5.00	0.87
PcNi	7.45	5.28	4.95	0.98
PcCo	7.46	5.07		0.77
PcFe	7.22	4.69	4.95	0.39
PcMn	7.26	4.16		−0.14

Determined by electron impact measurement (vertical ionization potentials) (Ref. [302]); [b]: calculated from equation $I_c \sim E_{1/2}^{ox} + 4.3$ (see Eq. III.10); [c]: from References [292], [303]; [d]: for the references and conditions see Table III.5.

The values of I_c calculated from Eq. III.10 are compared in Table III.4 with the experimental surface ionization potential (I_s). The use of the redox potentials to estimate I_c clearly overestimates the influence of the central metal ion. However, considering the drastic approximations made, the agreement between the experimental and calculated values is fairly good.

An exhaustive review of the oxidation and reduction potentials published in the literature is presented in Table III.5. Only the first redox potentials corresponding to the removal or the addition of one electron are given. Depending on M, the oxidation and the reduction potentials vary over more than 1 V. In all cases, the metallophthalocyanines are more easily reduced than the porphyrin homologues by approximately 0.4 V. The substitution of the benzene moieties by sulfonate groups makes the reduction potentials more positive by over 0.1 V [310, 311].

Table III.5 Oxidation and reduction potentials of metallophthalocyanines determined by cyclic voltammetry or polarography.

PcM (Ionic Radius)	E^{ox} vs SCE	Cond.	Ref.	E^{red}	Cond.	Ref.
PcH$_2$	0.9		[26]	−1.10		[26]
	1.10	a	[304]	−0.66	g	[310]
P$_c^{2-}$(PRA$^+$)$_2^1$			[310]	−1.24		[310]
PcLi$_2$	0.14	b	[305]	−1.44	b	[305]
PcNa$_2$				−1.06		[26]
PcMg	0.65		[26]	−0.96		[26]
(0.66)	0.70	c	[306]	−0.95	c	[306]
	0.59	b	[305]	−0.91	g	[310]
PcBa	0.46	k	[519]	−0.49	k	[519]
(1.34)						
PcCu	0.87		[26]	−0.84	g	[310]
(0.72)	0.98	a	[304]			
	1.00	a	[307]			
PcAg	0.28	b	[305]	−1.38	b	[305]
(0.89)						
PcZn	0.68	a	[304]	−0.90		[26]
(0.74)	0.78	c	[306]	−0.80	c	[306]
	0.83	b	[305]	−0.89	g	[310]
	0.68	d	[308]	−0.89	d	[308]
	0.68	a	[307]			
PcCd	0.54		[26]	−1.17		[26]
(0.97)						
PcHg	0.25	j	[518]	−1.30	j	[518]
(1.10)						
PcCr	0.52	d	[308]	−1.35	h	[308]
(0.89)						
PcMn	−0.14	e	[308]	−1.02		[26]
(0.80)	0.01	d	[308]	−0.90	b	[305]
	−0.26	b	[305]	−0.77	d	[308]
	−0.11	a	[307]			
PcFe	0.39	e	[308]	−1.05		[26]
(0.74)	0.38	f	[309]	−1.07	i	[309]
	0.70	d	[308]	−0.56	e	[308]
	0.19	a	[304]			
	0.16	a	[307]			

Table III.5 (continued)

PcM (Ionic Radius)	E^{ox} vs SCE	Cond.	Ref.	E^{red}	Cond.	Ref.
PcCo	0.77	a	[304]	−0.55		[26]
(0.72)	0.80	c	[306]	−0.20	c	[306]
	0.22	d	[308]	−0.43	b	[305]
	0.77	a	[307]	−0.37	g	[310]
PcNi	0.98		[26]	−0.70		[26]
(0.69)	1.05	a	[304]	−0.85	g	[310]
	1.1	a	[307]			
PcAlX	0.91		[26]	−0.66		[26]
	1.15	c	[306]	−0.50	c	[306]
	1.00	b	[305]	−0.		
PcGaCl	0.86	j	[518]	−0.73	j	[518]
PcInCl	0.84		[26]	−0.65		[152]
PcSi(O-amyl)₂			[519]	−0.54	k	[519]
PcPb	0.67		[26]	−0.72		[26]
(1.20)	0.52	b	[305]	−0.94	b	[305]

Conditions: a: 0.1 M tetra-n-butylammonium perchlorate, 1 chloronaphthalene; b: tetra-n-hexyl-ammonium perchlorate, N,N-dimethylacetamide, Pt electrode, 25 °C; c : tetra-n-butylammonium perchlorate, N,N-dimethylacetamide; d: perchlorate, pyridine; e: perchlorate, dimethylformamide; f: tetramethylammonium perchlorate, N,N-dimethylacetamide; g: tetrapropylammonium perchlorate, dimethylformamide; h: perchlorate, dimethylsulfoxide; i: tetraethylammonium perchlorate, pyridine; j: tetraethylammonium perchlorate, dimethylformamide; k: dimethylformamide; l: N-propylammonium cation.

Oxidation of PcM's may occur by removing electrons either from the π orbitals of the macrocyclic ring or from the metal:

$$PcM \rightleftarrows PcM^{+} \qquad \text{oxidation of the metal} \qquad (III.11)$$

$$PcM \rightleftarrows Pc^{+}M \qquad \text{oxidation of the ring} \qquad (III.12)$$

ESR studies have demonstrated that both cases are possible, depending on M [304]. In PcFe and PcCo the oxidation occurs first at the metal center; for PcCu, PcZn, and PcH₂ the macrocyclic ring is oxidized first, and for PcNi both metal and ring oxidations take place simultaneously [304].

PcM's can accept up to four electrons in the reduction process [310, 312]. The electrons are added in most cases into the lowest empty antibonding π orbital of the macrocycle [221], but it is the metal which is preferably reduced in the cases of PcCo [221, 310, 312–317], PcFe [308, 309], and PcMn [312].

Chemical reagents may be used to oxidize or reduce PcM's. PcCo reacts with iodine to give PcCo^{III}I, though PcNi is not oxidized under the same conditions [312,

318]. The dilithium salt of benzophenone reacts with PcM's in THF to yield the anions $(PcM)^{n-}$ with n varying from 1 to 4 [314, 319–322]. Although reduced PcM's are fairly stable in solution under an inert atmosphere, oxidized PcM's are relatively unstable and cannot be kept in solution over long periods of time at room temperature. PcCr, PcFe, PcCo, and PcZn react at 60–80 °C with $SOCl_2$ in nitrobenzene to yield the corresponding cation radicals; PcNi, PcH_2, and PcVO are unaffected by this treatment [318, 323]. The reactivity of PcM's towards NO follows the same trend as the reactivity towards $SOCl_2$ [324]. PcM's, when treated with strong acids lead to mixtures of products. PcFe reacts with HCl at 100 °C to give several derivatives including probably $PcFe^{II}$, HCl, and $PcFe^{III}Cl$ [318].

f Electron Spin Resonance Measurements (ESR)

ESR investigations of paramagnetic metallophthalocyanines have long been carried out [325, 326]. The amount of σ and π bonding between the central metal ion and the orbitals of the ligating nitrogen atoms can be deduced from these studies. In the σ system, the macrocycle acts as a donor towards the metal while in the π system, the phthalocyanine ring may act either as a π donor or a π acceptor. In the case of PcCu for example, antibonding molecular orbitals may be constructed from Cu(3d) and (4s) atomic orbitals and from the nitrogen (2s) and (2p) atomic orbitals [257, 258, 327–329]. The following states are thus obtained:

σ bonding:
$$\Psi^*_{B_{1g}} = a_1 \cdot \Psi_{d_{x^2-y^2}} - \frac{a'_1}{2} \cdot \Psi_{\sigma_N} \qquad (III.13)$$

in plane π bonding:
$$\Psi^*_{B_{2g}} = a_2 \cdot \Psi_{d_{xy}} - \frac{a'_2}{2} \cdot \Psi_{p_{x,y}} \qquad (III.14)$$

out-of-plane π bonding:
$$\Psi_{Eg} = \begin{bmatrix} a_3 \cdot \Psi_{d_{xz}} - \dfrac{a'_3}{\sqrt{2}} \cdot \Psi_{p_z} \\[2ex] a_3 \cdot \Psi_{d_{yz}} - \dfrac{a'_3}{\sqrt{2}} \cdot \Psi_{p_z} \end{bmatrix} \qquad (III.15)$$

ESR studies on isotopically pure $PcCu^{63}$ diluted in PcH_2 furnish the best determinations of the bonding parameters [258]. Knowing the ^{14}N and Cu hyperfine coupling tensors, a_1 and a_2 can be estimated. The calculation of a_3 requires in addition the value of the d-d electronic transition energy $\Delta E(B_{1g} \rightarrow E_g)$. The bonding parameters are thus found to be: $a_1^2 = 0.75$, $a_2^2 = 1$, and $a_3^2 = 0.51$ or 0.60 [258]. The amount of mixing is therefore very large for the σ bonding and for the out-of-plane π bonding, but there is no in-plane π bonding.

ESR measurements on most of the other paramagnetic PcM's have been carried out (Review: [136]). PcCo [332–334] and PcVO [335], in particular, have been extensively studied. In the case of PcAg, the electrons of the $d_{x^2-y^2}$ metal orbital have been shown to spend 46% of their time on the macrocyclic ring [136].

From measurements of the dependence of the susceptibility of PcCu (obtained by integration of the ESR signals) as a function of temperature it is possible to evaluate the magnitude of the intermolecular interactions between equivalent paramagnetic sites [330, 331]. As expected from the 4.72 Å metal-to-metal distance in β-PcCu, the interaction energy (expressed as the exchange integral J) is small, with $J \sim 10\,cm^{-1}$ [329].

Static magnetic properties of PcM's also have been determined [335]. Weak ferromagnetic interactions have been detected in PcMn monocrystals [336]. The magnetic transition from the ferro- to the paramagnetic state occurs at 8.6 K. Magnetic susceptibility measurements and proton magnetic resonance confirm the ferromagnetic transition occurring in PcMn [337].

Reduced and oxidized PcM's also give rise to ESR signals. ESR spectra of PcM^{+} have been reported [318, 338–340]. Charge-transfer complexes between PcH_2 and iodine are obtained by treating crystals of phthalocyanine at 100–120 °C in an iodine atmosphere [341]. A signal probably due to PcM^{+} is observed with a g value close to that of the free-electron ($g = 2.0036$). The line width is 5.03 gauss. ESR studies of $PcNiI_{1.0}$ monocrystals also have been reported [342]. The spectrum consists of a single line with $g_{\parallel} = 2.0075$ and $g_{\perp} = 2.0007$. The spin density corresponds at room temperature to 0.14 spins per macrocycle. The ESR intensity remains essentially constant from room temperature to 180 K where it starts to decrease gently. The line widths are angle dependent ($\Gamma = 3$–7 G). In a closely related macrocycle doped with iodine, the line width has been interpreted as being controlled by spin-lattice relaxation via polaron motion [342–344].

The paramagnetic metallophthalocyanines, PcCo and PcMn, when dissolved in oleum no longer show the characteristic signals due to the unpaired d electrons. Instead, ESR peaks with unresolved hyperfine structures are observed. This has been attributed to the formation of biradicalar species of the type $Pc^{+}M^{+}$, where electrons have been removed both from the macrocycle and from the metal [339]; the electron remaining on the macrocycle is responsible of the ESR peaks.

The successive reductions of PcM's may be followed by ESR [311, 320, 322, 340]. The lowest empty orbital is of symmetry e_g, it is doubly degenerate and can accommodate four electrons. For $PcMg^{n-}$ with n = 0, 2, and 4, the species are diamagnetic and no signal is observed. For n = 1 and 3, fairly narrow single lines are observed very close to the free-electron g value [320, 345]. In contrast, the first paramagnetic species observed by ESR for PcCu and PcCo is the dinegative ion (n = 2) [320]. This is a consequence of the spin coupling of the additional electrons with the unpaired electrons of the metals. In most cases the reduction is ligand-centered and the added electrons are predominantly π type. However, in $PcCr^{-}$ the electron has been found to be 73% d type [322].

By comparing the previous ESR results with the ones obtained with the porphyrin homologues, it has been concluded that the degree of covalency of the metal to nitrogen bond is significantly greater for PcM's [221].

III.2 Dark Electrical Properties

a Energy Band Structures

The energy band structure of a few PcM's has been calculated using principally extended Hückel methods [346–350]. Most PcM's crystallize in monoclinic lattices with two molecules per unit cell. As would be expected from the molecular structure, the size of the interaction between the constituent molecules, as measured by the exchange integral, is larger along the b axis than along the other two axes. The energy band structure of PcH_2 has been calculated within the tight binding approximation using Hückel-type molecular orbitals [348]. Configuration interaction (CI) was taken into account. The highest occupied molecular orbital (HOMO) which belongs to the irreducible representation a_u forms the valence band. This level is separated by about 0.5 eV from the nearest state and the effect of configuration interaction may be neglected. CI is, on the other hand, important for the lowest unoccupied molecular orbitals (LUMO's) which form the conduction band. Several levels are in the same energy range and interaction between bands of the same symmetry must be considered. In an isolated PcH_2 molecule, the lowest empty orbitals b_{2g}^* and b_{3g}^* are separated by only 0.09 eV and there will thus be significant mixing of the two levels in the solid state [348]. The calculated bandwidths of PcH_2 (β form) are given in Table III.6. The reference directions are the three directions of the reciprocal lattice a^{-1}, b^{-1}, c^{-1} in **k** space. The bandwiths are, as expected, strongly dependent upon the direction considered, varying from a few meV for directions parallel to a^{-1} and c^{-1} axes, to 30–100 meV along the b^{-1} axis. The corresponding mobilities have been computed within the constant-isotropic relaxation time or the constant-isotropic free path approximations [348]. The use of one or the other of the approximations leads to significantly different values of the mobilities, but only relative values must be discussed. Moreover, whatever the approximation used, the mean free time of the charge carriers, τ_c, must be known to calculate the mobility. A lower limit of τ_c may be estimated from the uncertainty principle. The product of τ_c with the bandwidth W must be greater than $\hbar$, giving:

$$\tau_c > \frac{\hbar}{W} \qquad\qquad (III.16)$$

This equation must necessarily be obeyed in order for the band model to be meaningful. Using this approach, it is possible to settle lower limits on the mobility components (Table III.6). In the b^{-1} direction the mobility thus calculated is of the order of 0.4 to 1 cm^2/V · s. The anisotropy with respect to the other directions is 5–10. The value of 1 cm^2/V · s is the lowest limit of the mobility compatible with a band model. Consequently, electron and hole transports in the b^{-1} direction should be describable by a simple band model, whereas, in the directions parallel to the axis a^{-1} and c^{-1} of the crystal, the scattering mechanisms should become predominant and a localized state model is presumably preferable.

Table III.6 Bandwidths and corresponding mobilities for β-PcH$_2$ obtained from tight-binding calculations with configurations interactions. Hückel molecular orbitals are used.
Only the larger bandwidth as a function of **k** in each direction is given. The numbers in parentheses are for band structure with configurations interactions. (From Ref. [348]).

Band	Direction (reciprocal lattice)	Bandwidth (meV)	Mobility[a] (cm^2/V · s)
hole	a^{-1}	1.05	
	b^{-1}	29.76	0.412
	c^{-1}	0.86	0.097
electron	a^{-1}	3.57 (4.64)	
(b$_{2g}$*)	b^{-1}	101.55 (82.58)	1.096
	c^{-1}	1.56 (2.39)	0.103
electron	a^{-1}	1.69 (2.76)	
(b$_{3g}$*)	b^{-1}	28.04 (49.15)	0.418
	c^{-1}	4.79 (5.90)	0.204

[a] constant free time approximation, vibronic overlap = 1.0. The values indicated are lower limits compatible with the uncertainty relation.

The mobilities of the charge carriers in PcM rarely exceed 1 cm^2/V · s. By comparison, the electron mobilities are 1350 cm^2/V · s for monocrystalline silicon, 4500 cm^2/V · s for InP, 8000 cm^2/V · s for GaAs, 340 cm^2/V · s for CdS [351]:
inorganic semiconductors have mobilities more than three orders of magnitude larger than those found in PcM.
The electronic structure of PcH$_2$ and its anion radicals have been calculated both by the Pariser-Parr-Pople method and by the unrestricted SCF method [347]. They have been used as basis functions for the band calculation in the tight binding approximation. The bandwidths were calculated to be of the order of 10 meV for electrons and 1 meV for holes [347]. Calculations including the contribution of the core orbitals have also been performed [349]. In all cases, the bandwith of PcH$_2$ in the b^{-1} direction is approximately an order of magnitude larger than the thermal energy kT at room temperature. The intermolecular interactions may therefore dominate over the various electron-phonon scattering mechanisms. A thorough examination of the applicability of the band model to PcH$_2$ has been reported [352].
Energy-band structure calculations have been carried out on PcPt [346]. The 5d orbitals of platinum contribute to the hole energy band. The bandwidths for **k** ∥ a^{-1} and **k** ∥ b^{-1} directions are 1 meV and 200 meV for electrons and 40 meV and 3.6 eV for holes. On account of the contribution of the metal orbitals, the hole bands are much broader than those found for PcH$_2$.

b Electrical Properties: Intrinsic Case

The intrinsic electrical properties of PcM's are very rarely observed. In most cases, the measurements are plagued by the effects of the ambient. Drastic precautions must be taken to avoid the influence of absorbed gases. The time t$_{ab}$, in seconds, which is required for the adsorption of a monolayer of gas on the surface of a material at room temperature is given by [353]:

$$t_{ab} = \frac{U_o}{\mathscr{P}_{st}} \, 2.63 \cdot 10^{-21} (P_{O_2})^{-1} \qquad \qquad (III.17)$$

U_o: number of available sites (for a surface $U_o \sim 10^{14}/cm^2$)
$\mathscr{P}_{st}$: sticking probability

The sticking probability, for an organic material, is of the order of 10^{-4}. Taking this approximate value, it is possible to calculate that, at 10^{-5} torr, only 4 min are needed to form a monomolecular layer of gas on the organic material; at 10^{-7} torr, 7 hours are necessary. Consequently, even under high vacuum conditions the surface properties of materials may be altered by the ambient. This is particularly true for junction studies where the very surface of the materials determine the electrical behavior of the contacts.

The semiconducting potentialities of PcM's were recognized early on and much study has been devoted to their electrical properties. The first systematic measurements were carried out in the late forties by Eley and Vartanyan [17, 354–356]. The primary goal of these studies was to determine the thermal activation energy for conduction by studying the temperature dependence of the conductivity. In most cases, an exponential relationship was found:

$$\sigma = \sigma_o \cdot \exp\left(-\frac{\Delta E}{2kT}\right) \qquad \qquad (III.18)$$

σ_o is the "intrinsic" conductivity and ΔE has been assigned as being the "energy gap" of the semiconductor. In fact, the purity of the molecular semiconductors is seldom controlled enough to obtain the intrinsic properties of the material. The term ΔE thus represents, in most cases, the energy needed to liberate charge carriers from traps or to ionize levels within the band gap. A relationship between ΔE and the energy of the absorption band in the visible region had been found [270, 357–359]. It will be seen in the next sections that this correlation is almost entirely fortuitous and that no information concerning the band structure of the materials can be drawn from these observations.

The determinations of electrical properties of molecular materials must be, as far as possible, carried out on monocrystals under high vacuum conditions in order to obtain the most reliable results. The dark electrical properties of monocrystals of β-PcH$_2$ have been investigated at a reduced pressure of 10^{-7} torr [360]. The temperature dependence of the dark conductivity has been determined. The conduction process is shown to be intrinsic only at temperatures exceeding 140 °C; at lower temperatures the conduction is influenced by an electron trap level located 0.32 eV below the conduction band edge. *The energy gap in the intrinsic domain is 2.0 eV.* Previous studies on β-PcH$_2$ [361] or β-PcCu [362], are in agreement with this value of the energy band gap. Further confirmation has been obtained from photophysical determinations [367]. The quantum efficiency of the carrier generation process under illumination is temperature dependent, the corresponding activation energy is

0.21 eV. This energy represents the energy difference between the bottom of the electron band and the singlet excited state level. For PcH_2 the singlet level is at 1.77 eV above the valence band, the energy gap is thus $1.77 + 0.21 = 1.98$ eV. This value is consistent with the dark thermal activation energy found from the temperature dependence of the dark conductivity. The variation of the position of the Fermi level as a function of temperature is shown in Fig. III.23. At low temperature, the number of free charge carriers is smaller than the density of trapped electrons, the thermal release of carriers from donor centers must be taken into account; the Fermi level is shifted towards the bottom of the conduction band. At sufficiently high temperatures the number of free charge carriers increases, the conduction becomes intrinsic, and E_F lies approximately at the middle of the band gap. In most experiments, values significantly lower than 2.0 eV are obtained for the thermal activation energy, reflecting, in most cases, an extrinsic charge transport process. The effect of successive sublimations on the dark conductivity of monocrystals of β-PcH_2 best illustrates the paramount importance of the purity. Crystals grown three times show a conductivity of $5 \cdot 10^{-16}\ \Omega^{-1}\ cm^{-1}$ and an activation energy of 1.9 ± 0.06 eV, if they are grown only once the corresponding values are $2 \cdot 10^{-13}\ \Omega^{-1}\ cm^{-1}$ and 1.65 eV, respectively [361].

Numerous measurements of the mobility of the charge carriers in PcM's have been carried out. "Hall" and "drift" mobilities are distinguished according to the method used to determine them. In Hall measurements, an electric field is applied to a sample placed in a perpendicular magnetic field. The current is deviated generating a Hall voltage in a direction perpendicular to both the current and the magnetic field. The corresponding Hall voltage is related to the mobility and the concentration of the charge carriers (see Ref. [15], page 107). This method has been extensively used for studying PcM's [363–365]. However, Hall measurements data are sometimes difficult to interpret. Anomalous Hall effects have been observed in the case of systems with narrow bands. The influence of surface states and the corrections which must be made for anisotropic materials are often difficult to estimate. Moreover, the Hall effect in intrinsic semiconductors can be accurately measured only if the electron and hole mobilities are significantly different from one another.

Drift mobilities are determined by an entirely independent method. e^-/h^+ pairs are generated at the surface of the semiconductor by irradiation, whilst simultaneously an electric field is applied across the material. The mobile charge carriers drift under the influence of this field and the transit time t_t necessary for the charges to go from

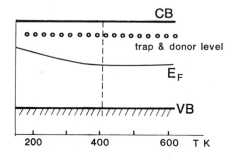

Fig. III.23. Variation of the Fermi level position in monocrystals of β-PcH_2 as a function of temperature. (After Ref. [363])

one side of the molecular semiconductor to the other is measured. The transit time is related to the corresponding drift mobility through:

$$t_t = \frac{d}{\mu_d \cdot \xi_a}$$

(III.19)

t_t: transit time
μ_d: drift mobility
d: thickness of the sample
ξ_a: electric field applied

This equation only applies in the absence of space-charge and trapping effects. The drift mobilities are therefore a measure of the overall transport properties of the materials as opposed to other methods which yield only local microscopic parameters. The values of the Hall and drift mobilities [361, 366–369] obtained for various PcM's are compared in Table III.7. For PcH$_2$, both methods give similar values for the mobilities, but, for PcCu, they differ by a factor of 10. However, it has been shown afterwards that the monocrystals of PcCu used for the Hall determinations contained a noticeable concentration of traps arising from the ambient [362]. With due attention to these problems, it seems, however, that metallated Pc's possess higher mobilities than PcH$_2$. In PcCu, the copper ion is axially coordinated to the aza-bridge atoms belonging to the macrocycles above and below it. This enhances the intermolecular coupling and, correspondingly, the mobility of the charge carriers.

The crystalline state has a drastic influence on the transport properties of PcM's. Thin films of PcH$_2$ demonstrate drift mobilities 10^2 to 10^3 lower than monocrystals. Structural disorders are expected to introduce only shallow trapping levels within the band gap. In polycrystalline materials, however, the number of displaced molecules is very large and deep traps may also be generated. Crystal imperfections in very thin films of metallophthalocyanines have been observed by means of high-resolution electron microscopy [520]. Planar defects and dislocations have been characterized.

Table III.7 Hall and drift mobilities of the charge carriers for metallophthalocyanines.

	Crystalline[a] state	Hall mobilities (cm^2/V · s)	Ref.	Drift mobilities (cm^2/V · s)	Ref.
β-PcH$_2$	SC[b]	0.1	[363]	$\mu_e \sim 1.2$; $\mu_h \sim 1.1$ (373 K)	[367]
		0.1–0.4	[364]	0.05–0.1	[369]
				$\mu_e \sim 0.43$–0.7;	
				$\mu_h \sim 0.24$–0.56 (295 K)	[361]
PcH$_2$	MC			10^{-2}–10^{-3}	[370]
β-PcCu	SC[b]	75	[363]	$\mu_e \sim 7$ (300 K)	[368]
		100	[365]		
PcCu	TF			10^{-3}–10^{-4}	[368]
PcPb	SC[b]			0.6–4.0	[369]

[a] SC: single crystal; MC: microcrystalline; TF: thin film. [b] For single crystals, the mobility is measured along the c'axis, which is perpendicular to the ab plane.

These latter may induce deep trap levels within the band gap. Thin films of PcM's are, on the other hand, more permeable to gases than monocrystals and they are consequently also less chemically pure. A model has been proposed to describe the electrical transport phenomena in "imperfect" crystals of PcCu [371]. The real crystal is approximated by a stacking of perfectly conducting disks, the perturbations brought about by the structural defects are computed knowing the way the molecule stacks break off within the crystal.

As to their dark electrical properties, PcM's behave as narrow band semiconductors, the bandwidth being of the order of the thermal energy kT. However, other descriptions are possible [10]. As usual in the case of molecular crystals, the applicability of the band model may be questioned. If the electronic exchange interactions dominate, the transport of the charge carriers is "coherent" and the band model is applicable; if electron-phonon interactions are predominant, the transport process is incoherent and a hopping model is preferable. The temperature dependence of the charge-carrier mobility permits an estimate of the magnitude and the nature of the electron-phonon interactions. It is even possible, in principle, to determine which of the phonon modes is coupled the most strongly with the electronic motion. The phonon modes are classified as two main types [372]:

— Lattice modes. These modes change the relative separation and orientation of molecules in the lattice. They are important for coherent transport. They are further divided into acoustic modes of lower frequencies and optical modes of higher frequency. In the acoustic modes the molecules in the unit cell move in phase while in the optical modes the molecules in the same unit cell have different phase relationship.

— Molecular or internal modes. These involve the in-plane stretching and out-of-plane bending modes.

In molecular crystals, the temperature dependence of the mobility is, in most cases, given by [10, 21]:

$$\mu = a \cdot T^{-n} \tag{III.20}$$

a: constant

The parameter n is generally of the order of unity. A model has been proposed for estimating n knowing the phonon mode involved in the scattering [10]. This model is valid for narrow-band systems ($W \leq kT$), but it neglects the temperature dependence of the electron overlap integrals (Table III.8). Other models have been developed in which the charge-carrier transport is considered to be determined by the change of the transfer integral induced by the change of the lattice spacing in the course of the lattice vibrations [21]. In that case almost-temperature-independent mobilities are found.

The temperature dependence of the drift mobility in single crystals of β-PcH$_2$ has been determined [366]. The experiments have been carried out between 290 and 600 K in high vacuum (10^{-7} torr). The determinations were made along the c' axis perpendicular to the ab plane of the monocrystals. A power law of the form of Eq. III.20 with n = 1.5 $\pm$ 0.2 for electrons and n = 1.3 $\pm$ 0.3 for holes has been found. These values suggest that the charge transport is controlled by lattice phonon scattering over the temperature range used (290–600 K) [366].

Table III.8 Temperature dependence of the mobility. The temperature dependence of the electron overlap integral has been neglected. (From Ref. [10]).

Origin of Scattering	μ^a (large bands)	μ^b (narrow bands)
Dislocation	$T^{-1/2}$	T^{-1}
Neutral impurity	T^0	T^{-1}
Lattice modes		
Acoustic		
One phonon	$T^{-3/2}$	T^{-2}
Two phonons		T^{-3}
Optical		
One phonon	$T^{1/2}\left[\exp\left(\dfrac{h\nu}{kT}\right)-1\right]$	$\left[\exp\left(\dfrac{h\nu}{kT}\right)-1\right]$
Two phonons		$\left[\exp\left(\dfrac{h\nu}{kT}\right)-1\right]^2$

[a] Standard results obtained with classical semiconductors; [b] Results for narrow band semiconductors with $W \leq kT$.

Hopping theories based on localized electronic state models may alternatively be used to describe the transport properties in molecular materials. The first hypothesis of most of the hopping theories is that the nuclear configurations of the reactants and of the products is not changed during the electron transfer, i.e. the Franck-Condon principle must be obeyed. An energy E_{reo} is required to change the nuclear coordinates such that isoenergetic electron transfer may take place. This reorganization energy arises from two sources [377]:

$$E_{reo} = E_{in} + E_{out} \tag{III.21}$$

E_{in} is the inner shell reorganization energy corresponding to the intramolecular processes; E_{out} is the outer sphere reorganization energy (intermolecular processes). *From the relative orders of magnitude of the characteristic transfer times involved in hopping models, the intramolecular processes are expected to predominate during the charge migration. By contrast, in the domain of applicability of the band model, the coherent charge-transport process will probably be more perturbed by the relaxation modes of the lattice (intermolecular processes).* In PcM's, if the moving charge is metal centered, the electronic transfer may be represented as:

$$PcM^{I}, PcM^{II} \rightleftarrows PcM^{II}, PcM^{I} \tag{III.22}$$

In this example, the internal reorganization energy is the energy required to inter-convert the large metal-to-ligand distance when the metal is in its lower oxidation state to a smaller value when the metal is in its oxidized form. The frequency of vibration of Pc-M bonds is of the order of 150–200 cm^{-1} [201]; this frequency shifts to higher values according to M = Zn > Pd > Pt > Cu > Fe > Co > Ni. This provides an order of magnitude value for the corresponding reorganization energy. Interconversion between ligand and metal-oxidized electronic tautomers recently has been shown to be important in determining the charge-transport properties of iodine doped (tetrabenzoporphyrinato) nickel(II) [378].

The compressibility of molecular molecular crystals is generally fairly high. Under the effect of high pressure, the $\pi - \pi$ orbital overlap between the constituting molecules is increased yielding higher mobilities of the charge carriers. In fact, the conductivity of PcM's can reach $10 \, \Omega^{-1} \, cm^{-1}$ under several hundred kilobars of pressure [384, 385].

c Determination of the Trapping Levels

In the intrinsic domain, the energy gap of PcM's is approximately 2 eV, and any doping, fortuitous or not, will lead to thermal activation energies lower than this value. This yields a good test for the presence or the absence of dopants in the material. Thin films of PcCu treated at 200 °C under a stream of nitrogen demonstrate, for example, an activation energy E_{act} of 1.12 eV [373]. The same film treated at room temperature in vacuum (10^{-6} torr) shows an activation energy of 1.44 eV [374, 375]. The value of E_{act} corresponding to the intrinsic domain (1.98 eV) is obtained only when the films are heated at temperatures higher than 200 °C under high vacuum [376]. Only these drastic conditions permit the release of gases absorbed in the bulk or at the surface of the materials. In most cases, only minute amounts of impurities suffice to greatly perturb the electrical properties of molecular semiconductors. Only a few physico-chemical methods are available for characterizing the nature and the extent of the doping. Space-Charge Limited Current (SCLC) measurements permit the determination of the density of traps and their energy distribution within the band gap. PcM's have been much studied in this manner. However, in subsequent discussion in this section, only materials possessing an activation energy of 2.0 eV at high temperatures will be considered. The results obtained from SCLC measurements on PcCu and PcH$_2$ are shown in Table III.9. E_{act} seems not to depend on the crystalline state of the PcM's (α or β thin films, single crystals) nor on the presence of a complexed central metal ion. On the other hand, the depth of the traps in single crystals is strongly influenced by the nature of M. The density of traps N_t erratically varies from sample to sample and it is difficult to correlate N_t with any molecular properties of the organic material. In single crystals, the SCLC measurements have been interpreted by postulating a single discrete level of trapping. The chemical nature of this trapping level is unknown but its depth within the band gap is clearly related to the nature of the metal ion of the PcM. This discrete trapping level in monocrystals is transformed into an exponential distribution of traps in the polycrystalline material. The environment of an impurity is nonuniform in a disordered material and the corresponding trapping state is therefore smeared out in energy, leading to a distribution of traps instead of a discrete level.

It is possible to determine, from the current-voltage curves, the voltage which corresponds to the transition from ohmic behavior to the space-charge-limited current region. It is, however, useful to check if, in the ohmic region, the current is really inversely proportional to the thickness of the organic material. In fact, Ohm's law may be obeyed at low voltage while the current is space-charge limited, but in this case a d^{-3} relationship is expected. The thickness dependence of the current

Table III.9 Space charge limited current studies on phthalocyanine derivatives. Determinations of the density of traps (N_t), the depths of traps for discrete levels (E_t) and the type of trap-level distribution within the band gap (h(E)).

	E_{act} [a]	N_t [b] (cm^{-3})	E_t (eV) [c] below CB	Type of distribution	Ref.
PcH$_2$					
β-SC [d]	2.00 (137 °C)	N_t^e:7·10^{16}	0.32	discrete	[360]
		N_t^h:2·10^7	0.32		
β-SC		N_t^e:5·10^{19}	0.38	discrete	[379]
		N_t^h:2·10^{12}	0.38		
PcCu					
β-SC	2.04 (88 °C)	N_t^e:1.7·10^{13}	0.88	discrete	[362]
		−3.2·10^{14} [g]			
β-TF [e]		N_t^e:6·10^{15}		exponential [f]	[376]
(heat treatment)					
270 °C)					
α-TF	1.98	N_t^e:6·10^{14}		exponential [f]	[376]
(heat treatment)					
200 °C)					

[a] Activation energy in the intrinsic domain; the break point temperature is given in parentheses;
[b] N_t^e: density of traps for electrons, N_t^h: density of traps for holes; [c] trap depth below conduction band; [d] SC: single crystals; [e] TF: thin films;

[f] $h(E) = \dfrac{H}{kT_c} \cdot \exp\left(-\dfrac{E_g - E}{kT_c}\right)$ H: density of traps
   E_g: energy gap
   T_c: temperature characteristic of the distribution;

[g] depending on the transition voltage considered in the SCLC.

density has been determined for β-PcH$_2$ monocrystals [379]. In the ohmic region, the slope is −1, the current is inversely proportional to the thickness; in the SCLC region the slope is −3. It seems, therefore, that the number of injected charge carriers from the electrodes is less than the number of intrinsic charge carriers in the ohmic domain. It is noteworthy that the term "intrinsic charge carriers" refers only to the charge carriers present in the material before injection. These "intrinsic charge carriers" might well be generated from donors or acceptors. With these precautions in mind, it is possible to estimate the number of intrinsic charge carriers in the material from the transition voltage. *For single crystals of PcH$_2$, the density of the intrinsic carriers is then found to be in the order of 10^6–10^7 cm^{-3}* [380].

The density of states within the valence band, N_v, has been determined in PcM's by studying the temperature dependence of the photoconductivity [373]. A value of 10^{21} cm^{-3} has been found; this corresponds approximately to one state per molecule of PcM.

Detailed SCLC measurements on thin films of PcMg have been carried out [381]. Treatment of the thin films at 169 °C in vacuum with an applied bias of 10 V leads to a room temperature conductivity 10^4 smaller (10^{-15} Ω^{-1} cm^{-1}) than in the preannealed state. The SCLC measurements have been interpreted by assuming uniform distribution of trap levels within the band gap. In this model, a trap concentration of 2.09 · 10^{17}/cm^3 · eV is found.

Table III.10   Thermally stimulated current (TSC) measurements on monocrystals of PcH$_2$ and PcPb. (From Ref. [382]).

	N$_t$ (cm^{-3})	E$_t$ (eV)
PcPb (SC)	1.2–2.0 · 10^{14}	E$_t^e$: 0.36; 0.38^a
		E$_t^h$: 0.30; 0.36^b
PcH$_2$ (SC)	1.0–7.0 · 10^{13}	E$_t^e$: 0.35; 0.37^a
		E$_t^h$: 0.30^b

N$_t$: density of traps; E$_t$: depth of traps; SC: single crystal.
^{a} below conduction band; ^{b} above valence band.

SCLC measurements also permit an order of magnitude estimate of the mobility of the charge carriers. The mobility of the carriers in thin films of PcCu is thus calculated to be 0.02 cm^2/V · s [373]. This value differs markedly from the drift mobilities previously determined (10^{-3}–10^{-4} cm^2/V · s).

Thermally stimulated current (TSC) measurements have been performed on monocrystals of PcH$_2$ and PcPb [382, 521], and on PcCu powder dispersed in polyurethane [383] (Table III.10). In this method, the traps are filled at low temperature by irradiation and the current produced by emptying the traps is measured as the temperature is raised. TSC studies allow a more direct determination of the trap concentration and the associated distribution in energy within the band gap. By comparing the results shown in Tables III.9 and III.10, it is seen that the values of the depths of traps obtained from SCLC and TSC measurements are quite in agreement. TSC measurements have also demonstrated that when PcCu is incorporated into a binder, the interface between the PcCu particles and the insulating binder is the main source of traps [383].

d Doping of PcM by O$_2$

O$_2$ is an almost unavoidable dopant of molecular semiconductors. The redox properties of O$_2$ in solution are well known [386, 387]. The superoxo species O$_2^-$ is formed by the monoelectronic reduction process:

$$O_2 + e^- \rightleftarrows O_2^- \qquad E_{1/2} = -0.75 \text{ V vs SCE} \qquad \text{(III.23)}$$

O$_2^-$, in presence of protons or other acceptors, can dismutate into O$_2$ and O$_2^{2-}$:

$$O_2^- + H^+ \rightleftarrows HO_2 \qquad \text{(III.24)}$$

$$2 \, HO_2 \rightleftarrows H_2O_2 + O_2 \qquad \text{(III.25)}$$

O$_2^-$ may alternatively be further reduced into the peroxo-species:

$$O_2^- + e^- + H^+ \rightleftarrows HO_2^- \qquad E_{1/2} = 2.0 \text{ V vs SCE} \qquad \text{(III.26)}$$

The main equilibrium which must be taken into account when O_2 is incorporated into the lattice of PcM's involves a monoelectronic transfer from PcM to O_2:

$$PcM, O_2 \rightleftarrows PcM^+, O_2^- \tag{III.27}$$

O_2 is, however, only a weak monoelectronic oxidant in solution and it should not be able to oxidize PcM's. The equilibrium represented by Eq. III.27 must be largely displaced to the left side. This may be very different in the solid state. The binding of O_2 by PcM will favour the electronic transfer between them. A chlorophyll-O_2 complex has been detected in the solid state [388]. The binding energy of O_2 with Chl_a and Chl_b is 1.4 eV and 0.63 eV, respectively. This energy maintains the two reactants close to each other, allowing the electronic transfer to occur.

The additional possibility exists that the complex PcM, O_2 may act as an electron trap:

$$PcM, O_2 + e^- \rightleftarrows PcM, O_2^- \tag{III.28}$$

Hydrogen offers less chemical versatility than O_2. The bi-electronic oxidation process is of course well known and usually serves as a reference potential:

$$H_2 \rightleftarrows 2 H^+ + 2 e^- \tag{III.29}$$

The monoelectronic transfer necessitates much more energy (>3 eV) and is far less common:

$$H_2 \rightleftarrows H_2^+ + e^- \tag{III.30}$$

The effect of ambients on the properties of PcM's has been extensively studied by ESR. It has been known for a long time that "diamagnetic" PcM's (M = H_2, Zn, Ni, Pt, Pd) show an ESR signal [389]. In all cases, the ESR peak corresponds closely to the free electron g value. The intensity of the signal depends on the conditions of preparation and purification of the sample [389–396, 522]. Repeated crystallization from concentrated sulfuric acid decreases the intensity of the ESR signal [136]. Successive sublimations also reduce the overall concentration of spins and the residues of sublimation contain a high concentration of spins [391]. Oxygen and hydrogen have an equally drastic effect on the overall spin concentration [393]. The effect of ambients have been examined for a single crystal of PcH_2 crushed into powder and heated under vacuum (10^{-4} torr) from 25 to 500 °C. This treatment does not increase the concentration of spin. Under a pure atmosphere of O_2, the spin density remains constant up to 200 °C and then increases linearly with the temperature. In an atmosphere of hydrogen, the spin concentration decreases at 300 °C to reach a constant level after 6 hours. These experiments show that

(i) O_2 is associated, at least in part, with the ESR signal,

(ii) heating under vacuum does not change the concentration of the paramagnetic species,

(iii) H_2 at high temperature reduces the spin density but without completely eliminating it. Two types of defects probably coexist in the material. The first one is sensitive to O_2 and H_2 and may be formed reversibly. It is reasonably related to the complex PcM,O_2. H_2 decreases the concentration of PcM,O_2 by reducing the oxygen to form water, reaction catalyzed by PcM [435]. The corresponding ESR peak is probably $PcM^{\dot{+}}$ formed by charge transfer within the complex. ESR studies have shown, however, that in the case of O_2 chemisorption on $\gamma\text{-}Al_2O_3$, PcCo no electron transfer from the oxygen to the metal complex occurs [407]. This effect may be attributed to the coordination properties of $\gamma\text{-}Al_2O_3$.

The second type of defect is more difficult to identify though it probably arises from decomposition products of PcM. It has been shown that the density of spins in the material is dependent upon the way the sublimation has been carried out [391]. High spin concentrations occur when PcH_2 is sublimed above 500 °C and the crystals are grown below 400 °C, low spin concentrations are obtained when the crystals are grown at 450 °C. The impurities associated with the ESR signal seems therefore to sublime at a slightly different temperature compared to PcM's. PcH_2 pyrolyzed at 400 °C also shows a spin concentration increase [394]. Paramagnetism is therefore related to some thermal degradation process of the PcM's. Samples irradiated with a ^{60}Co source demonstrate a similar increase in the spin density [391].

The second type defect has therefore the following ensemble of properties:
— it is O_2- and H_2-insensitive
— it probably arises from the thermal degradation of the PcM's
— it sublimes at temperatures close to PcM's, and therefore possesses a molecular weight of the same order of magnitude as PcM's and must be neutral.

The second type defect will be abbreviated as $P^{\cdot}$. The formula shown in Fig. III.24 meets all the above-mentioned requirements for $P^{\cdot}$ and is a reasonable tentative structure.

In the light of the previous observations, the effect of ambients on the charge-transport properties of PcM's may be more easily understood. The temperature dependence of the conductivity has been seen to take the form:

$$\sigma = \sigma_0 \cdot \exp\left(-\frac{\Delta E}{2kT}\right) \tag{III.31}$$

By plotting $\log \sigma$ vs $(1/T)$ it is possible to determine the thermal activation energy ΔE. For PcCu single crystals, no discontinuity in the curve of $\log \sigma$ vs $(1/T)$ is observed when the experiments are carried out under oxygen or under hydrogen (Fig. III.25). The activation energy found under H_2, 2.14 eV, agrees satisfactorily with expected

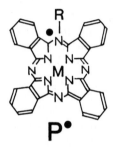

Fig. III.24. Tentative formula of the species responsible for the second type of defects in metallophthalocyanines (paramagnetic defects O_2 and H_2 insensitive)

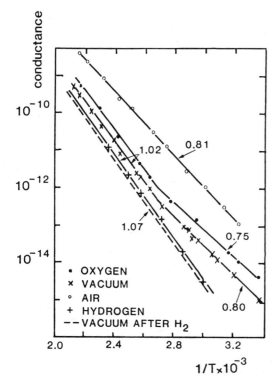

Fig. III.25. Plots of log σ vs (1/T) for a monocrystal of PcCu under different ambients. The values indicated must be multiplied by 2 to obtain the activation energies of the charge transport. [Reproduced with permission of (363)]

"band gap" of PcM's. The activation energy under O_2 is lowered to 1.62 eV and the conductivity in the whole temperature range studied (40–180 °C) is clearly extrinsic. The value of 1.62 eV must correspond to the energy needed to liberate free charge carriers from the oxygen complex:

$$PcM, O_2 \rightleftarrows PcM^{\cdot +}, O_2^{\bar{\ }} \rightleftarrows \text{free charge carriers} \qquad (III.32)$$

The value of 2.14 eV may be assigned to the intrinsic generation of the charge carriers:

$$PcM \rightleftarrows \text{free charge carriers} \qquad (III.33)$$

Under vacuum or in air, a discontinuity in the log σ vs (1/T) plot appears at approximately 100 °C indicating a transition between an intrinsic transport regime at high temperatures and an extrinsic one at low temperatures [363].

The influence of the ambient on the concentration of traps has been studied by the SCLC method. Studies on β-PcH_2 single crystals [379] and PcCu thin films [373] have been reported (Table III.11). O_2 causes the Fermi level to move closer to the valence band and therefore acts as a dopant with an increase in its concentration inducing a net augmentation of the density of free charge carriers. H_2 has just the opposite effect [373], E_F is displaced towards the middle of the band gap and the electrical behavior of the molecular material tends towards intrinsic regime.

Table III.11 Determinations of the density of traps (N_t) and their distribution by SCLC determinations under various ambients. (From Ref. [373] and [379]).

	Conditions	N_t (cm^{-3})	E_t (eV) below CB
β-PcH$_2$	vacuum	$5 \cdot 10^{19}$	0.38
(SC)	$O_2{}^a$	10^{14}	0.82^b
	$H_2{}^a$	10^{13}	0.95^b
PcCu	$N_2{}^c$	$3.3 \cdot 10^{18e}$	exp.^{d}
(TF)	$H_2{}^c$	$3.3 \cdot 10^{18e}$	exp.^{d}
	$O_2{}^c$	$4.0 \cdot 10^{18e}$	exp.^{d}

SC: single crystal, TF: thin film, CB: conduction band.
^{a} After stabilization 24 hours in the ambient, the process is reversible; ^{b} vacuum levels still exist; ^{b} two annealings have been carried out: 5 min at 200 °C, 15 min at 200 °C in the ambient; ^{d} exponential distribution of traps; ^{e} density of traps in a slice kT_c at the band edge.

The kinetics of the thermal diffusion of O_2 inside monocrystals of PcPb have been measured [397]. The diffusion constant at 242 °C is $7.2 \cdot 10^{-9}$ cm^2 s^{-1} with an activation energy of 1.4 eV. O_2 will diffuse over 50 μm in approximately one hour.

The nature of the central ion does influence the PcM-O_2 association constant. Temperature-programed desorption studies [398] have shown that PcCu and PcFe reversibly absorb O_2 while, under the same conditions, PcH$_2$ does not interact with O_2. The molecule of O_2 is therefore very probably directly bound to the central metal ion, and the magnitude of the binding constant is related to the ability of M to coordinate extra axial ligands.

e Doping of PcM by other Doping Agents

Many doping agents may be used to oxidize or reduce metallophthalocyanines. The oxidation potentials of most of the PcM's lie between $+0.6$ and $+1.0$ vs SCE. For significant charge transfer between the dopant and the PcM to occur, the reduction potential of the dopant, E^{red}_{dopant}, must be such that:

$$E^{red}_{dopant} - E^{ox}_{PcM} \geq -0.25 \text{ V} \tag{III.34}$$

Most of the usual oxidants satisfy this condition: halogens, benzoquinone, anthraquinone derivatives etc. [399, 400].

The influence of halogens on the conductivity of single crystals [401] or thin films [402] of PcM's has long been recognized. Doping of PcM's by halogenes reduces the overall resistivity by many orders of magnitude. By comparison with O_2 ($E^{red} =$

$= -0.56$ V vs NHE), Br_2 and I_2 are far stronger oxidizing agents [400]. In aqueous solutions, the corresponding reduction potentials are:

$$Br_2 + 2\,e^- \rightleftarrows 2\,Br^- \qquad E^{red} = 1.06 \text{ V vs NHE*} \qquad \text{(III.35)}$$

$$I_2 + 2\,e^- \rightleftarrows 2\,I^- \qquad E^{red} = 0.636 \text{ V vs NHE*} \qquad \text{(III.36)}$$

Doping of single crystals of PcH_2 at 100–120 °C in an atmosphere of I_2 has been followed by ESR [341]. The peak observed, with a g value of 2.0036 and a linewidth of 5.03 G, has been attributed to the species $PcM^{+\cdot}, I_3^-$.

Thin films of PcM's may be doped by I_2 vapour or solutions. The corresponding compressed-pellet conductivities are of the order of 0.06–4.2 Ω^{-1} cm^{-1} depending on M [404, 405]. Resonance Raman Spectroscopy studies have shown that the iodine is in the form of I_3^- [404]. The thermal activation energies are approximately 50-fold less than for the undoped complexes.

Single crystals of $PcNiI_{1.0}$ have been prepared [342, 406]. The crystalline structure deduced from X-ray measurements is very different from the structures classically found for undoped PcM's (Fig. III.26). The PcM's are stacked in columns parallel to the c axis, the interplanar Ni-Ni separation is 3.244 Å, successive phthalocyanine rings within a stack are staggered by 39.5°. The iodine chains are located in channels defined by the benzo groups of the phthalocyanines. Partial oxidation has no measurable effect on the geometry and the dimensions of the phthalocyanine ring. $PcNiI_{1.0}$ crystals show a single ESR signal with $g_{||} = 2.0075$ and $g_{\perp} = 2.0007$. These values

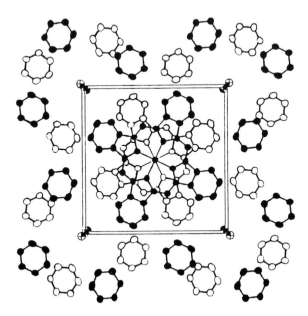

Fig. III.26. Crystal structure of $PcNiI_{1.0}$ viewed along the stacking direction. [Reproduced with permission of (342)]

* The redox potential vs NHE may be converted into the corresponding value vs SCE by adding 0.24 V [403].

are in agreement with a ligand-centered oxidation since the corresponding g values for PcNi^{III}, the metal oxized species, are $g_{||} = 2.11$ and $g_{\perp} = 2.29$ [339, 342]. The peak-to-peak linewidth is 5–7 G at room temperature. The spin susceptibility at room temperature corresponds to 0.14 spin par macrocycle. The conductivity along the needle axis is 260–750 Ω^{-1} cm^{-1} at 300 K and corresponds to a mean free path of a carrier along the stacking direction of 1.0–2.3 intermolecular spacings (3.3–8.2 Å). By comparison, under the same conditions, the mean free paths in TTF-TCNQ and in $K_2Pt(CN)_2Br_{0.3} \cdot 3\,H_2O$ are 0.4 and 0.6 intermolecular spacings, respectively [342]. The temperature dependence of the conductivity is shown in Fig. III.27 for a monocrystal of NiPcI$_{1.0}$. The conductivity below ambient temperature rises to a maximum corresponding to a temperature T$_m$, then the conductivity abruptly decreases. T$_m$ varies between 55 K and 100 K from crystal to crystal. In the region above T$_m$, the electrical properties of the crystals are entirely reversible with temperature changes, the temperature dependence of the resistivity is given by the relationship [9, 408]:

$$\frac{\varrho(T)}{\varrho(T_1)} = a + b\left(\frac{T}{T_1}\right)^n \qquad (III.37)$$

a, b: constants
T$_1$: 295 K
ϱ: resistivity

A least-square fit of the data yields n = 1.9 $\pm$ 0.2. Charge transfer complexes have been classified into three groups depending on the form of the conductivity-

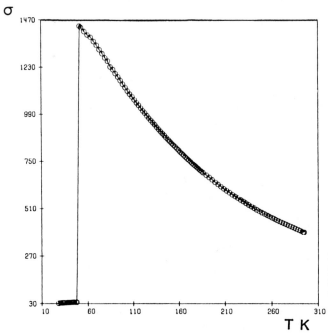

Fig. III.27. Temperature dependence of the conductivity for a PcNiI$_{1.0}$ crystal along the c axis. (After Ref. [342])

temperature relationship observed [9]. It is the class III system which demonstrates a sharp maximum of conductivity at low temperature. TTF-TCNQ is a representative example of this class. A relationship of the form of Eq. III.37 is generally used for class III compounds. The dominant scattering mechanism should be deducible from the parameter n. However, a good deal of the temperature dependence of the conductivity may arise from the thermal expansion of the lattice rather than from the dominating scattering process [409]. The temperature dependence of the conductivity at constant lattice parameter rather than at constant pressure is more easily correlated with a precise scattering mechanism. The value of n for many other charge-transfer complexes range from 2.0–2.4 [8, 9]. For TTF-TCNQ $n \sim 2.3$, such a value may arise from collective many-body effects (charge-density wave fluctuations), electron-electron scattering between electrons on TCNQ, and between holes on TTF or interaction of electrons with vibrational modes of TCNQ. Depending on the temperature range studied one of these contributions is predominant. It is reasonable to think that the electrical properties of $PcNiI_{1.0}$ will be influenced by these same scattering mechanisms.

Bridged polymers of formula $[PcMF]_x$ (M = Al, Ga, Cr) have been doped with I_2 and Br_2 [177, 410, 411]. Adducts of different stoichiometries $[(PcMF) I_y]_x$ with y ranging from 0 to 3.4 have been isolated. Doping of μ-oxo-bridged polymeric systems has also been achieved [168, 181] (Table III.12). Their conductivity is increased by more than 9 orders of magnitude by bromination or iodination. The interplanar spacing within a column has been estimated from X-ray powder diffraction studies. It seems there is a correlation between the inter-ring distance and the maximum conductivity obtained after doping, larger interplanar spacings leading to smaller conductivities.

Table III.12 Transport properties of bridge-stacked metallophthalocyanine polymers. (From Ref. [176, 181, 411]).

	σ^a $(\Omega^{-1}\,cm^{-1})$	E^b_{act} (eV)	Interplanar[c] spacing (Å)
$[PcAlF]_x$	$< 10^{-7}$		3.66
$[(PcAlF)I_{3.4}]_x$	0.59	0.03	
$[PcGaF]_x$	$< 10^{-9}$		3.86
$[(PcGaF)I_{2.1}]_x$	0.15	0.04	
$[PcCrF]_x$			3.87
$[(PcCrF)I_{3.2}]_x$	0.62		
$[PcSiO]_x$	$3 \cdot 10^{-8}$		3.32
$[(PcSiO)I_{1.55}]_x$	1.4	0.04	3.32
$[(PcSiO)Br_{1.00}]_x$	$6 \cdot 10^{-2}$		
$[PcGeO]_x$	$< 10^{-8}$		3.50
$[(PcGeO)I_{1.80}]_x$	$3 \cdot 10^{-2}$	0.08	3.50
$[PcSnO]_x$	$< 10^{-8}$		3.83–3.95
$[(PcSnO)I_{1.2}]$	10^{-6}		

[a] Polycrystalline samples, conductivities at 300 K on pressed pellets; [b] thermal activation energy for the conduction; [c] estimated from X-ray powder diffraction data. Only the conductivities for maximally doped samples have been indicated.

Quinone derivatives and related compounds (Fig. III.28) may also be used as dopants since they have suitable redox potentials. However, the incorporation of compounds such as bromanil, chloranil, or TCNQ into solid phases of PcM's does not change the overall conductivity of the material to any great extent [176, 412]. It has been suggested that these oxidants lead to nonconducting integrated stack-charge-transfer complexes $(A^+I_A^- A^+I_A^- A^+I_A^-)$ rather than segregated structures $(A^+A^+A^+ I_A^-I_A^-I_A^-)$ [176, 412]. The μ-bridged polymers, which cannot form integrated stack complexes, are indeed efficiently doped in the presence of quinone derivatives (Table III.13). The doping is achieved by stirring the polymers with solutions of the dopants.

Doping has also been achieved by spraying an organic solution of the dopant onto the surface of thin films of PcM's [402, 413–418]. Addition of o-chloranil onto the surface of PcH$_2$ films leads to a 10^7 increase in the steady-state photoconductivity

Table III.13 Electrical conductivity data for polycrystalline samples of μ-bridged phthalocyanine polymers in presence of dopants. (From Ref. [176]).

	σ^a $(\Omega^{-1} \, cm^{-1})$	E^b_{act} (eV)
[PcSiO]$_x$	$3 \cdot 10^{-8}$	
I$_{1.55}$	1.4	0.04
K$_{1.0}$	$2 \cdot 10^{-5}$	
(TCNQ)$_{0.5}$	$2.8 \cdot 10^{-3}$	0.09
(Fl)$_{0.23}$	$7.2 \cdot 10^{-4}$	0.13
(Chlr)$_{0.37}$	$6.9 \cdot 10^{-4}$	0.13
(Brl)$_{0.84}$	$5.8 \cdot 10^{-4}$	0.15
(DDQ)$_{1.00}$	$2.1 \cdot 10^{-2}$	0.08

Fl: fluoranil; Chlr: p-chloranil; Brl: bromanil; DDQ: dichlorodicyanoquinone.

Doping is achieved by stirring the polymer in solutions of the dopants except for K doping where the polymer is reacted with potassium vapor in a sealed tube.
[a] At 300 K, polycrystalline samples; [b] thermal activation energy.

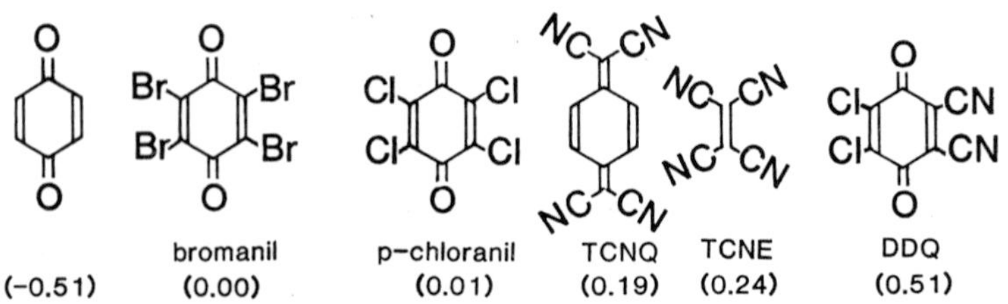

	bromanil	p-chloranil	TCNQ	TCNE	DDQ
	(−0.51)	(0.00)	(0.01)	(0.19)	(0.24)	(0.51)

Fig. III.28. Quinone derivatives which may be used as p-type dopants for metallophthalocyanines. The corresponding reduction potentials are given in parentheses (vs. SCE)

[402, 418]. The addition of o-chloranil induces a rise in the concentration of unpaired spins. Upon irradiation the spin concentration diminishes, probably as a consequence of the formation of the dianion of o-chloranil [402, 418]:

$$PcM, \text{o-Chlr} \rightleftarrows PcM^{\dot{+}}, \text{o-Chlr}^{\dot{-}} \tag{III.38}$$

$$PcM^*, \text{o-Chlr}^{\dot{-}} \rightleftarrows PcM^{\dot{+}}, \text{o-Chlr}^{2-} \tag{III.39}$$

PcM's may be "auto-doped" by using mixtures of two different metallophthalo-cyanines [26, 419]. A theoretical estimate of the energy levels introduced within the band gap may be provided knowing the redox potentials of the host and guest phthalocyanines (Fig. III.29). PcCu and PcNi, for example, form only acceptor levels within the band gap of host PcH_2 crystals, PcCd and PcFe act only as donors. In most cases, the donor and acceptor levels are close to the band edges; only PcMn and PcFe have filled orbitals lying deep within the forbidden gap. The mixtures of PcM's are made by dissolving PcH_2 and the appropriate PcM in concentrated sulfuric acid and by reprecipitating the homogeneous mixture by pouring the solution into ice water. Mixed PcM thin films have also been obtained by cosublimation of two PcM's under high vacuum [420]. It has been shown that the incorporation of PcZn, PcCo, PcFe, or PcNi has no detectable influence on the electric properties of PcCu. These results must however be taken with care since it is fairly difficult to keep constant the density of impurities introduced during the preparation processes. Nonetheless, it seems clear than the mixed PcM's do not contain deep trap levels. Such is not the case for PcVO incorporated into PcCu [420]. PcVO is not planar so the energetic perturbation caused to the PcCu lattice is fairly significant. The admixture indeed shows the presence of impurity centers 1.04 eV below the conduction band edge

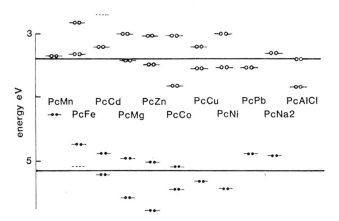

Fig. III.29. Energy levels introduced by host PcM's into guest PcH_2. The valence and conduction band edges represented are those of PcH_2. The energy levels induced by host PcM's are calculated from Eqs. I.58 and I.59:
$I_c \sim E_{1/2}^{ox} + 4.3$
$A_c \sim E_{1/2}^{red} + 4.3$
(After Ref. [26, 419])

Table III.14 Electrical properties of thin films of mixtures of PcCu and PcVO. Before each measurement the thin films were annealed at 180 °C for 15 h under vacuum ($5 \cdot 10^{-5}$ torr) to remove adsorbed gases. (After Ref. [420]).

		σ^a $(\Omega^{-1}\,cm^{-1})$	E_{act}^b (eV)	μ^c $cm^2/V \cdot s$
PcCu	β	$7 \cdot 10^{-16}$	1.98	17
	α	$2 \cdot 10^{-15}$	1.96	3.5
PcVO	α	$2.5 \cdot 10^{-11}$	1.34	2.4
PcVO:PcCu α (1:9)		$1.1 \cdot 10^{-10}$	1.06	$9 \cdot 10^{-3}$
PcVO:PcCu α (1:1)		$4.7 \cdot 10^{-11}$	1.02	$8 \cdot 10^{-4}$

[a] At 300 K; [b] thermal activation energy from the plots of the logarithm of the conductivities against $1/T$; [c] calculated drift mobilities; for the admixtures the number of donor centers is equated with the number of PcVO molecules (see Ref. [420]).

[420] (Table III.14). The thermal activation energies E_{act} for α-PcCu, β-PcCu, and α-PcVO are approximately equal to the intrinsic energy gap. In the mixed polycrystalline thin films PcVO-PcCu the corresponding activation energy is 1.0 eV. This value may be assigned as being the energy difference between the donor level and the conduction band edge. PcVO therefore acts as a dopant towards PcCu and leads to a strong increase of the conductivity of PcCu by increasing the number of charge carriers. The transport properties in admixtures of PcCu and PcVO are complicated, however, by the formation of clusters of pure PcCu and PcVO.

III.3 Photovoltaic Effect and Solar Cells

a Photoelectrical Properties

Early determinations of the photoconductive properties of PcM's were made without paying sufficient attention to the effects of the ambients [356, 359, 421–423]. It was long thought, for example, that the photoconductive properties of the α- and β-crystalline forms of PcH_2 were drastically different [424], though it was subsequently demonstrated that this difference arose in fact from the greater ability of the α form to bind O_2. The influence of O_2 is even more spectacular on the photoelectrical properties of PcM's than on their dark properties [425]. The photoconduction was first thought to be strictly correlated with the absorption properties of the PcM's, higher photocurrents being observed in the visible region from irradiation in the $S_0 \rightarrow S_1$ absorption band. However, the magnitude of the photocurrent is highly dependent upon the amount of O_2 present in the organic material [268, 426–428]. The photoconductivity seems also to be correlated with the axial coordinating ability of the central metal ion. The ability of PcM's to participate in axial coordination decreases in the order: PcZn > PcCu > PcNi [268]. PcNi does not show a high photoconductivity in air and remains unchanged in vacuo. The photoresponse of PcZn in the visible is diminished 40-fold after 24 hours at 10^{-5} torr [268]. The quantum yield of generation of free charge carriers from the photoexcited singlet state must consequently be rather low:

$$PcM \xrightarrow[S_0 \to S_1]{h\nu} PcM^* \xrightarrow{\times} \text{free carriers} \tag{III.40}$$

The major photochemical pathway towards the formation of free carriers is therefore extrinsic and involves a guest impurity, very probably O_2:

$$PcM, O_2 \xrightarrow[S_0 \to S_1]{h\nu} PcM^*, O_2 \to PcM^+, O_2^- \rightleftarrows \text{free carriers} \tag{III.41}$$

The most important intermediate involved in the primary ionization steps is presumably PcM^+, O_2^-, though some more energy is needed to generate the free carriers.

The photoconductive properties of single crystals of PcPb doped with various amounts of O_2 have been examined [427]. O_2 is incorporated into the single crystals by heating them at 250 °C under variable pressures of pure oxygen. When PcPb is doped with O_2, the photocurrent is multiplied by a factor of more than 100. Similar results are found in the case of single crystals or thin films of PcCu [429, 523]. For monocrystals the photocurrent is found to be thermally activated with an activation energy of 0.41 eV in the range of temperature 30–200 °C. The magnitude of the activation energy is wavelength independent from 648 to 1200 nm. The photocurrent is increased by a factor of 40 in the presence of O_2. Surface photocurrents are not responsible for this increase since the use of guard rings does not significantly change these results, so the O_2 molecules must diffuse into the bulk material [429]. Several conclusions may be drawn at this point. First, since PcM^+, O_2^- is probably the primary ionized intermediate, the thermal activation energy of photoconduction must correspond to the equation:

$$PcM^+, O_2^- \xrightarrow{0.4\,eV} \text{free charges} \tag{III.42}$$

An energy of 0.4 eV is needed to separate the opposite charges. Second, since the photocurrent activation energy is wavelength independent, the same intermediate is formed whatever the photoexcited state primarily formed. Third, energy migration to the PcM, O_2 centers is probable since the concentration of O_2 is in all case fairly low.

The photocarrier generation processes involved in thin films of PcZn exposed to air have been investigated [524]. The Onsager distance corresponding to the distance between the initially formed charged pairs is found to be 9 Å. Such a separation may be associated with the mean intermolecular distance between neighboring phthalocyanine molecules. It may also be associated with the mean separation between initially formed PcM^+ and O_2^-.

In undoped PcH_2, the photocurrent activation energy is 0.21 eV [367]. The energy difference between doped and undoped materials, $0.41 - 0.21 = 0.2$ eV, must correspond to the difference in energy between the singlet excited state PcM^* and

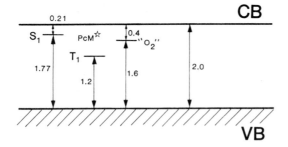

Fig. III.30. Tentative representation of the principal levels involved in the transport properties of metallophthalocyanines in the dark or under illumination. Depending on the nature of the central metal, slight deviations from the energy shown may occur

the ionized intermediate $PcM^{+\cdot}, O_2^{-\cdot}$. The approximate band scheme shown in Fig. III.30 summarizes most of the previous observations. The energy state of PcM* is calculated from the absorption properties of the materials. The energy level of $PcM^{+\cdot}, O_2^{-}$ is obtained from the thermal activation energies of conduction either in the dark or under illumination [363, 429]. The I–V characteristics of PcM's as a function of the temperature have allowed the determination of the intrinsic band gap (2.0 eV) [363].

Photons of lower energy than those associated with the $S_0 \rightarrow S_1$ absorption band also induce a photoconductivity effect [426, 428, 430]. Metallated phthalocyanines indeed show weak absorption bands in the near IR region (1000–1100 nm). These bands have been attributed to the spin-forbidden singlet-triplet transitions $S_0 \rightarrow T_1$ [264, 265]. Since no $S_0 \rightarrow T_1$ peak is detectable in PcH_2, it is the spin-orbit coupling induced by the metallic center which influences the intensity of the near IR bands. The photoconductivity associated with the $S_0 \rightarrow T_1$ transition is less sensitive to O_2 than the photoconductivity associated with the $S_0 \rightarrow S_1$ band [428]. Triplet excited states have long lifetimes and they can generally migrate over longer distances than singlet excited states. Thus, the probability that the triplet excited state encounters an O_2 molecule is increased and minute amounts of O_2 suffice to yield charge-transfer complexes efficiently. Added to this, it seems that triplet excited states are better precursors to free charge carriers than singlets even in solution. Quenching of the excited states of chlorophyll and porphyrin derivatives by quinones or nitroaromatic compounds has been studied by flash kinetic experiments (see the references cited in [43]). The conclusion of these extensive studies is that ionic photoproducts are only observed when quenching proceeds from a triplet (T_1) excited state, but not when it proceeds from a singlet excited state (S_1):

$$Porp. \xrightarrow{h\nu} Porp^*(S_1) \rightarrow Porp^*(T_1)$$

$$+A \qquad\qquad +A$$
$$\downarrow \qquad\qquad\qquad \downarrow$$
$$Porp^{+\cdot}, A^{-\cdot}(S) \quad Porp^{+\cdot}, A^{-\cdot}(T)$$

$$Porp^{+\cdot} + A^{-\cdot}$$

ionic photoproducts   (Scheme III.43)

Quenching of singlet excited states by electron transfer yields singlet radical pairs and spin-allowed recombination may occur before dissociation to the free ionic photoproducts. In the case of triplet excited states, the primary ionized intermediate has the character of a triplet so the spin-forbidden electron-hole recombination rate is decreased and ionic photoproducts are observable [431–434].

Hydrogen ambients also influence the photoconduction properties of PcM's [428]. Monocrystals of PcH_2 and PcCu have been prepared by sublimation either in a stream of pure N_2 or in a stream of a $N_2:H_2$ mixture (9:1). PcM's catalyze the formation of water from H_2 and O_2 [435], and, consequently, the growing of crystals in a reducing atmosphere ensures that little O_2 remains in the material. Under these conditions ESR measurements indicate a weaker density of paramagnetic centers, but a significant quantity remains. No photoconductivity is observed in the $S_0 \rightarrow S_1$ transition, but photoconductive properties still arise from irradiation in the $S_0 \rightarrow T_1$ absorption band. These results are not significantly changed when air is readmitted. Single crystals grown in N_2 show this same behavior only when they are previously annealed at 200 °C for 12 h under high vacuum (10^{-6} torr). These results confirm that at least two kinds of paramagnetic defects coexist in PcM monocrystals. The magnitude of the photoconductivity associated with the $S_0 \rightarrow S_1$ absorption band is clearly correlated with the O_2-dependent defects, it is less evident in the case of the $S_0 \rightarrow T_1$ transition. The sensitivity of PcM's toward dopants (O_2, NH_3, H_2S, etc.) is related to their degree of purity; the purest materials are markedly less sensitive to the dopants [436].

Single crystals of PcMn and PcFe have drastically different behaviors because, in the presence of gases (O_2, N_2, a.o.), their photoconductivity is depressed [437]. This effect is reversed by evacuation, and gassing/degassing cycles may be repeated indefinitely. PcMn and PcFe are fairly easily oxidized and they can form ionized charge-transfer complexes with O_2. These species may in turn reduce the mobility of the charge carriers by acting as trapping states.

Surface photovoltage measurements on polycrystalline PcNi [353, 438] indicate that O_2 is absorbed in two different ways at the surface of the films. One of the adducts is reversibly bound and can be removed by evacuating the ambient; the second one is only released by heating at 160 °C under vacuum. This last species causes an order of magnitude increase of the surface photovoltage, while the first one leads, on the contrary, to an overall decrease in the surface photovoltage. The effect of o-chloranil on the surface photovoltages of thin films of PcH_2 has also been reported [439], monolayer coverages result in the inversion of the sign of the photovoltage.

Pulsed techniques have been used to determine the wavelength dependence of photoconductivity (action spectra) of vacuum sublimed β-PcH_2 thin films in the range 360–740 nm [440]. It is concluded that the carrier generation process proceeds via the lowest excited singlet state and that it is a bulk phenomenon. The quantum efficiency of the carrier generation is affected by the amount of absorbed O_2.

The charge-carrier photogeneration process has been further studied by electric-field-induced fluorescence-quenching measurements [441–443, 525]. The rate of generation of free charge carriers from the excited state of PcM is assumed to be dependent upon the magnitude of the electric field in the material. The lifetime of the excited state and the efficiency of generating charge carriers are correlated:

$$PcM \xrightarrow{\ h\nu\ } PcM^* \xrightarrow{\ k(\xi)\ } \text{free carriers} \tag{III.44}$$

ξ: electric field

Table III.15   Charge-carrier photogeneration efficiency. φ_{pe}, of x-PcH$_2$ dispersed in polyvinylacetate as a function of doping by metallophthalocyanines. (From Ref. [442]).

Dopant[a]	φ_{pe} (%)	Fluo.[b] efficiency	σ $(10^{-7}\ \Omega^{-1}\ cm^{-1})$
none	28	100	0.5
PcMg	35	90	0.23
PcCd	24	88	0.23
PcFe	26	81	0.28
PcMn	36	77	0.4
PcCo	16	55	0.22
PcPb	39	49	0.55

The doping is achieved by dissolving PcH$_2$ and the dopant PcM in concentrated sulfuric acid followed by reprecipitation by H$_2$O.
[a] Concentration of dopant: 5000 ppm; [b] fluorescence efficiency in arbitrary units.

Electric-field-induced fluorescence-quenching measurements as a function of applied field, light intensity, and fluorescence wavelength have been performed on PcH$_2$ thin films or on x-PcH$_2$-particle dispersions in polyvinyl acetate (PVA) [441–443, 525]. They confirm that the first excited state is involved in the carrier photogeneration process. The charge-carrier photogeneration efficiencies, φ_{pe}, for x-PcH$_2$ doped with metallophthalocyanines have been determined [442] (Table III.15). PcPb seems to be a promising dopant since it increases the charge-carrier photogeneration efficiency while it decreases the resistance of the molecular semiconductor layer.

At sufficiently high illumination, reaction between two excited states may occur. Fluorescence-quenching measurements for solutions of PcH$_2$ in 1-chloronaphthalene at 77 K have indicated that singlet-triplet fusion might take place. The role of singlet-triplet fusion on the photogeneration of charge carriers has been discussed [444].

b Photovoltaic Effect: Generalities

The photovoltaic effect associated with molecular semiconductors was first detected at the end of the forties [445]. Shortly after this first discovery, several groups described molecular solar cells based on metallophthalocyanine semiconductors [446–448]. At that time, the detailed chemical events occurring at the interfaces to give rise to the photovoltaic effect were unknown. Surprisingly, despite a great deal of work on molecular solar cells, the understanding of the chemistry and the photochemistry arising at the junctions has made little progress. Several fundamental questions have not been answered. What is the mechanism of the formation of the space-charge region? What are the ionizable impurities giving rise to the space-charge region? By how much does the space-charge region extend into the molecular semiconductor? What role in the electrical properties of Schottky contacts is played by the oxide layers on the metallic electrode? Only fragmentary answers have been provided. It might be thought that inorganic semiconductor devices are more thoroughly understood.

However, it is only recently that even the simple n-Si/Au Schottky contact has been analyzed in depth [449, 450, 526–528]. The mechanism of formation of the space-charge region in the dark will be first examined. Then, in the light of these results, an attempt will be made to rationalize the photovoltaic properties of PcM-based devices.

c Junction Studies in the Dark

Most of the studies on molecular semiconductors have been made using Schottky-type devices. Studies on polycrystalline films of PcM, $M = H_2$, Cu, Ni, Mn showed that a small amount of a liquid polar impurity at the interface was essential to obtain a rectification effect in the dark [451]. It was postulated at that time that an ionic space-charge barrier in the vicinity of the least noble electrode was necessary in order to obtain a significant rectification ratio. Various chemicals were tested for their ability to give rise to this ionic space-charge barrier. From the tested agents sulfuric acid, oxalic acid, acetic acid, and dilute hydrochloric acid, the first was the most efficient "activating agent". These early qualitative observations have not been followed by more thorough determinations.

A huge influence of O_2 on the junction properties of PcM's is expected. Figure III.31 shows a current/voltage plot of a Au/PcZn/Al cell made without breaking the vacuum at any stage (10^{-7}–10^{-8} torr) [452]. *No rectification effect is observed.* The I–V curve is almost perfectly symmetrical, although the two metallic electrodes have different work functions ($\Phi_{Au} = 5.1$ eV, $\Phi_{Al} = 4.28$ eV) [453]. In the low-voltage range studied (± 2 V), the plots of log I vs log V give a straight line with a slope of 1.2–1.5, i.e.:

$$I \sim V^n \qquad\qquad (III.45)$$

$$n = 1.2\text{–}1.5$$

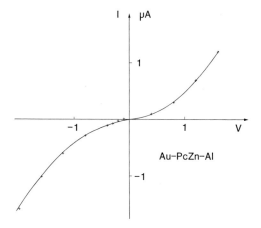

Au–PcZn–Al

Fig. III.31. Current-voltage plots for organic solar cells of the type Au/PcZn/Al made and studied entirely under vacuum (10^{-7}–10^{-8} torr). The thickness is of the PcM is 2.0 μm. (After Ref. [452])

The two electrodes are thus ohmic and do not limit the conduction. It was seen previously that the current is proportional to the voltage when the number of charges injected from the electrodes is less than the density of the charge carriers generated within the organic material. But the same relationship is found at low voltages for space-charge-limited currents. The two mechanisms may only be differentiated by determining the thickness dependence of the current.

At higher voltage ($|V| > 3$–5 V), the currents flowing through the Au/PcZn/Al device become space-charge limited. The log I–log V plots remain linear but the parameter n rises to approximately 3 [452]. The carriers are trapped by local states exponentially distributed in energy within the band gap. The electrical behavior of Au/PcNi/Al devices is almost identical at both negative and positive high voltages. The electric field corresponding to the higher voltage is of the order of $5 \cdot 10^4$ V $\cdot$ cm^{-1} [452]. .

M_1/PcH$_2$/M_2 devices (M = Al, Au, Pb) made under ultra-high vacuum conditions had been previously reported [454]. In this case, contrary to the above results, a rectification effect was observed. The dark current is lower for a given voltage when the non-substrate electrode, M_2, is positive. This probably indicates a nonsymmetrical distribution of traps in vicinity of the two metallic electrodes. M_1, which is in contact with the glass substrate, must present less structural irregularities than M_2, which is deposited onto the PcM polycrystalline film.

The behavior of the cells is drastically different in the presence of oxygen (Fig. III.32). The O$_2$ doping is achieved by leaving the PcZn thin films in air for 10 min before a semi-transparent overlayer of aluminum is deposited. All subsequent studies are carried out without breaking the vacuum (10^{-7} torr). This brief exposure to

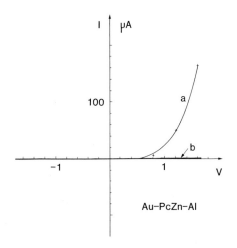

Fig. III.32. Current-voltage plots for Au/PcZn/Al devices. Measurements are made under vacuum (10^{-7} torr).
a: PcZn is first sublimed on Au and left 10 min in air before the layer of Al is deposited. All subsequent studies are made under vacuum (10^{-7} torr).
b: The depositions of PcZn and Al and the determinations of the characteristics are all achieved without breaking the vacuum at any stage. (After Ref. [452])

air is sufficient to give rise to a very large rectifying effect (r_{rf} = 16,000 at ± 1.6 V) as compared with the cell entirely made under high vacuum (r_{rf} = 1.2). The sensitivity to O_2 has been found to be highly dependent upon the nature of the central metal ion. For PcNi, for example, O_2 doping must be carried out at 150 °C under an atmosphere of pure O_2 (15 h) to demonstrate the same rectifying effect (Fig. III.33). At room temperature, noticeable effects are obtained only after several weeks of exposure. For comparison, a PcNi thin film was treated at 150 °C under argon, left 10 min in air and then studied under vacuum (Fig. III.33). Although the "apparent" resistance of the thin layer is markedly lowered, no rectification effect is observed. It was once assumed that rectification is due to the presence of a thin layer of Al_2O_3 at the interface between PcM and the aluminium electrode. To test this hypothesis, Al was deposited first and left 5 min in air to be superficially oxidized. Then, overlayers of PcZn and Au were successively evaporated under vacuum. Without breaking the vacuum, the I–V characteristic in the dark was measured. No rectification was observed (r_{rf} = 0.9 at ±1.6 V) [452]. The same results have been obtained for Al/PcCu/Au devices [459, 460]. When aluminum is first deposited and left in air to be superficially oxidized, only a small rectifying effect is noticed. To show how the rectification really arises, a sandwich cell Au/PcZn/Au was prepared where the organic layer had been left 5 min in air before deposition of the gold over-layer [452]. *Although symmetrical and non-oxidable electrodes were used, a strong rectification effect was observed* (r_{rf} = 6100 at ±1.5 V). It is therefore evident that a rectification effect is obtained whenever an unsymmetrical concentration of O_2 is present at the interface with the two metallic electrodes. The superficial metal-oxide layers do not directly intervene in the diode behavior of the sandwich cell.

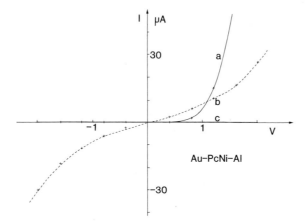

Fig. III.33. Current-voltage characteristics for Au/PcNi/Al devices.
a: PcNi O_2-doped at 150 °C, 15 h, 1 atm. of O_2.
b: PcNi treated under argon at 150 °C, 15 h.
In cases a) and b), after the deposition of the aluminum electrode, the I–V plots are drawn under vacuum (10^{-7} torr).
c: Device made and studied entirely under vacuum. In all cases the thickness of the organic layer is 2.0 µm. (After Ref. [452])

There have been other studies demonstrating the dramatic effect of exposure of the cells to air on their electrical behavior [455–461]. Doped and undoped Al/PcMg/Ag diodes have been prepared [456]. The undoped cells are made and studied in 10^{-5} torr vacuum without contact with the atmosphere. The doped cells are exposed to air for 5 min and heated at 50 °C in 10^{-5} torr to remove the excess dopant. The rectification coefficient in undoped cells does not exceed 2($V \leq 1.5$ V), while the doped cells show rectification ratios reaching 40. The same behavior is observed for Au/PcMg/Al devices [455]. The effects of O_2 on the dark rectifying properties of junctions are very difficult to reproduce accurately, since the concentration of O_2 in the polycrystalline thin films is hard to control. The state of purity of the material, the size of the micro-crystals, the temperature, the humidity, all influence the rate of migration of O_2 in the PcM thin films.

The I–V characteristics of M_1/PcM/M_2 devices have been the object of various publications [453, 454, 456, 466–468]. Knowing the current-voltage relationship it is possible, in principle, to determine the mechanism of charge transport (Fig. III.34). The main mechanisms of transports in insulators are briefly recalled (see also Chapter II).

— The Schottky emission mechanism. The carrier transport is limited by a thermo-ionic emission process, the interfacial energy barrier ΔE_{MI} has to be overcome. When both electrodes are identical, the current-voltage relationship is given by [462]:

$$\log I \sim V^{1/2} \qquad\qquad (III.46)$$

— The tunnel mechanism. For very thin insulating layers, charge carriers may go from one electrode to the other by tunneling. Depending on the relative magnitude of the applied voltage and of the barrier heights at the interface, several cases are possible [463, 464].

— The Frenkel-Poole emission mechanism. This is due to the field-enhanced thermal excitation of trapped electrons into the conduction band [465]. The I–V relationship is identical to that of the Schottky emission mechanism, except that the depth of the trap potential replaces the barrier height in the equation.

— The SCLC mechanism. Charge carriers are injected from the electrodes into the organic material. The corresponding I–V relationships have been established previously. At high voltages a log I–log V relationship is expected, while at low voltages the current is directly proportional to the voltage.

— The classical Schockley equation may be applied to standard Schottky junctions, it predicts a linear relationship between log I and V [95].

The Schottky emission is the predominant conduction mechanism for Al-Al_xO_y/PcCu/Al_xO_y-Al devices [456]; the log I vs $V^{1/2}$ plots give straight lines in the potential range of 1–6 V. Schottky emission occurs through the insulating aluminum layers, and the slopes of the log I–$V^{1/2}$ curves allow their thickness to be determined. The values found (25–90 Å) are in agreement with those found for aluminum thin films subjected for a short time to the ambient atmosphere. At higher voltages a space-charge-limited current mechanism becomes predominant. For the air-exposed devices NESA*/x-PcH_2 in PVA/Al [466], Al/PcMg/Ag [454], and Al/PcMg/Ag [453], the

---

* NESA is the abbreviation for the registered mark NESATRON. It is a conductive transparent glass made of doped indium oxide.

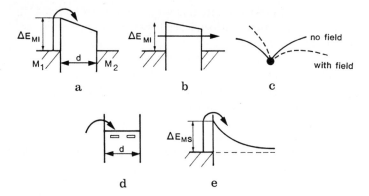

Fig. III.34. Basic conduction mechanisms for sandwiches metal/insulator/metal with their expected I–V dependence.

a: Schottky emission

$$\Delta = \Delta_1 - \Delta_2$$

— Symmetrical case $\Delta = 0$

$$\log J \sim V^{1/2}$$

— Unsymmetrical case $M_1 \neq M_2$ high voltage:

$$J \sim a \cdot T^2 \cdot \exp\left(-\frac{b}{T} + \frac{c \cdot \Delta}{d \cdot T}\right) \cdot \left[1 - \exp\left(-\frac{e \cdot V}{kT}\right)\right] \qquad \text{(Ref. [462])}$$

b: Tunneling
   — Unsymmetrical case $M_1 \neq M_2$

$$V \sim 0 \qquad\qquad J \sim V \quad \text{(ohmic)}$$

$$0 < V < \frac{\Delta E_{MI}}{e} \qquad J \sim a \cdot (V + b \cdot V^3)$$

$$V > \frac{\Delta E_{MI}}{e} \qquad J \sim V^2 \cdot \exp\left(-\frac{a}{V}\right)$$

(Ref. [463, 464])

c: Frenkel-Pool emission.
Same relationships as Schottky emission, but ΔE_{MI} is replaced by the depth of the trap potential. (Ref. [465]).

d: Space-charge limited currents

$$\text{low voltage} \quad J \sim \frac{V}{d^{2b+1}}$$

$$\text{high voltage} \quad J \sim \frac{V^{1+b}}{d^{2b+1}}$$

(Ref. [105–110]).

e: Classical Schottky contact

$$J \sim \left[\exp\left(\frac{q \cdot V}{kT}\right) - 1\right]$$

$$\log J \sim V$$

(Ref. [95])

electrical data may be fitted to a Schockley equation at very low voltages (0 to 0.4 V). At higher voltages (0.4 to 2 V), a quadratic dependence between the current density and the applied voltage is observed, indicating SCLC effects. In all three cases a rectifying effect was observed. In/PcH$_2$/Au and In/PcZn/Au devices have been O$_2$ doped by leaving the samples 1–3 days in dry air [467]. The rectification ratio is higher in the case of PcZn ($r_{rf} = 10^3$) than for PcH$_2$ ($r_{rf} = 200$). However, both cells demonstrate very similar I–V characteristics. A transition around 0.3 V between a region where the current varies exponentially with voltage to a region where the current is space-charge limited is observed. The electrical behavior of Al/PcH$_2$/Au and Al/PcZn/Au cells is perturbed by the superficial oxidation of the aluminum electrode, the forward-biased I–V curve adheres to a relation of the type:

$$I \sim a \cdot \exp (b \cdot V^{1/2}) \qquad\qquad (III.47)$$

a and b: constants

This has been interpreted as evidence for a predominant Frenkel-Poole emission process (field-assisted thermal detrapping) or Schottky emission mechanism (field-assisted thermoionic emission over a surface barrier) arising through the thin insulating Al$_2$O$_3$ layer. The electrical studies of thin layers of microcrystals of x-PcH$_2$ (diameter 800 Å) dispersed in various polymers: polycarbonate, polyvinylacetate (PVA), or polyvinylcarbazole (PVK) [468] have also been determined. Typically, in In/xPcH$_2$ in PVA/NESA cells, an ohmic region ($I \sim V^{1.1}$) is followed by a superquadratic dependence of current on voltage ($I \sim V^{4.7}$) in the potential range 200–600 mV.

The interface between thin films of PcM and metallic electrodes has been examined by transmission electron microscopy [469] in order to better understand the junction properties. Au/PcZn/Au sandwiches have been observed transversally by embedding the devices into an epoxy resin. Electron microscope observations reveal that the interfaces are sharp and well defined. The organic layer, however, sometimes shows pinholes which are filled with metal. This suggests that the metal, Au or Ag, can migrate on the PcM surface to fill the pinholes even at room temperature. There is, on the other hand, no interdiffusion of Au into PcZn.

From all these I–V determinations in the dark, several conclusions may already be drawn. First, O$_2$ is essential for the establishment of a space-charge region; the rectification ratio is directly related to the presence of O$_2$. While aluminum electrodes are superficially oxidized in air, the insulating Al$_2$O$_3$ is not responsible for the rectification effect. Devices with two similar noble-metallic electrodes such as Au/PcM/Au, give rise to a rectification effect under conditions where an unsymmetrical distribution of O$_2$ within the organic material is generated. Second, a space-charge layer is definitely formed within the organic material. The depletion region is probably formed by ionization of PcM,O$_2$ complexes through the equilibrium:

$$PcM,O_2/metal \rightarrow PcM,O_2^-/metal (+) \qquad\qquad (III.48)$$

However, the ionized charges are presumably firmly trapped, and this can induce characteristics distinct from usual Schottky devices. Capacitance measurements can help to elucidate this mechanism.

Whenever the number of free charge carriers in the molecular material is sufficient ($n > 10^{16}$ cm^{-3}), classical interpretations may be used to rationalize the capacitance measurements. The capacitance varies with the applied voltage with a $1/C^2$ vs V dependence. The width of the depletion layer l_{sc} is then obtained from the relation II.118, for n-doped materials:

$$l_{sc} = \sqrt{\frac{2 \cdot \varkappa_s \cdot (V_{bi} \pm V)}{q \cdot N_d}}$$

(III.49)

N_d: density of ionized donors
$\varkappa_s$: dielectric constant of the semiconductor
V_{bi}: built-in potential

On the hand, for very low carrier concentrations ($n \ll 10^{16}$ cm^{-3}), the capacitance is voltage independent, its value is only related to the geometric parameters of the cells: thickness of the organic layer, shape, overall dielectric constant, etc. (geometric capacitance). In PcM's, both voltage-dependent and voltage-independent capacitances have been found depending on the frequency of the probing voltage, the temperature, and the mode of preparation of the sample [456–458, 468]. The usual assumption made to interpret inorganic device measurements is that the number of trapped charges is less than the number of free carriers. The reverse is certainly true for most

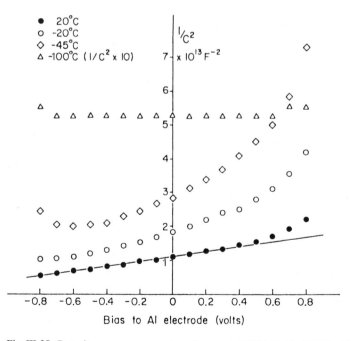

Fig. III.35. Capacitance measurements on the system NESA/PcMg/Al. The $1/C^2$ vs V plots are shown for different temperatures. The frequency of the applied field is 1 Hz. [Reproduced with permission of (470)]

molecular semiconductors [470]. Most capacitance measurements have been performed with high-frequency probing voltages (10^4 Hz or higher) on samples maintained at room temperature. To detect the capacitance corresponding to the space-charge region, the oscillating probe voltage must be able to expand and contract the Schottky layer at the frequency of the measurements [470]. If the carriers are immobilized by trapping, they cannot respond to the high-frequency probe voltage. Capacitance measurements must therefore be made at very low frequencies (10^{-2} — 100 Hz). The trapped charges may additionally be "mobilized" by light or by rising the measurement temperatures [470]. In NESA/PcMg/Al devices, at low frequencies (1 Hz), and at −100 °C, the capacitance is independent of the voltage; it becomes voltage dependent only at 20 °C (Fig. III.35) [470]. A sigmoidal curve is obtained for the variation of the capacitance with temperature (Fig. III.36) [470]. The geometric capacitance is measured at low temperatures; within the high-temperature range the characteristics of the space-charge region can be determined. The temperature T_{sc} at which the capacitance drops to half its depletion-layer value is frequency dependent, increasing with the frequency of the probing voltage. The measure of T_{sc} allows an estimate of the thermal activation energy needed to detrap the charge carriers. This activation energy is highly dependent upon the conditions of preparation of the cells and it seems difficult to obtain reproducible values. The values of 0.42 eV and 0.26 eV have been obtained for two different preparations. These difficulties are not really surprising, since the concentration and the distribution of O_2, which is most probably involved in space-charge region formation, are not controlled. The analyses of the $1/C^2$ vs V plots for the NESA/PcMg/Al cell give, at 20 °C, barrier widths varying from 60 to 210 Å and concentrations of ionized impurities varying

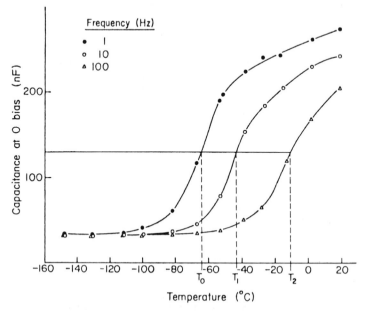

Fig. III.36. Capacitance vs temperature curves for the cell NESA/PcMg/Al. The frequency of the applied voltage is parametrically varied. [Reproduced with permission of (470)]

between $2 \cdot 10^{18}$ and 10^{19} per cm^3. Such densities correspond to one acceptor for 10^2 or 10^3 molecules of PcM [470]. NESA/x-PcH$_2$/In cells have also been characterized by low-frequency capacitance measurements [471]. The determinations have been performed both in the dark and under illumination. The space-charge density in the depletion region is determined either by shallow traps or by deep acceptor levels depending on the bias conditions. The capacitance-voltage relationship $(1/C^2$ vs V) is described by two intersecting lines; the corresponding trap densities are $6.3 \cdot 10^{16}$ cm^{-3} for the shallow levels and $2 \cdot 10^{16}$ cm^{-3} for the deep acceptor levels. Irradiation markedly increases the number of filled shallow traps and, correspondingly, the space-charge density. The experiments were performed on cells exposed to ambient atmosphere for 3 months. The deep acceptor traps may be tentatively assigned to the complex PcM,O$_2$, while the shallow traps are probably associated with structural defects. Capacitance measurements on NESA/PcZn/Al devices confirm that O$_2$ is involved in the Schottky-barrier formation [529]. The space-charge density is seen to be inhomogeneous close to the aluminum contact. The electrical behavior is not the result of the creation of an Al$_2$O$_3$ layer [529].

Capacitance measurements on Al/PcCu/Al entirely prepared in a vacuum of 10^{-5} torr have been described [472]. The temperature dependence of capacitance (at 1 and 5 kHz) has been determined (Fig. III.37). In the model of Simmons et al. [128] (see Chap. II), the capacitance at low temperature and high frequencies is defined both by the capacitance of the bulk C$_b$ and by the capacitance of the two Schottky barriers C$_j$:

$$C_p \sim \left[\frac{2}{C_j} + \frac{1}{C_b}\right]^{-1} \qquad \text{(III.50)}$$

C$_b$: bulk capacitance
C$_j$: junction capacitance

When the temperature is raised, the impurities start to ionize and the bulk resistance becomes negligible compared to the junction resistances. At high temperatures and low frequencies the capacitance of the system is therefore governed only be the capacitances of the two Schottky barriers:

$$C_p \sim \frac{C_j}{2} \qquad \text{(III.51)}$$

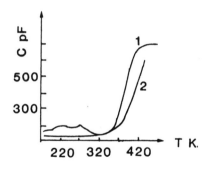

Fig. III.37. Capacitance vs temperature curves for the cell Al/PcCu/Al prepared under vacuum (10^{-5} torr) (1) at 1 kHz; (2) at 5 kHz. (After Ref. [472])

The depth of the trapping levels, E_t, may be determined from the temperature and frequency dependence of the capacitance. In the case of the Al/PcCu/Al cell, E_t is found to be equal to 0.54 eV. The chemical nature of this level is difficult to identify since any intermediate contact with air has been avoided in these experiments. The capacitance of the junction, C_j, may be easily related to the width of the Schottky barrier [474, 475], yielding a value of 1000 Å, which is almost 10 times larger than the one previously found in presence of air.

Capacitance measurements on single crystals of PcPb have also been described [473]. Both the temperature and the frequency dependences of the capacitance are highly influenced by the presence of O_2 (Fig. III.38). O_2 doping is achieved by baking the crystals at 240 °C in oxygen. The capacitance of the junction, C_j, decreases from 1940 pF to 280 pF by O_2 doping. This corresponds to an increase in the concentration of the ionizable impurity centers and, thus, correspondingly to a decrease of the width of the space-charge region.

The I–V and C–V characteristics of sandwiches of PcH$_2$ have also been studied by capacitor discharge measurements [476, 477]. There is, however, some controversy over this method of measurement [478, 479].

In conclusion, both I–V and C–V studies agree with the fact that O_2 is implicated in the formation of the space-charge region. A rectification effect is observed whenever there is an asymmetrical distribution of O_2 in the organic material sandwiched between the two metallic electrodes. The ability of the metal to axially coordinate extra ligands is an important parameter determining the amount of O_2 bound, but the crystalline state, the purity of the material, or the state of surface are also important factors. It may be now wondered whether the dark rectifying behavior is correlated with the photovoltaic properties of the cells.

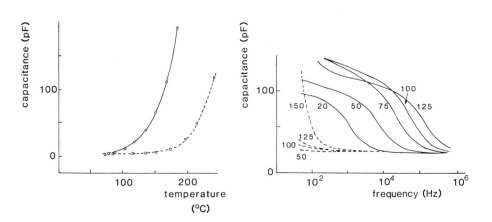

Fig. III.38. Temperature dependence and frequency dependence of the capacitance of single crystals of undoped or O_2-doped PcPb. Doping is achieved by a treatment at 240 °C in O_2 for 8 hrs, the capacitances are measured at 1 kHz.
Full lines: O_2-doped; Dotted lines: undoped. (After Ref. [473])

d Junction Studies under Illumination

Under irradiation, solar cells are usually characterized by the open-circuit photo-voltage V_{oc} and the short-circuit photocurrent I_{sc} (see Chapter II). For classical Schottky barriers, V_{oc} cannot exceed the built-in potential V_{bi} (Fig. III.39a)

$$q \cdot V_{bi} = \Phi_M - \Phi_{PcM} \tag{III.52}$$

Φ_M, Φ_{PcM}: work functions of the metal and of the PcM, respectively

$$\boxed{q \cdot V_{oc} \leq \Phi_M - \Phi_{PcM}} \tag{III.53}$$

In the case where the molecular material is considered as an insulator (Fig. III.39b), no charge transfer between the organic layer and the electrodes may take place and the two metallic electrodes M_1 and M_2 adjust their Fermi levels respectively to each

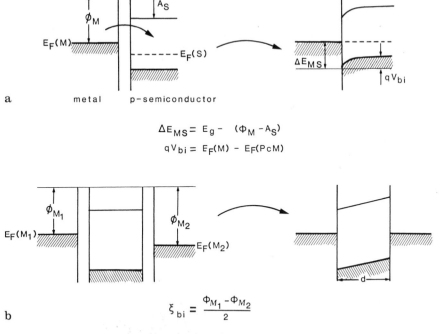

Fig. III. 39. Energy schemes of Schottky-type devices.
a: Semiconductor/metal contact, the built-in potential is determined by ionized impurities.
b: Metal/insulator/metal contacts. The two electrodes are "rectifying", there are no ionizable impurities within the organic material, the electrical field is constant throughout the organic layer.

other through the external circuit. A constant electrical field ξ_{bi} is generated in the organic layer such that [92, 93, 454, 480]:

$$\xi_{bi} = \frac{\Phi_{M_1} - \Phi_{M_2}}{d} \tag{III.54}$$

Φ_{M_1}, Φ_{M_2}: work function of the two metallic electrodes
d: thickness of the organic layer

In this case, V_{oc} cannot exceed the potential difference between the two metallic electrodes:

$$\boxed{q \cdot V_{oc} \leq \Phi_{M_1} - \Phi_{M_2}} \tag{III.55}$$

It is therefore relatively easy to predict the upper limit of V_{oc} under strong irradiation conditions. Figure III.40 shows the approximate band diagram for typical PcM's relative to the work functions of the most usual metallic electrodes. The position of the Fermi level within the band gap depends on the doping level. Heavy doping with acceptor impurities causes the Fermi level $E_F(PcM)$ almost to coincide with the valence-band edge situated in the region 4.95–5.2 eV depending on M. In the absence of dopant $E_F(PcM)$ should lie very near the middle of the gap at 4 eV. This last value has been confirmed experimentally by contact potentials determinations [481].

The equivalent electrical circuit of a solar cell has been represented in Fig. II.23. For negligible shunt and series resistances it has been shown that:

$$I_{sc} = I_{ph} \tag{III.56}$$

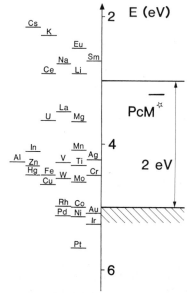

Fig. III.40. Energy levels of the valence and conduction bands of PcH$_2$ relative to the work functions of metals.
• Metal work functions: from Ref. [453].
• Surface ionization potentials of PcM's: from Ref. [292] and [303].
M = H$_2$: 5.2; Zn: 5.0; Cu: 5.0; Ni: 4.95; Fe: 4.95

and

$$V_{oc} = A_o \cdot \left(\frac{kT}{q}\right) \cdot \ln\left(\frac{I_{sc}}{I_{oo}} + 1\right)$$ (III.57)

A_o: junction perfection factor
I_{oo}: preexponential factor of the dark current

The dependence of the steady-state photocurrent on the light intensity has been thoroughly studied [482–484]. In the simplest case, the photocurrent is proportional to the number of absorbed photons:

$$I_{ph} = e \cdot \mathcal{G} \cdot N_{abs}$$ (III.58)

$\mathcal{G}$:   gain factor
N_{abs}: number of absorbed photons

The gain factor $\mathcal{G}$ is the number of carriers passing through the outer circuit divided by the number of photons absorbed. Since the number of absorbed photons is proportional to the incident photon flux Φ_o, it can be seen that:

$$I_{sc} \sim e \cdot G \cdot \Phi_o$$ (III.59)

Different modes of recombination of charge carriers induce different dependences of the steady-state photocurrent with light intensity. If free charge carriers recombine through a monomolecular process:

$$n \rightarrow deactivation$$ (III.60)

then the photocurrent is still proportional to the incident photon flux. If, on the other hand, bimolecular processes occur:

$$n + n \rightarrow deactivation$$ (III.61)

then the photocurrent is no longer directly proportional to the incident flux, but:

$$I_{ph} \sim \Phi_o^{1/2}$$ (III.62)

Various other types of recombination mechanisms have been envisaged and it has generally been found that [488]:

$$I_{ph} \sim \Phi_o^m$$ (III.63)

with $0.5 < m < 1$.
The parameter m is only rarely lower than 0.5 though a model has been elaborated which gives $m = 0.33$ [489, 490]. This unusual power-law dependence is found by

assuming that the carriers are generated at the surface and that they are destroyed in the bulk in a second order process. Superlinear photoconductivity (m > 2) indicative of two-photon processes is also possible [15].

However, in the case of low conducting materials, the intensity of the current flowing through the molecular material may no longer be limited by the number of absorbed photons but may become space-charge limited [485–487]. In this case, the previous relations are of course not applicable, and equations of the type of Eq. II.79 must be used.

A relationship correlating I_{ph} with the exciton diffusion length and the probability of interfacial charge transfer has been established previously (Eq. II.102). This equation, however, does not take into account the presence of a space-charge region. In the case of Schottky contacts, a more suitable equation may be derived [457, 491]. Two terms must be distinguished depending on whether the carriers are photogenerated within or outside the space-charge region. The number of photocarriers produced within dx at a distance x from the interface is proportional to:

$$\eta_{cc} \cdot \Phi_o \cdot \alpha_a \cdot \exp\left(-\alpha_a \cdot x\right) dx \tag{III.64}$$

η_{cc}: quantum yield for charge-carrier generation
Φ_o: incident photon flux
α_a: absorption coefficient

The total number of carriers produced in the depletion region n_{sc} is, assuming that η_{cc} is constant in this region:

$$n_{sc} = \eta_{cc} \cdot \Phi_o \cdot \int_0^{l_{sc}} \alpha_a \cdot \exp\left(-\alpha_a \cdot x\right) dx \tag{III.65}$$

l_{sc}: width of the space-charge region

$$n_{sc} = \eta_{cc} \cdot \Phi_o \cdot \left[1 - \exp\left(-\alpha_a \cdot l_{sc}\right)\right] \tag{III.66}$$

If the photon is absorbed outside the depletion region at a distance x from the interface, a fraction proportional to:

$$\exp\left(-\frac{x - l_{sc}}{l_{dif}}\right) \tag{III.67}$$

l_{dif}: carrier diffusion length

will diffuse to the space-charge layer. By integration over the overall thickness of the organic material it is found that:

$$n_b = \eta_{cc} \cdot \Phi_o \cdot \left[\frac{\alpha_a}{\alpha_a + 1/l_{dif}} \cdot \exp\frac{l_{sc}}{l_{dif}}\right] \cdot \{\exp\left[-(\alpha_a + 1/l_{dif}) \cdot l_{sc}\right]$$
$$- \exp\left[-(\alpha_a + 1/l_{dif}) \cdot d\right]\} \tag{III.68}$$

d: thickness of the organic material

In the case where $\alpha_a \cdot l_{sc} < 1$, the overall short-circuit photocurrent $n_{sc} + n_b$ may be approximated by:

$$I_{sc} \sim q \cdot \eta_{cc} \cdot \Phi_o \cdot \left\{ \alpha_a \cdot l_{sc} + \left(\frac{\alpha_a}{\alpha_a + 1/l_{dif}} \right) \cdot \exp\left(-\alpha_a \cdot l_{sc} \right) \right\} \qquad (III.69)$$

A very similar equation would be obtained if exciton migration up to the space-charge region were considered rather than charge-carrier migration [468, 490]. The exciton diffusion length must then be substituted for the charge-carrier migration length.

The effect of O_2 on the photovoltaic behavior of $M_1/PcM/M_2$ devices is, as expected, drastic. In absence of O_2, a Au/PcNi/Al system shows an ohmic behavior in the dark and no photovoltaic effect ($V_{oc} = 0$; $I_{sc} \sim 0$). When the metallo-organic layer is treated at 150 °C in an atmosphere of pure dioxygen, a strong rectification is observed in the dark, and a large photovoltaic effect arises ($V_{oc} = 680$ mV; $I_{sc} = 5.4$ µA/cm^2; 100 mW/cm^2 white light) (Fig. III.41).

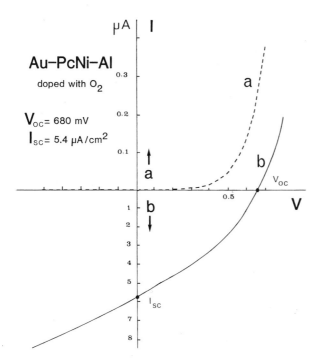

Fig. III.41. Electrical properties of O_2-doped Au/PcNi/Al devices.
a: In the dark.
b: Under illumination (100 mW/cm^2; white light).
Doping is achieved by treating the PcNi layer at 150 °C for 15 hrs in a pure atmosphere of O_2. The electrical characterizations are made under vacuum (10^{-5} torr). (After Ref. [452])

Further experiments have been reported to elucidate the chemical mechanism underlying this photovoltaic effect [452]. Two series of experiments have been described. In a first one, Au/PcZn/M sandwiches have been realized by successive vacuum sublimations. The current-tension (I–V) characteristics in the dark and under illumination are measured in situ. The vacuum (10^{-7}–10^{-8} torr) is not broken at any stage of the experiments. Irradiation is carried out through the last deposited semi-transparent electrode M. The nature of M is varied as to span a wide range of work junctions from gold ($\Phi_{Au} = 5.47$ eV) to samarium ($\Phi_{Sm} = 2.7$ eV). These results are compared with those obtained for O_2-exposed molecular solar cells. The conditions of doping by O_2 have been strictly defined and standardized to all the cells. The organic layer is left 1 hour at air and then treated at 150 °C for 15 hrs at 10^{-2} torr. This treatment is necessary to standardize the amount of O_2 remaining in the organic layer. It has been checked that the photoelectrical properties of the organic layer is not altered by this heat treatment. The cell is again exposed to air for 10 min and the last metallic overlayer is sublimed. The cells are then transferred into a jar where further studies are made under vacuum (10^{-4}–10^{-5} torr).

The electrical properties of the Au/PcZn/M devices entirely made under vacuum (conditions a) or exposed to air (conditions b) are shown in Table III.16. The rectification ratio (r_{rf}) at ± 0.5 V has been used to characterize the dark electrical behavior of the sandwiches. The open-circuit photovoltage V_{oc} and the short-circuit photocurrent I_{sc} are utilized to characterize the devices under illumination.

In a classical scheme, the space-charge region is formed by ionization of impurities present in the semiconductor near one of the interface. The space-charge region is detected in the I–V plots by giving rise to a rectification effect. For all air-exposed devices, a high rectification ratio varying from 20 to 80 at ± 0.5 V is observed; r_{rf} rises rapidly with the voltage considered; at ± 1 V, it reaches several thousands. On the contrary, no rectification effect in the dark is noticed for the cells made entirely under vacuum ($r_{rf} \sim 1$). From examination of Table III.16, it is clear that there is no straightforward relationship between the dark junction properties and the magnitude of the photovoltaic effect. In absence of O_2 (conditions a), a strong photovoltage is found, while the rectification ratios are close to unity. Additionally, there is no obvious relationship between the work function of the semi-transparent metallic electrode Φ_M and V_{oc}. On the contrary, in presence of O_2 (conditions b), both a strong rectifying effect and a strong photovoltaic effect are observed; V_{oc} increases when the work function of the metallic electrode is lowered. This behavior is expected for Schottky junctions between p-type semiconductors and metals. It consequently seems that at least two different mechanisms give rise to the photovoltaic effect. One of them is correlated to Φ_M and to the presence of O_2, the second one is independent of both of these parameters.

It is possible to tentatively assign these two contributions to chemical mechanisms. The semi-transparent electrode through which irradiation is performed is always negative; in consequence, the O_2-independent mechanism must involve some electronic transfer process between the metallophthalocyanine photochemically excited, PcM*, and the electrode:

$$\text{PcZn*/electrode} \rightarrow \text{PcZn}^+\text{/electrode} \qquad (III.70)$$

Table III.16 Electrical properties of Au/PcZn/M devices in the dark and under illumination (from Ref. [452]).

Metal	$\Phi_M{}^c$ (eV)	Conditions a			Conditions b		
		r_{rf} (±0.5 V)	V_{oc} (mV)	I_{sc} (μA/cm^2)	r_{rf} (±0.5 V)	V_{oc} (mV)	I_{sc} (μA/cm^2)
Au	5.47	1.0	450	0.6	1.2	20	3.2
Cu	4.59	1.0	0	0	3.	80	4.7
Cr	4.50	1.1	450	0.3	19.0	228	51
Al	4.41	1.0	420	0.3	82	640	144
In	4.12	0.5	100	0.015	39	300	57
Sm	2.7	1.0	230	0.2	?^{d}	1640	72

Thickness used: PcZn: 2 μm; Au and M: 300 Å with exception of In and Sm (e = 600 Å); Abbreviations: r_{rf}: Rectification ratio in the dark (±0.5 V); V_{oc}: open-circuit photovoltage; I_{sc}: short-circuit photocurrent.

  Conditions a: cells entirely made and studied under vacuum, without breaking the vacuum at any stage (10^{-7}–10^{-8} torr); irradiation conditions: 10 mW/cm^2, white light.

  Conditions b: the PcZn layer is first deposited on the gold counterelectrode left 1 h at air and treated at 150 °C under vacuum (10^{-2} torr) for 15 hrs, the organic layer is again left 10 min at air and the second metallic electrode is deposited; the studied are made under vacuum (10^{-4}–10^{-5} torr); irradiation conditions: 100 mW/cm^2 white light.

^{c}: From references [453].

^{d}: Too instable to be measured accurately.

On the other hand, the O_2-dependent contribution should occur through a two-step mechanism. In a first step, the ionization of O_2 gives rise to a space-charge region near one of the interfaces:

$$PcZn, O_2/\text{electrode} \rightarrow PcZn, O_2^-/\text{electrode} \qquad (III.71)$$

In a second step, the reverse reaction takes place under illumination:

$$PcZn^*, O_2^- / \text{electrode} \rightarrow PcZn, O_2 / \text{electrode} \qquad (III.72)$$

It is noteworthy that, in the first case, the metallic electrode acts as an acceptor of electrons, while the formation of the space-charge region in the second type mechanism arises by transferring electrons from the metal to the molecular material. This is in agreement with the different dependences of V_{oc}'s with Φ_M found in the two cases.

Under illumination, the log I–log V plots are still linear both under reverse- and forward-bias conditions. For O_2-doped Au/PcNi/Al devices, in the forward direction, the relationship $I \sim V^n$ is found with $n \sim$ 4–5, the same value of n was found in the dark, indicating a space-charge limit. Under reverse-bias conditions, the parameter n lies between 1 and 2, the trapping levels are probably all filled, and the conduction is governed by the trap-filled limit regime.

The Schottky barrier parameters may be characterized in more detail by two independent methods: the determination of the capacitance and the measure of the photovoltaic action spectra. Capacitance measurements at low frequency have allowed an estimation of the width of the space-charge region (100–500 Å) [457, 466,

468, 470, 471]. This is probably an order-of-magnitude estimate of the distance over which the O_2 molecules are ionized to form O_2^-. Photovoltaic action spectra provide further information about the interfacial active layer [457, 466, 468, 490]. Action spectra are obtained by illuminating the front or the back electrode by monochromatic light and by measuring the corresponding short-circuit current generated (Fig. III.42). A strong correlation between the photovoltaic current and the absorption spectrum of the films is observed when the cell is irradiated through the semitransparent rectifying electrode (Al). In contrast, if the cell is irradiated through the ohmic electrode, the magnitude of the photocurrent is inversely proportional to the absorption coefficient of the PcM. This demonstrates that the ohmic interface is ineffective in creating charge carriers and that photons must reach the other interface through the organic layer to give rise to a photocurrent. However, most of the photons are absorbed in the bulk organic material without yielding charge carriers. An estimate of the width of the interfacial "active layer" may be calculated by fitting the data with the theoretical equations that relate I_{sc} to the absorption coefficient of PcM (Eq. III.69). A space-charge region of the order of 100–500 Å is found in this way [457, 466, 490]. Other calculations have been made taking into account exciton diffusion up to the space-charge region [468], yielding a value of 300–500 Å for the exciton diffusion length. These values must be treated with caution, however, since action spectra are rather insensitive to the choice of the parameters l_{sc} and l_{dif} (Eq. III.69) especially when the approximation $\alpha_a \cdot l_{sc} \ll 1$ is valid [490]. Nevertheless, all measurements are in accordance with an "active layer" of a few hundreds Å, extending into the molecular material from the rectifying electrode. Only this region is effective in generating the carriers, photons absorbed further within the bulk do not contribute to the photocurrent.

For none of the studied devices is the short-circuit photocurrent directly proportional to the intensity of the incident light [457, 466, 468, 490]. A power-law dependence of the form:

$$I_{sc} \sim \Phi_o^n \qquad\qquad (III.73)$$

Φ_o: incident photon flux

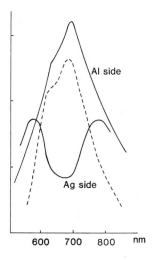

Fig. III.42. Photovoltaic action spectra of Ag/PcMg/Al cells. Full lines: photovoltaic action spectra by irradiation through the aluminum or the silver electrode.
Dotted line: absorption spectra of PcMg. The PcMg film thickness is 1500 Å and after fabrication the films were exposed to air. (After Ref. [457])

is observed with n varying from 0.3 to 0.8. One explanation for this suggests that at high light intensities, the photocurrent may be limited by space-charge effects. If this were the case, the density of traps and their distribution in energy within the band gap determine the magnitude of I_{sc}, so when the thickness of the organic material is varied, a d^{-3} dependence would be expected [485, 487]. But for NESA/x-PcH$_2$ in PVA/In devices, the photocurrent is found to be inversely proportional to the cell thickness in the range 0.2 μm to 4.0 μm [468]. Under the irradiation conditions used (1 mW/cm^2), this seems to exclude SCLC limitations. The other explanation suggests that the quantum yield for charge-carrier generation, η_{cc} in Eq. III.69, may decrease when the incident photon flux increases. η_{cc} is strongly field dependent [82–84, 492], principally because of the importance of geminate recombination in molecular materials. Under high irradiation conditions, the electric field within the space-charge region may be markedly decreased leading to a drop in the charge-carrier generation efficiency [466]. In NESA/x-PcH$_2$/In devices, the quantum efficiency of charge-carrier generation for photons entirely absorbed within the first few hundred angströms from the interface, in the space-charge region, has been calculated [468]. At 0.1 mW/cm^2 of transmitted 600 nm light, assuming that 60% of the photons are absorbed within the barrier region, the quantum efficiency is 73%; at 2 mW/cm^2, the quantum efficiency drops to 30%. Previous determinations on NESA/x-PcH$_2$/Al cells indicate similar values, $\eta_{cc} = 44\%$ at 6 μW/cm^2 and $\eta_{cc} = 62\%$ at 0.6 μW/cm^2 for 670 nm irradiation [466].

Transient action spectra for x-PcH$_2$/Al solar cells have been recently determined [499, 500]. The width of the space-charge region is varied by applying an external voltage. If the barrier width is larger than the exciton diffusion length, the corresponding action spectra of the photovoltaic effect must be correspondingly modified. If, on the other hand, the effective light collection region is determined by the exciton diffusion length, the external voltage should not affect the action spectra. It is found experimentally that the photovoltaic action spectra are modified significantly by external applied voltages; this suggests that the first hypothesis is correct.

e Effect of Doping on the Performances of Molecular Solar Cells

The effect of O$_2$ on the photovoltaic properties of molecular solar cells has been studied in detail in the previous section. Many attempts have been made to use additional doping agents to improve the electrical performances of molecular solar cells. Once again, it must be emphasized that in most cases the doping was made in addition to O$_2$ doping, no precautions being taken to avoid the presence of O$_2$ in the materials. The importance of such doping has long been recognized. PcMg pressed powders have been coated with a film of air-oxidized tetramethyl-p-phenylenediamine by evaporating an acetone solution of the amine [446–448]. In the absence of dopant, no photovoltage under irradiation was observed. Surface doping induces a photovoltage of 200 mV under strong illumination conditions (500 W lamp). Similar effects have been described for other devices [456, 493, 494]. Based on the same principle, solar cells consisting of two different organic semiconductors have been described [495–498]. Sandwiches such as NESA/PcM/organic dye/In have been realized (Fig. III.43). The organic dye is deposited on the PcM thin film by the spin-coating

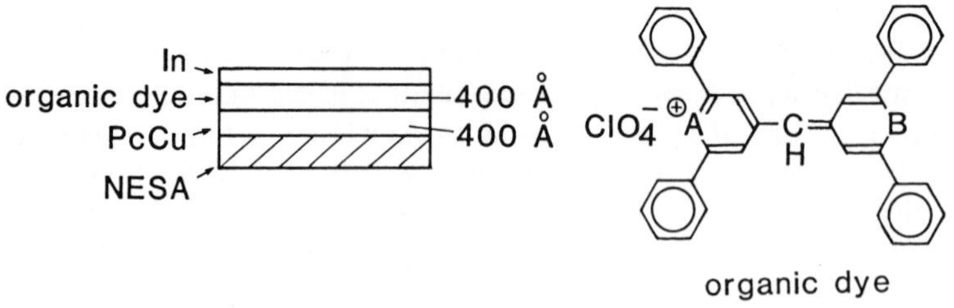

organic dye

Fig. III.43. Organic solar cells made up from two different dyes. (After Ref. [495]).

Substituents		V_{oc}[a]	I_{sc}[a]	Conversion[a] efficiency
A	B	(V)	mA/cm^2	(%)
O	O	0.55	0.9	0.29
O	S	0.45	1.6	0.43
S	Se	0.31	1.5	0.28
Se	Se	0.24	2.0	0.28
O	Se	0.42	0.4	0.10

[a] Air Mass 2 irradiation, Kodak 600 slide projector with glass and water filters (75 mW/cm^2).

Table III.17   Effect of additives on the photovoltaic properties of NESA/x-PcH$_2$ in MK/In cells. (MK = Makrolon 5705, a polycarbonate from Bayer). (From Ref. [468]).

Incident flux (mW/cm^2)[a]	Dopant								
	None			Coumarine 6			TNF		
	V_{oc}	I_{sc}	eff. (%)	V_{oc}	I_{sc}	eff. (%)	V_{oc}	I_{sc}	eff. (%)
5.2	0.45	122	0.35	0.45	396	1.17	0.42	420	1.13
0.54	0.42	21	0.55	0.42	66	1.81	0.38	84	1.96
0.135	0.39	7	0.64	0.38	25	2.3	0.35	27	2.30

The additives correspond to a loading of 14% by weight; solar stimulated irradiation; organic film thickness 1.8 μm; V_{oc} in volts; I_{sc} in μA/cm^2.
[a] Flux transmitted through the upper semitransparent electrode.

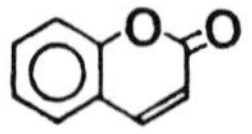

coumarine 6

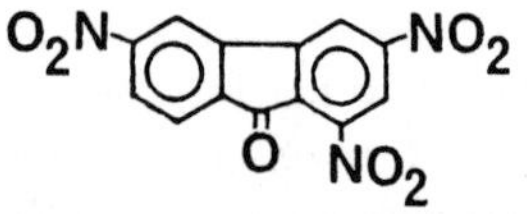

2,4,7-trinitrofluorenone (TNF)

method. High open-circuit photovoltages and high short-circuit photocurrents are obtained when the dye has a higher oxidation potential than the corresponding PcM. The energy conversion efficiencies of these devices are remarkably good and reach 0.5% for white light under natural condition of irradiation (70 mW/cm²) [495]. It has been assumed that the organic dye yields a rectifying interface with the metallophthalocyanine; however, the mechanism involved has not been further characterized.

In NESA/x-PcH₂,PVA/Al cells, a strong electron acceptor, 2,4,7-trinitrofluorenone (TNF), has been added to the organic layer [466]. Doping significantly increases the carrier generation efficiency η_{cc} from 20% in absence of dopant to 33% (monochromatic light, 670 nm, 50 µW/cm²). More thorough studies have been carried out on NESA/x-PcH₂/In cells [468] (Table III.17). The incorporation of 2,4,7-trinitrofluorenone or coumarine causes a large increase in the energetic conversion efficiencies of the solar cells. The open-circuit photovoltage does not significantly vary on doping, but the corresponding short-circuit photocurrent is markedly increased. At a solar input power of 0.14 mW/cm² (for light transmitted through the front semitransparent metallic electrode) the power conversion efficiency of the doped cells approach 2.3%. An increase of the input solar power to 5 mW/cm² causes the efficiency to drop to 1.1%.

f Solar Energy Conversion Efficiencies of Molecular Solar Cells

Much work has gone into increasing the energy conversion efficiencies of molecular solar cells. But it is very difficult to make a comparison between the results found in the literature because of the vastly different experimental conditions used. The power conversion efficiency of a cell is defined by the equation:

$$\chi_{pc} = \frac{P_{out}(max)}{P_{in}} \, 100 \qquad (III.74)$$

where $P_{out}(max)$ is the maximum power extractable from the cell and P_{in} is the power of the incident irradiation. The power conversion efficiency may be expressed as a function of the fill factor FF, the short-circuit photocurrent I_{sc}, and the open-circuit photovoltage:

$$\chi_{pc} = \frac{I_{sc} \cdot V_{oc} \cdot FF}{P_{in}} \qquad (III.75)$$

χ_{pc} is expressed under various experimental conditions and the following parameters must be considered to obtain comparable results:
— Is monochromatic or white light irradiation utilized?
— Is light transmitted through the semitransparent electrode considered in the calculation of the solar cell efficiency?
— What is the incident photon flux?
The efficiencies of the various molecular solar cells under "standard conditions" must be calculated. The results found in the literature will be, as far as possible, converted into "standardized yields" obtained under the following conditions:
— No correction for the transparency of the semitransparent metallic electrode is taken.

Table III.18 Power conversion efficiencies of the best organic solar cells based on metallophthalo-cyanine semiconductors published in the literature. (From Ref. [501]).

System	Ref.	Efficiencies originally described (%) and corresponding exp. conditions	Standardized yields[a] %
Ag/PcMg/Al	[456]	$1.5 \cdot 10^{-3}$; no FF white light; 0.5 mW/cm^2	$1.2 \cdot 10^{-4}$
Ag/PcMg/Al	[457]	10^{-3}; white light	?
NESA/x-PcH$_2$/Al	[466]	0.02; monochromatic light (670 nm); 100 mW/cm^2; corrected for transmission	$1.7 \cdot 10^{-3}$
Au/PcH$_2$/Al	[490]	$2.5 \cdot 10^{-4}$; monochromatic light (6328 Å); 67 mW/cm^2	$2.1 \cdot 10^{-5}$
Au/PcZn/In	[490]	0.023; same conditions	$1.9 \cdot 10^{-3}$
?/(PcAl)$_2$Se/?	[497]	3; unfiltered light; 1–100 lux	?
?/(PcV)$_2$O/?			
Au/PcZn/Cr	[501]		$7.6 \cdot 10^{-3}$
Au/PcZn/In	[501]		$5.9 \cdot 10^{-3}$
Au/PcMg/Cr	[501]		$1.2 \cdot 10^{-2}$
Au/PcMg/In-Al	[501]		$1.9 \cdot 10^{-2}$
?/PcInCl/?	[502]	0.5; 10 mW/cm^2	?
NESA/xPcH$_2$/In	[468]	1.2%; solar simulated; corrected for transmission	$\sim 1.2 \cdot 10^{-2}$

[a] See text for the definition of the standard conditions.

— White light irradiation is used (Halogen lamp equipped with a water filter).
— The white light irradiation is fixed at 100 mW/cm^2.

A comparison of the main systems exhibiting the higher conversion efficiencies described in the literature is shown in Table III.18 [501]. The best power conversion efficiency does not exceed $2 \cdot 10^{-2}\%$. Higher yields have been described using (PcAl)$_2$Se or (PcV)$_2$O; however, in these cases the experimental conditions were not sufficiently detailed to allow the calculation of the "standardized yields". The influence of the complexed metal ions on the performances of the organic solar cells strictly follow their O$_2$-binding ability. By order of decreasing efficiency, it is found [503]:

$$PcMg > PcBe > PcFe > PcCu > PcCo \qquad (III.76)$$

Metalloporphyrin complexes show a very similar correlation [504] (TPP: tetra-phenylporphyrin):

$$TPPMg > TPPZn > TPPCu > TPPH_2 > TPPCo \qquad (III.77)$$

This gives further evidence, if necessary, of the role of O$_2$ on the photovoltaic efficiencies of the solar cells.

The energy conversion efficiencies of metallophthalocyanine-based devices are fairly low especially when compared to merocyanine-based solar cells. Under standard conditions (solar-simulated irradiation, 78 mW/cm^2) the energy conversion efficiency of merocyanine devices is 0.7% with no correction for electrode transmission (V_{oc} = 1.2 V; I_{sc} = 1.8 mA/cm^2; FF = 0.25) [92–94]. There is no straightforward way to rationalize such a dramatic difference, the exceptionally high interactions between the merocyanine molecules in the solid state [505] probably provide part of the explanation. However, the unusual stability of the PcM's, their large chemical versatility, as well as their very low cost of production designate them as very good candidates for future molecular solar cells.

The power conversion efficiency of molecular solar cells is still limited by the electrical field dependence of the quantum efficiency of charge-carrier generation. This limitation is clearly correlated with the very low overall conductivity of the molecular materials and the space-charge limitations which correspondingly arise. These problems may be in part solved by finding proper doping agents able to give rise to space-charge regions extending deeply within the organic material and favoring the migration of the charge carriers. The eventual advantages brought by using molecular material for making efficient molecular solar cells more than justify the challenge.

IV Polyacetylene

Polyacetylene, abbreviated as $(CH)_x$, consists of a one-dimensional chain of conjugated double bonds; it is the simplest possible conjugated organic polymer. Polyacetylene was first obtained in 1929 by Champetier from the polymerization of acetylene [530, 531]. However, the development of the catalysts necessary for the polymerization has taken place only since 1957. Organo-magnesium systems [530, 532], Ziegler-Natta catalysts [533, 534], or complexes of nickel or cobalt [535–537] have been used to initiate the polymerization. In the early work, only poorly conducting materials were obtained ($\sigma < 10^{-9}\ \Omega^{-1}\ cm^{-1}$), and the conductivities were highly sample dependent since the purity of the polymer was not really controlled [538–539]. In all the materials studied, a strong paramagnetism was noticed [538–546]. Further studies showed that the paramagnetic centers corresponded to a fundamental state of the molecule and that there was no direct relationship between the charge carriers responsible for the electrical conductivity and these paramagnetic centers [542].

The physico-chemical properties of $(CH)_x$ were not further elucidated until Shirakawa et al. developed and improved the techniques for preparing the polymer [547–549]. For the first time the polyacetylene was obtained in the form of lustrous silvery films. The problem of the configuration of the chains was also studied in detail. Two isomers of $(CH)_x$ are of importance; the cis (or cis-transoid) and the all-trans (Fig. IV.1). A variant, the trans-cisoid may also occur. Shirakawa et al. determined the ratio of the cis to the trans isomer and the conditions under which the cis is converted into the trans isomer.

The advent of such nice-looking films generated interest in polymeric organic semiconductors. Heeger and McDiarmid [550–552] achieved the doping of the polycrystalline films of $(CH)_x$ with oxidizing agents such as iodine or AsF_5. Doping induced orders-of-magnitude increase in the conductivity of the material. Moreover,

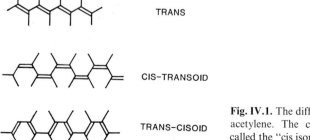

TRANS

CIS–TRANSOID

TRANS–CISOID

Fig. IV.1. The different possible isomers of polyacetylene. The cis-transoïd isomer is usually called the "cis isomer" for short

the concentration of dopants could be varied such that the overall conductivity of the films could be varied continuously from the insulator to the semiconductor range and also from the semiconductor to the metallic domain.

Since then, an increasing number of publications have appeared in the literature describing in great detail the various physico-chemical properties of $(CH)_x$. The possibility of n-type or p-type doping, the compensation effect, the realization of organic solar cells or batteries have all been the object of intensive researches. At the same time, the "soliton" formalism was developed to provide a highly successful rationalization of the electrical properties of $(CH)_x$.

IV.1 Synthesis and Physico-Chemical Properties of Polyacetylene

a Synthesis

Polyacetylene, like most conjugated polymers, is highly insoluble. The standard techniques of purification and characterization are therefore inappropriate. Determinations of the molecular weight, the polydispersity or the amount of cross-linking can only be achieved with great difficulty [553]. Much effort, then, has gone into finding highly effective catalysts able to form pure materials directly (Table IV.1). In all cases, however, the polymerization of acetylene is accompanied by secondary reactions, giving rise to cyclic oligomers such as benzene [567–570].

Table IV.1   The main catalysts used to polymerize acetylene.

Catalysts	Cocatalysts	Ref.
NiX_2 [a]	$P\Phi_3$	[537]
	$P\Phi_3$-$NaBH_4$	[554]
$NiCl_2$	$P(C_4H_9)_3$	[535, 536]
$NiBr_2$	$P(C_3H_7)_3$	[535, 536]
$Ni(CO)_2$	$P\Phi_3$	[555]
$CoCl_2$	$P\Phi_3$-$NaBH_4$	[554]
$Co(NO_3)_2$	$NaBH_4$	[535]
$Co(acac)_3$	$Al(C_2H_5)_3$	[556, 557]
$TiCl_4$	$Al(C_2H_5)_3$	[538]
	$Al(i-C_4H_9)_3$	[539]
	$Li(n-C_4H_9)$	[539]
	$Zn(C_2H_5)_2$	[539]
$Ti(CH_2-C_6H_5)_4$		[558]
$Ti(OR)_4$	AlR'_3	[533, 547]
$Ti(OC_3H_7)_4$	$Li(C_5H_{11})$	[534]
	$Al(C_2H_5)_3$	[534]
$Ti(OC_4H_9)_4$	$Al(C_6H_{13})_3$	[534]
	$Al(C_2H_5)_3$	[538, 560, 562]
$Ti_2(C_5H_4)(C_5H_5)_3$ [c]		[559]
$FeCl_3$	$\Phi MgBr$	[530]
$Fe(acac)_2$ [b]	AlR_3	[563]
$Fe(dmg)_2$ [d]	$AlEt_3$	[564]

Table IV.1   (continued)

Catalysts	Cocatalysts	Ref.
$CoCl_2$	$P\Phi_3$-$NaBH_4$	[554]
$Co(NO_3)_2$	$NaBH_4$	[535]
$Co(acac)_3$[b]	$Al(C_2H_5)_3$	[556, 557]
$TiCl_4$	$Al(C_2H_5)_3$	[538]
	$Al(i-C_4H_9)_3$	[539]
	$Li(n-C_4H_9)$	[539]
	$Zn(C_2H_5)_2$	[539]
$Ti(CH_2-C_6H_5)_4$		[558]
$Ti(OR)_4$	AlR_3'	[533, 547]
$Ti(OC_3H_7)_4$	$Li(C_5H_{11})$	[534]
	$Al(C_2H_5)_3$	[534]
$Ti(OC_4H_9)_4$	$Al(C_6H_{13})_3$	[534]
	$Al(C_2H_5)_3$	[538, 560–562]
$Ti_2(C_5H_4)(C_5H_5)_3$[c]		[559]
$FeCl_3$	$\Phi MgBr$	[530]
$Fe(acac)_2$[b]	AlR_3	[563]
$Fe(dmg)_2$[d]	$AlEt_3$	[564]
$MoCl_5$		[565]
	Φ_4Sn	[558]
$MoCl_4$		[565]
$Mo(CO)_3$		[566]
WCl_6		[565]
WCl_4	Φ_4Sn	[558]

[a]: X = Cl, Br, SCN, NO_3; [b]: acac: acetylacetonate; [c]: μ-(η^1:η^5-cyclo-pentadienyl)-tris (η-cyclopentadienyl) dititane; [d]: dmg: dimethylgly-oxime.

The transition metals used as catalyst for the polymerization of acetylene belong mostly to the first row of the periodic table: Ni, Co, Ti, Fe; Mo and W are the only exceptions. Cocatalysts for "activating" the transition metals are often employed [571]; they act by polarizing the metal complexes or by yielding an appropriate valence state of the metal. The catalysts or the catalytic systems are soluble and the polymerization of acetylene occurs in homogeneous phases or at the interface between the catalyst dissolved in a solvent and a gaseous phase containing acetylene. The crystallinity and the morphology of the polymer films are highly dependent upon the conditions of the reaction: the concentration of the reactants, the nature of the solvent, the pressure used, the temperature, etc. [561]. Two types of catalyst have been the most widely utilized: the Ziegler-Natta catalyst [533] is composed of titanium tetrabutoxide and triethylaluminum; the Luttinger catalyst results from the reaction of cobalt nitrate with a reductant, usually $NaBH_4$ [572, 573]. Electron micrographs have recently revealed that the two catalysts give very similar materials [574].

The mode of action of the Luttinger catalyst is not known in details. It is very probable that the Co(+I), (+II), and (+III)-ions are all involved in the polymerization process: the cobaltous complex LCo^{II} is presumably transformed into LCo^{I} species by reacting with the reducing agent ($NaBH_4$). Addition of acetylene with the reduced cobaltous complex might, in a two-electron transfer process, yield LCo^{III} complexes. The other chemical reactions involved may only be guessed at.

In the case of the Ziegler-Natta catalyst, two fairly different mechanisms have been invoked. The first one is directly transposed from the mechanism inferred for the polymerization of ethylene [562, 571, 575]. The "active complex", which is at the origin of the polymerization process, involves reduced titanium ($+III$) species. Several forms of the active complexes have been postulated [562, 576–578]. In all cases, the growth of the polymer chain takes place by repeated insertion of the monomer into an organo-metallic bond linking titanium with one carbon of the alkyl group (Fig. IV.2). One of the coordinate sites of titanium is vacant and is used to bind the acetylene molecule. The insertion of acetylene within the Ti-alkyl bond occurs through a Möbius-type transition state [562, 571] and yields the cis-isomer.

The second mechanism is based upon the methathesis reaction [553, 566, 579] (Fig. IV.3). Successive cycloadditions of acetylene monomer lead to a ladder-type polymer. At some stage during the polymerization, the cyclic intermediate opens and yields cis-polyacetylene. Such a mechanism is, however, highly improbable.

The conditions of polymerization using Ziegler-Natta catalysts strongly affect the morphology, the crystallinity, and the cis-to-trans content of the material obtained. The temperature of polymerization is the most important parameter which determines the cis-to-trans ratio. Films prepared at low temperatures ($-78\ °C$) contain 93 to 95% of the cis-isomer, whereas at room temperature values as high as 40% of the trans isomer are found [549]. The cis-trans ratio may be easily determined by infrared spectroscopy [549] or solid-state ^{13}C NMR [580]. The efficiency of the Ziegler-Natta catalyst is strongly modified by varying the Al/Ti ratio [561, 576]. In most cases, an Al/Ti ratio of between 3 and 4 has been adopted [549, 581, 582]. The concentration of the catalyst plays an important role in determining the mor-

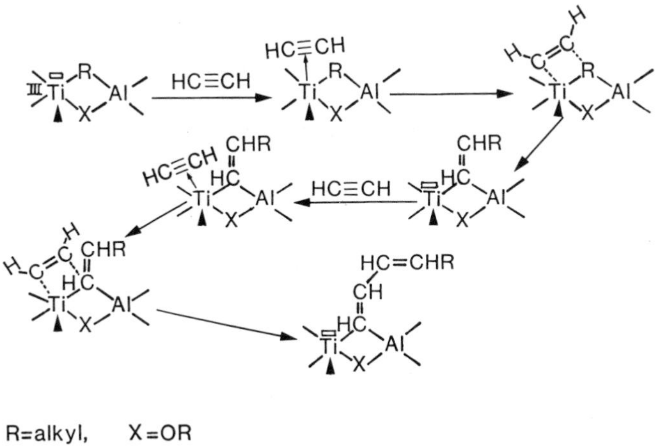

R=alkyl,   X=OR

Fig. IV.2. Schematic representation of the mechanism involved in the polymerization of acetylene using Ziegler-Natta catalysts

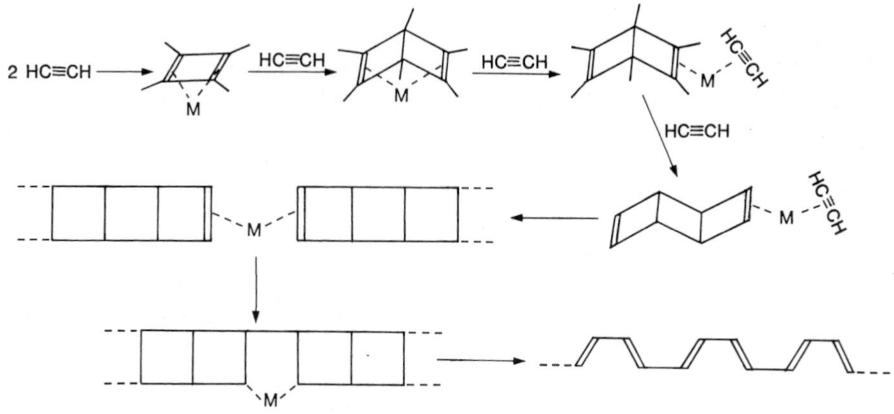

Fig. IV.3. Methathesis mechanism of acetylene polymerization. (After Ref. [553] and [566])

phology of the materials prepared, high concentrations of catalyst producing films and low concentrations producing powders [583]. The choice of the solvent is also important. Toluene is the most widely employed, but n-hexane, n-hexadecane, diethylether [549], and toluene-anisole mixtures [583] have also been used to dissolve or disperse the Ziegler-Natta catalyst. In toluene, at temperatures higher than 80 °C, no polymer is formed [584], whereas in n-hexadecane the polymerization occurs up to 200 °C [549]. This effect is probably due to the difference in volatility of the two solvents producing a greater or lesser dilution of the acetylene gas. By varying the pressure it is possible to prepare films of different thicknesses, from 50 Å [585] at low pressures up to a few millimeters [586].

A few other ways of obtaining polyacetylene have been reported. Polymerization of the adduct of hexafluorobut-2-yne and a derivative of cyclooctatetraene catalyzed by transition-metal complexes leads to a substituted homologue of $(CH)_x$ (Fig. IV.4). This polymer decomposes spontaneously on standing in the dark to give polyacetylene and 1,2-bis-(trifluoromethyl)benzene [587]. Elimination of hydrogen chloride from poly-(vinylchloride) has also been said to yield polyacetylene [588, 589]. Thermally-induced [590–592] or radiation-induced [593, 594] liquid and solid-phase polymerizations of acetylene have been reported.

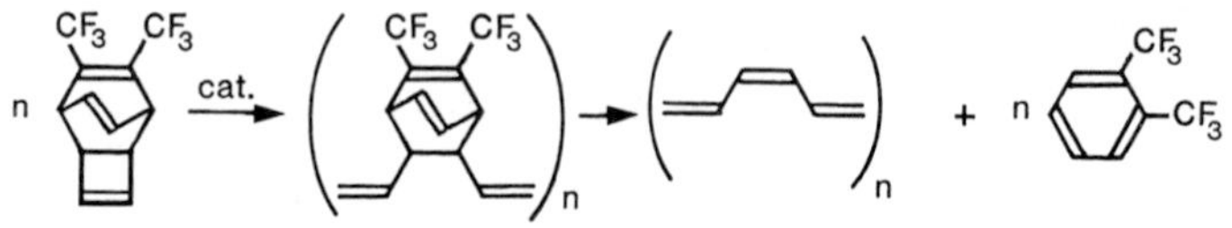

Fig. IV.4. Synthesis of polyacetylene starting from the adduct of a derivative of cyclooctatetraene and hexafluorobut-2-yne. (After Ref. [580])

b Morphology

As mentioned in the preceding section, by varying the experimental conditions it is possible to change the morphology of the polyacetylene from powder to gel or film. This great variety is reflected in the bulk densities obtained; they vary from 0.02 to 0.5 g/cm³ for the gels [583, 585, 595] and from 0.2 to 0.6 g/cm³ for the films [583]. The density of the fibers constituting the films may be obtained by the flotation technique and is found to be of the order of 1.2 g/cm³ [585, 596, 597], so the polymer fibers only fill one third of the total volume of the films.

The fibrillar structure associated with the low bulk density corresponds to a large surface area of 60 m² · g⁻¹ as evaluated by absorption of nitrogen gas (BET technique [585]).

Depending on the methods of preparation, the films may present some inhomogeneities. The most commonly used method of polymerization consists of wetting the inside wall of a glass reactor with the catalyst solution and then admitting the acetylene gas inside the reactor. The side of the film in contact with the glass has then a bright, silvery aspect, whereas the other side is dull [586]. Electron microscopy studies confirm this difference of aspect (Fig. IV.5). The dull surface consists of randomly oriented fibers a few tenths of a micron long with a diameter in the range 50 to 400 Å [583, 585]. The most abundant fibrils have a diameter of 200 Å [547–549,

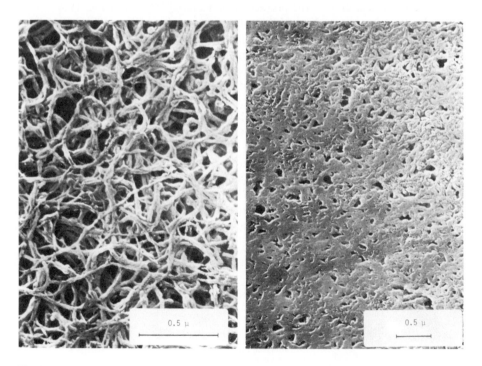

Fig. IV.5. Electron micrographs of polyacetylene films obtained with a Ziegler-Natta catalyst.
a: Dull surface (exposed to the gas)
b: Shiny surface (in contact with the glass)
[Reproduced with permission of (583)]

585]. The shiny surface also presents a fibrillar structure but the fibrils have a larger diameter (600–1000 Å) and they are more tightly packed. The dull surface is thought to be representative of the bulk morphology [583, 585].

The gels obtained with lower concentrations of catalyst are foam-like materials. Despite this very different aspect, they also have a fibrillar structure [505]. The apparent fibril diameter is somewhat larger, ca. 600 to 800 Å, than that observed for films [583]. Gels and films have therefore roughly the same morphology; they only differ in the filling fraction of polymer within the samples [583].

Wegner et al. did not observe such fibrillar structure [573]. However, recent studies have shown that a proper washing of the films makes the fibrillar morphology of the films clearly apparent, whatever the conditions under which they were prepared [574].

Globular structures have also been reported. Such structures appear in particular parts of the films and may be the result of local heating related to the exothermic character of the polymerization [585].

Polyacetylene films are highly crystalline. By the same methods as these used for polyethylene [598], it is possible using X-ray diffraction data, to estimate the crystallinity of the films; a value of 81 % has been found [599]. Crystallinities as high as 90 % have recently been reported [600]. Further X-ray observations strongly suggest that the polymeric chains of $(CH)_x$ are extended parallel to the fibril axis [583]. A very schematic respresentation of the polyacetylene material is given in Fig. IV.6.

Much effort has been devoted for obtaining oriented films of polyacetylene. Alignment of the fibrils would result in the alignment of the polymeric chains and a highly anisotropic material should be prepared. Small areas of oriented fibrils may be obtained spontaneously in some samples [601] or in the vicinity of fractures in the films [585]. These fractures are observed in smooth films prepared at a high acetylene gas pressure (1 atm.). They are formed during the washing procedure or during the thermal contractions which accompany temperature changes [585]. Large oriented samples have been obtained by polymerization under a shear gradient [602, 603]. Rolling [628] or stretching [583, 604] methods may also be used to produce oriented samples. Only partial orientations are obtained as shown by electron microscopy (Fig. IV.7). The degree of stretch orientation is related to the mechanical properties of the films. These are in turn influenced by the cis to trans ratio of the material, for the elongation-to-break decreases rapidly with an increase of the trans content so that the trans-isomer cannot be oriented by stretching. The cis isomer is,

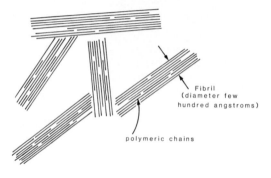

Fibril
(diameter few
hundred angstroms)

polymeric chains

Fig. IV.6. Schematic representation of polyacetylene showing its fibrillar morphology and the orientation of the polymeric chains within the fibrils

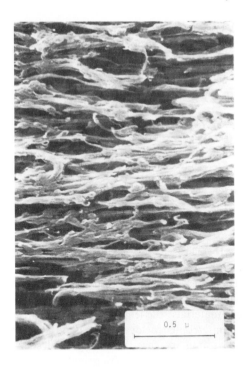

Fig. IV.7. Electron micrograph of a highly stretch-aligned polyacetylene film. [Reproduced with permission of (583)]

on the other hand, elongated by a factor of 3 under a tensile strength of 4 kg/mm^2 [583].

c Molecular Weight and Length of the Conjugated Sequences

Polyacetylene is totally insoluble, so no direct determination of its molecular weight is possible. Two indirect methods can be resorted to. The first requires chemical transformation of the polymer into a soluble material, the second uses a radioactive labelling process.

The chlorination or hydrogenation of polyacetylene yields a polymer which is partly soluble [628]. The chlorination applied to a polyacetylene freshly prepared with the Luttinger catalyst at $T \leq -30$ °C is complete and the polymer is fully soluble [573, 605]. The insoluble fraction develops with storage time even at low temperature (-30 °C). The insoluble fraction reflects the spontaneous formation of cross-links [573]. The average molecular weight obtained in that way is approximately 5900 for the chlorinated material; this corresponds to a degree of polymerization of about 100 [573]. The temperature of polymerization seems to have little ·influence on the degree of polymerization [607].

A significantly higher molecular weight is obtained by radioactive labelling ($\overline{M} = 22,000$, $\overline{N} \sim 1500$) [606]. This discrepancy might be attributed to some differences in the quality of the two materials used. However, it is more probably associated with the difficulty of using the labelling technique whenever cross-linking processes are operative.

The termination of the growth of the polymer chains occurs because of secondary chemical reactions or because the film limits diffusion of monomer through to the catalytic center. The secondary chemical reactions include irreversible degradations of the catalytic system [562]:

— reduction by termination:

$$Ti^{III}R + Ti^{III}R' \rightarrow 2\ Ti^{II} + R_{-H} + R'H$$

— reduction by alkylation:

$$Ti^{III}R + AlEt_3 \rightarrow Ti^{III}R(Et) \rightarrow Ti^I + RH + H_2C=CH_2$$

— oxidative coupling:

$$2\ Ti^{III} + HC\equiv CH \rightarrow Ti^{IV} -C_2H_2-Ti^{IV}$$

(Ti^{III} is the abbreviation for species shown in Fig. IV.2 or for closely related adducts).

However, the length of the polymeric chain is not the factor that dominates the physico-chemical properties of polyacetylene materials; the length of the conjugated sequences, i.e. the average length of the uninterrupted delocalized π-bond system is, in most cases, the crucial parameter. Cross-links increase the average molecular weight of the material, while they decrease the length of the conjugated sequences (Fig. IV.8).

Branching, termination, or cross-link reactions all give rise to sp^3-hybridized carbon atoms. The content of sp^3 carbon atoms within the polyenic chain may be evaluated by ^{13}C NMR, and proportions varying from a negligible amount to a few percent have been found [580, 608].

d Stability of Polyacetylene and Effect of O₂

The thermal stability of polyacetylene under vacuum and in air as well as its stability toward radiation has been the object of numerous publications.

The $(CH)_x$ films have an apparent good thermal stability under vacuum. Thermograms of both cis and trans isomers show an exothermic peak at 325 °C, at which temperature the decomposition is rapid [586]; this is followed by an endothermic degradation process occurring around 420 °C. However, beside these decomposition reactions, which may be macroscopically detected, local degradations of the conju-

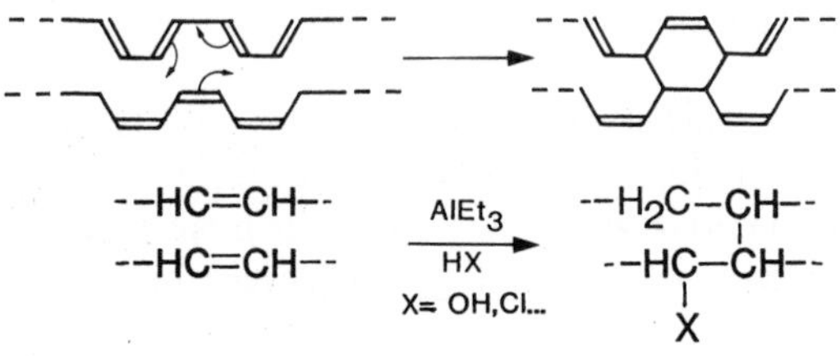

--HC=CH--
--HC=CH--

AlEt₃
————
HX
X= OH,Cl...

--H₂C—CH--
       |
--HC—CH--
     |
     X

Fig. IV.8. Cross-linking reaction and cycloaddition as examples of side-reactions which possibly occur during the polymerization of acetylene using Ziegler-Natta catalysts. (After Ref. [573])

gated chains may occur at low temperatures. Already at temperatures lower than 200 °C, ESR measurements show that some degradation processes take place [609]. Raman spectroscopy demonstrates that the mean length of the conjugated sequences decreases when the material is heated to temperatures higher than 140 °C [610]. Correspondingly, the proportion of sp^3 carbon atoms is increased [609].

Undoped $(CH)_x$ also has a remarkably good stability under irradiation. Electron beam or γ radiation as well as UV light do not significantly affect the conductivity and the thermal characteristics of $(CH)_x$ [611]. This does not, nevertheless, preclude local chain breakages or cross-linkings [612].

Even if careful precautions are taken, polyacetylene films are more or less briefly exposed to air during their history; even films handled in a glove box may demonstrate properties affected by the presence of minute amounts of O_2. This point is important since most of the potential applications of $(CH)_x$ depend on an eventual protection of the material from the ambient atmosphere [613].

For a long time, $(CH)_x$ has been recognized to be sensitive to the presence of air [582, 614]. After a few days in air, the films become brittle and they loose their characteristic silvery luster [586]. At the same time, the crystallinity of the material is decreased [615, 616] along with its overall conductivity [614, 617]. Exposure to air of $(CH)_x$ films is also detected by ESR, both the intensity of the signal and the corresponding linewidth are increased [562, 618–620]. Evidence that O_2 causes chemical modifications has been found by IR [616]; in particular, peaks corresponding to C–O bonds appear [586, 611]. This has been confirmed by high-resolution NMR [621]. Such effects are only detectable for air exposures over long periods of time.

The variation of the conductivity of $(CH)_x$ with the time of O_2 exposure casts further light on the mechanisms of degradation [614, 622, 623] (Fig. IV.9). Two different phases may be distinguished over approximately the first 10 min as the conductivity increases and then abruptly drops. The position of the maximum depends on the temperature and on the cis-trans ratio. A concentration of O_2 corresponding to 0.14 O_2 molecule per double bond of the polymer is found at the maximum [614].

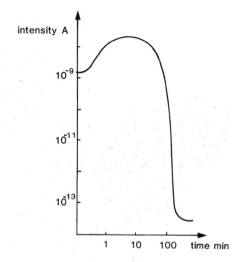

Fig. IV.9. Effect of O_2 exposure on the current observed by applying an external voltage of 0.5 V on a sample containing 60% of trans-$(CH)_x$ and 40% of cis-$(CH)_x$ (t = 70 °C). (After Ref. [614])

The simplest kinetic scheme rationalizing these observations is represented below:

$$(CH)_x + O_2 \overset{k_1}{\underset{k_2}{\rightleftharpoons}} (CH)_x, \; O_2 \overset{k_3}{\longrightarrow} \text{degradation products} \qquad (IV.1)$$

In a first step, O_2 is reversibly complexed to the polymeric chain; a significant charge transfer from $(CH)_x$ to O_2 probably occurs. Such complexation processes are already known with electron-rich systems [624, 625] and with polyolefinic compounds [626]. O_2 acts as an acceptor of electrons and is therefore a p-type dopant. The reversibility, at least partial, of the complexation has been assessed by conductivity experiments [617, 622], photoconductivity [612], and ESR [618] studies. However, irreversible effects are visible by ESR even at low concentration of O_2 and for short exposure: the residual Ti^{III} coming from the catalyst is oxidized to Ti^{IV} species and a strong increase of the concentration of the paramagnetic centers arising from the polymeric backbone is noticed [618, 620] (Fig. IV.10). After times of exposure as short as 1 min, the tensile properties of $(CH)_x$ are considerably altered [604]. These effects are probably due to minute amounts of O_2 in the material (as low as 10^{-4}). For short exposure times, the O_2 is presumably bound only at the surface of the fibrils [619, 627]; the diffusion of O_2 into the fibrils is probably fairly slow.

For longer times of O_2 exposure, irreversible chemical reactions become predominant. A strong decrease of the crystallinity is observed [616]. O_2 also induces the isomerization of the cis-$(CH)_x$ into the trans form [610]. It has been shown that the paramagnetic centers always present in the trans form of $(CH)_x$ rapidly react with O_2 [614].

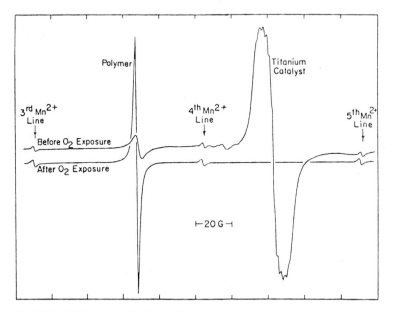

Fig. IV.10. ESR spectra of polyacetylene obtained from the Ziegler-Natta catalyst before and after a short exposure to O_2. [Reproduced with permission of (618)]

e Isomers

Six different configurations of polyacetylene may be reasonably considered (Fig. IV.11) [629]. These six configurations may be grouped in two classes: the all-trans structures noted A_1, A_2, A_3 and the cis-trans structures noted B_1, B_2, and B_3. Two of these six configurations possess equidistant carbon atoms; but only the trans isomer may lead to entirely symmetrical structures. Experimentally, two of these structures have been observed with certainty: the trans-configuration (A_1), which is the thermodynamically stable isomer, and the cis-transoid isomer (B_1), which is the pristine form resulting from the polymerization of acetylene with Ziegler-Natta catalysts. There are reports that the trans-cisoid (B_2) isomer is present in materials synthesized with the Luttinger catalyst [630].

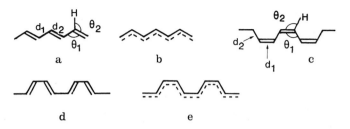

Fig. IV.11. The various configurations of the polymeric backbone of polyacetylene with the corresponding notation.

a: trans alternant $d_1 \neq d_2$ (A_1)
b: trans equidistant $d_1 = d_2$ (A_2, A_3)[a]
c: cis transoid $d_1 < d_2$ (B_1)
d: trans cisoid $d_1 > d_2$ (B_2)
e: trans equidistant $d_1 = d_2$ (B_3)
[a]: The two structures A_2 and A_3 differ by the fact that in the second one bond order alternation still subsists leading to a symmetry broken solution. (After ref. [629])

Table IV.2   Relative stabilities of polyacetylene isomers. Energies (kcal/mole) are given relative to structure A_1 and per C_2H_2 unit. (After Ref. [629]).

Structure	a	b	c	d
A_1	0.0	0.0	0.0	0.0
A_2	7.3	8.4	—	4.3
A_3	5.2	—	0.4	—
B_1	1.9	−2.6	19.8	1.9
B_2	2.1	−0.1	20.7	4.3
B_3	7.2	6.9	20.0	6.6

[a]: from ref. [629], ab initio crystal orbital method with optimization of geometry; [b]: from ref. [631] as a) with a fixed geometry; [c]: from refs. [632, 633], as a) with a different basis set and a fixed geometry; [d]: from ref. [634], tight-binding LCAO-SCF-MO method with optimization of geometry.

Much theoretical work has gone into determining the relative stabilities of the different isomers [629–635] (Table IV.2). Calculations performed with optimization of the geometries are preferable since the experimental values are not known with certainty. Under this condition, the ab initio crystal method and the tight-binding LCAO-SCF-MO method give very similar results. The most stable configurations are, in all cases, the alternant structures (A_1, B_1, and B_2); among them, the alternate all-trans isomer is the most stable. The regular structures with equidistant carbon atoms are destabilized by more than 5 kcal/mole with respect to the "localized bonds" configurations. As in the case of butadiene [636], the trans isomer is more stable by 2.5 kcal/mole than the cis isomer.

Very simple calculations made at the end of the fifties [637, 638] already indicated that consecutive bonds of polyenic backbones are not equal: they "alternate" between the distances d_1 and d_2 (Table IV.3). Using a very simplified relationship for the calculation of the resonance integrals:

$$\gamma = -a \cdot \exp\left(-\frac{d}{b}\right) \tag{IV.2}$$

a and b: constants

The degree of alternance $\Delta d = d_1 - d_2$ was found to be of the order of 0.04 Å. Ab initio calculations give Δd around 0.1 Å [629], in agreement with the experimental value found for trans-$(CH)_x$ as determined by analysis of X-ray scattering data [639]. No experimental data have been reported for the cis-transoid isomer.

The cis and the trans isomers have very different magnetic and optical properties. When cis-$(CH)_x$ is isomerized, the material becomes brittle [604]. The transformation may be easily followed by IR or Raman spectroscopy. The IR [547, 549, 630, 642, 643] and Raman [610, 644–648] absorption spectra of both isomers have been studied in detail. Precise assignments of the absorption bands have been reported [547, 645, 648].

Table IV.3 Theoretical distances between three consecutive C atoms calculated for polyacetylene and comparison with the experimental values avalaible.

Structure	d_1 (Å)	d_2 (Å)	Δd^a (Å)	Ref.
Calculated Values				
A_1	1.423	1.387	0.036	[638]
	1.435	1.350	0.085	[634]
	1.476	1.327	0.149	[629]
B_1	1.350	1.445	0.095	[634]
	1.328	1.480	0.152	[629]
B_2	1.489	1.324	0.165	[629]
Experimental Values				
trans-$(CH)_x$	1.455	1.385	0.07	[639][b]
alkanes	1.54			[640]
alkenes		1.35	0.19	[640]
butadiene	1.483	1.337	0.146	[637]

[a]: $\Delta d = d_1 - d_2$; [b]: determined by analysis of X-ray scattering data.

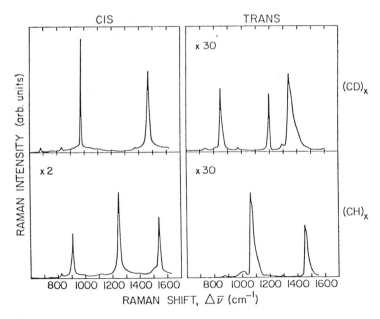

Fig. IV.12. Raman spectra of cis and trans isomers of normal and deuterated polyacetylene. T = 77 K. Excitation: 6000 Å. [Reproduced with permission of (645)]

The Raman spectra of the cis and trans isomers of normal and deuterated $(CH)_x$ are shown in Fig. IV.12 [645].

In Raman spectra of trans $(CH)_x$, the profile and the line position of the ⟩C=C⟨ stretch modes depend on the exciting laser wavelength [648]. Long wavelength photons excite only the sequences of conjugation which are sufficiently long [637] (Fig. IV.13).

Moreover, the frequency of vibration of the ⟩C=C⟨ mode is dependent upon the length of the conjugation sequence [646]. For a sequence of 100 double bonds, the frequency of vibration is about 1455 cm^{-1}; it increases to 1541 cm^{-1} for five conjugated double bonds [646].

As previously mentioned, at low temperatures, it is the cis-transoid isomer which is first formed during polymerization. The subsequent isomerization into the thermodynamically stable trans isomer has been much studied and may be achieved by the use of high temperatures, by doping, or by rolling, though only preliminary results have been published for this last method [649].

The thermal isomerization may be detected by Differential Scanning Calorimetry (DSC)[549]. An exothermic peak occurring at 145 °C may be unambiguously attributed to the isomerization process. The enthalpy change associated with the isomerization of a 100% cis sequence to a 100% trans is 1.85 kcal/mole [549, 650]. Pure trans-polyacetylene may be obtained with the minimum of degradation by treating the cis polymer around 150–160 °C for half an hour. However, in all cases, the thermal isomerization is accompanied by a decrease of the length of the conjugated sequences [549].

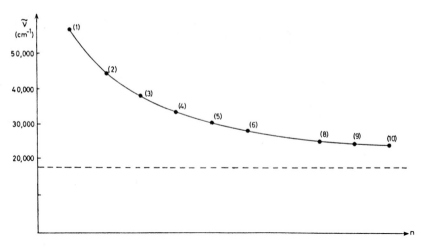

Fig. IV.13. Frequency of maximum absorption of all-trans α,ω-dimethylpolyenes, $CH_3-(CH=CH)_n-CH_3$, plotted against the chain length (solvent: ether or ethanol). [Reproduced with permission of (637)]

Isomerization may also be induced by doping. The incorporation of AsF_5 [647], iodine [643, 651], sodium [651], or oxygen [610] into cis samples cause the cis to trans isomerization. At room temperature the trans isomer is formed at 95% for a concentration of dopant of about 10% [651].

The mechanism of the isomerization process is not known in detail. However, based upon the fact that pure cis-$(CH)_x$ is diamagnetic, while the trans isomer contains paramagnetic centers, a plausible mechanism of isomerization has been proposed in which intermediate biradicals are formed [652] (Fig. IV.14).

f Crystalline Structures

The crystalline structure of cis-$(CH)_x$ has been extensively studied. X-ray diffraction experiments [596, 654] have been performed on randomly oriented films prepared

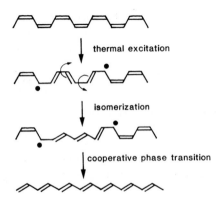

Fig. IV.14. Tentative mechanism of isomerization of cis-polyacetylene into the trans-isomer. (After Ref. [652])

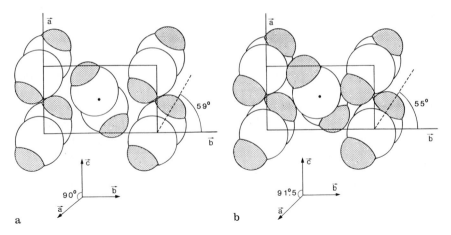

Fig. IV.15. Crystalline structures obtained for the cis and trans isomers of polyacetylene (most probable structures).
a: Cis-(CH)$_x$ (from Ref. [596, 653]); the a and b axes have been inverted compared to references [596] and [653].
a = 4.46 Å; b = 7.68 Å; c = 4.38 Å;
b: Trans-(CH)$_x$ (from Ref. [601, 639]).
a = 4.24 Å; b = 7.32 Å; c = 2.46 Å.
The polymeric chains are aligned with the c axis

using a Ziegler-Natta catalyst. Electron diffraction studies have been made on randomly oriented materials [572] (Luttinger catalyst) or oriented films [653] (Ziegler-Natta catalyst). The crystal-cell dimensions were found, in all cases, to be very similar. There is, however, some controversy about the direction of the polymeric chain within the cell; some studies have been interpreted by postulating that the chains are aligned along the b direction [596], some others along the c axis [653]. This last interpretation is no doubt the most convincing (Fig. IV.15). The packing shown in Fig. IV.15 is in accordance with NMR results although small discrepancies still exist between the second moment experimentally obtained [655, 656] and the one calculated from the diffraction data [596, 653].

Trans-polyacetylene has been also the object of numerous investigations by X-ray diffraction [639, 657] or electron diffraction [602, 603]. The most probable structure has been reported by Baughman et al. (Fig. IV.15) [657]. Although it has been disputed [603] further studies [601, 639] confirmed these results. NMR second-moment determinations have been carried out [655]. For the trans-(CH)$_x$ obtained with the Ziegler-Natta catalyst, the NMR results rule out the alternative structure postulated [603], whereas the agreement is satisfactory with the structure shown in Fig. IV.15.

The isomerization of polyacetylene from the cis to the trans form involves a significant expansion of the polymeric chain along the c axis from 4.38 Å to 4.92 Å. This is accompanied by a contraction in the two other directions. Experimentally, it has been shown that an elongation of the samples is correlated with the doping induced isomerization process [651].

IV.2 Theoretical Properties

Polyacetylene is a very attractive material for determining theoretical properties of one-dimensional polyconjugated systems. However, experimentally, the intrinsic properties of this polymer are rarely observed. Numerous inhomogeneities within the material induce the presence of charge carrier traps. It is known, for example, that a fairly high concentration of residual catalyst remains in the polymer especially on the bright side of the films [658]. Structural traps may also arise from inhomogeneities of chain lengths or crystallinities. More importantly, during the preparation of the samples, the polymer is contaminated with O_2. The presence of dioxygen drastically influences the transport properties of the material. Even when the manipulations are performed in a glove box, it does not preclude the absorption of significant amounts of O_2 within the polymer. There is therefore a large difference between the theoretically simple behavior of $(CH)_x$ and the experimentally determined macroscopic properties of the "undoped"* material. The transposition of the theoretical expectations into the experimental domain must therefore be made with care.

a Origin of the Band Gap

The optical absorption spectra of both the cis and the trans isomer of polyacetylene give information about the band structure of the material and especially about the magnitude of the band gap. The absorption coefficients first show a slow increase around 1.0 eV; they rise sharply at 1.4 eV up to a maximum at 1.9 eV for trans-$(CH)_x$ and 2.1 eV for the cis isomer (Fig. IV.16). At the maxima of absorption, the extinction coefficients are of the order of 10^5–10^6 cm^{-1} [643]. Such behavior is close to that expected for a direct band gap semiconductor. What is the origin of the band gap thus detected?

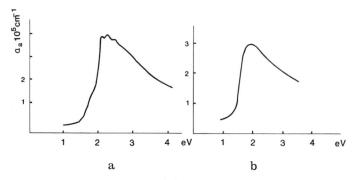

Fig. IV.16. Absorption spectra of cis (a) and trans (b) polyacetylene. (After Ref. [643])

---

* "Undoped" material corresponds to a sample which has not been intentionally doped.

In a Hückel approximation, the orbital energies of a regular linear polyene $C_{2N}H_{2N+2}$, with all equal carbon-to-carbon bonds, are given by [637]:

$$E_j = 2\gamma \cdot \cos \frac{j \cdot \pi}{2N + 1}$$

(IV.3)

$$j = 1, \dots, 2N$$

γ: resonance integral

The lowest N energy levels are occupied; the energy difference between the highest occupied molecular orbital (HOMO) and the lowest unoccupied molecular orbital (LUMO) is then:

$$E = -4\gamma \cdot \sin \frac{\pi}{2(2N + 1)}$$

(IV.4)

For a polyene having an infinite number of carbon atoms ($N \to \infty$), there is no gap between the HOMO and the LUMO; the material should behave as a metal. It is worth recalling that this result is obtained when no bond alternation is postulated.

If the double bonds are considered to be localized in some way, a finite value for the band gap is found even when $N \to \infty$. Bond alternation implies that two resonance integrals, γ_1 and γ_2, corresponding to the C-C lengths d_1 and d_2, must be considered:

$$\gamma_2 \leq \gamma_1 \leq 0$$

(IV.5)

In a Hückel approximation, the orbital energy levels of an infinite cyclic even polyene $C_{4n+2}H_{4n+2}$ are given by:

$$E_j = \pm \left[\gamma_1^2 + \gamma_2^2 + 2\gamma_1 \cdot \gamma_2 \cdot \cos \frac{2 \cdot j \cdot \pi}{2n + 1} \right]^{1/2}$$

(IV.6)

$$j = 0, \pm 1, \pm 2, \dots, \pm n$$

The energy levels are split into bonding ($\gamma_1 + \gamma_2 \leq E \leq \gamma_2 - \gamma_1$) and antibonding ($\gamma_1 - \gamma_2 \leq E \leq -\gamma_1 - \gamma_2$) bands. A band gap proportional to the difference of the resonance integrals ($\gamma_1 - \gamma_2$) is created.

Bond alternation thus induces the creation of a band gap between occupied and unoccupied orbitals. Kuhn already pointed out in 1948 that bond length alternation opens a gap near the Fermi level of long trans-polyenic systems [659]. However, this origin of the band gap has been disputed. If the interelectronic correlation energy is taken into account, it has been shown [546] that collective excitations may give rise to a band gap of about 2.0 eV. More sophisticated calculations have lately demonstrated that these types of excitations in fact have higher energies [660, 661] and that they contribute only slightly to the band-gap value. These calculations are highly dependent on the type of correlations used [662]. Extended Hückel-type calculations give basically the same results (Fig. IV.17) [663].

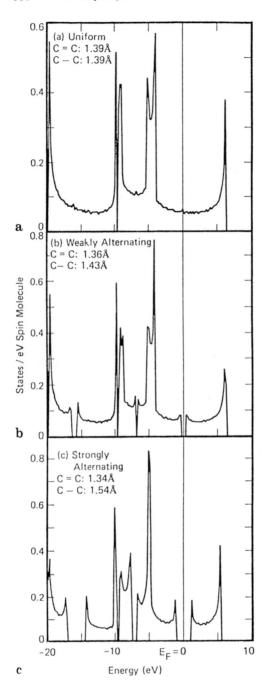

Fig. IV.17. Extended Hückel-type calculations on trans-polyacetylene. The densities of states versus energy are represented for three cases:

a: No alternation ($E_g = 0$)

b: Weakly alternating bond lengths ($E_g = 0.8$ eV).

c: Strongly alternating bond lengths ($E_g = 2.3$ eV). [Reproduced with permission of (663)]

Further discussions of the origin of the band gap in polyenic systems are available [546, 637, 643, 662, 664–666].

The problem of bond alternation is closely related to the Peierls theorem [667]. The energy gained by the bond-alternated structure with respect to the regular one is interpreted in terms of mixing of ground-state levels with low-lying excited states. In one-dimensional systems, the introduction of a distortion with a periodicity of r atoms reduces the translational symmetry. The total energy diverges as $n \cdot \ln n$ when $n \rightarrow \infty$; the regular configuration is not stable. The gain in energy is important if a break in the energy band appears near the Fermi level and if the periodicity is small. The condensation energy of the dimerization of a regular chain has been calculated [668].

b Band Structure

Numerous calculations have been directed toward the determination of the band structure of polyacetylene [550, 631, 634, 643, 663, 664, 669–678]. The representative results are shown in Table IV.4. Two types of calculations are compared: semi-empirical [664, 669] (valence-effective hamiltonian technique) and SCF ab initio Hartree Fock [631]. In this last case, 120(CH) units have been considered. It seems that the semi-empirical method is the only one which fairly accurately reproduces the experimental results. The experimental ionization energy of trans-$(CH)_x$, 4.5 eV [677], is close to the calculated value (4.7 eV). The calculated band gap of trans-$(CH)_x$ is found to be 1.4 eV; this value well agrees with the optical absorption threshold (1.4 eV) [643, 673] determined from the absorption spectra. However, in this case, the maximum absorption peak corresponds to higher energies (1.8–1.9 eV) [643, 674, 675]. This may be attributed to indirect interband transition effects in which the

Table IV.4 Calculated band-structure parameters of the three isomers of polyacetylene.

	Isomers			Ref.
	trans	cis-transoid	trans-cisoid	
I_c	4.7	4.8	4.7	[664][a, b, c]
I_g	6.6	6.7	6.6	[669][a, b]
	5.3	4.6	4.5	[631][d, e]
A_g	4.5	3.6	3.3	[631][d, e]
ΔW_{VB}	6.5	6.4	6.5	[664][a, b]
	6.4	7.6	7.5	[631][d]
ΔW_{CB}	8.2	9.9	10.1	[631][d]
E_g	1.4	1.5	1.3	[664][a, b]
	9.7	8.2	7.8	[631][d]

I_c: crystal ionization energy; I_g: gas-phase ionization energy; A_g: gas-phase electron affinity; ΔW_{VB}, ΔW_{CB}: maximum bandwidths for the valence band and conduction band, respectively; E_g: band gap. All values are expressed in eV.
[a]: Calculations use the valence effective hamiltonian technique; [b]: the following bond lengths have been taken $d_1 = 1.346$ Å, $d_2 = 1.446$ Å; [c]: a correction of 1.9 eV corresponding to the polarization energy of the lattice has been taken into account; [d]: SCF ab initio Hartree Fock calculations; 120 (CH) units have been considered; [e]: potentials expressed relative to the Fermi level.

electronic transition is correlated in some way with a vibration of the lattice. The indirect interband transition requires the absorption or the emission of a phonon prior to the electronic excitation. Such an effect could arise from interchain couplings; these have been neglected in the previous calculations. The bandwidth in the direction of the chain axis ($\Delta W_{||}$) is calculated to be of the order of 6–10 eV. This corresponds to a resonance integral value of 2–2.5 eV since:

$$\Delta W_{||} = 2 \cdot Z_c \cdot |\gamma_{||}| \qquad\qquad (IV.7)$$

$\gamma_{||}$: resonance integral in the chain axis
Z_c: coordination number

The transverse resonance integral ($\gamma_\perp$) which will give rise to interchain coupling is much less; but it may be sufficiently large to promote indirect transitions in the optical spectra. Following this interpretation it is calculated that the transverse resonance integral should be approximately 0.1 eV [643].

Optical spectra may alternatively be interpreted in a very different way. In the preceding discussion, the electron-hole electrostatic attraction was neglected. The excitonic approach, on the contrary, describes a linear polyene in terms of fully localized ethylenic molecular orbitals [666, 679] or Wannier functions [662]; therefore, very strong e^-/h^+ interactions are postulated. For even "trans"-polyene $C_{2N}H_{2N+2}$ of symmetry C_{2h}, the fundamental π electronic state has a symmetry A_g. The lowest allowed transition is an excitation between the totally symmetric A_g ground state to the B_u excited states, whereas the transitions to the A_g ones are forbidden. For a long polyene, the energy difference between A_g and B_u tends towards about 2 eV [662, 666, 679]. Very similar results are obtained for odd trans-polyenes of C_{2v} symmetry [679]. In the excitonic model, the difference in energy between the valence band and the conduction band is 4 eV [662] or 5.5 eV [666] depending on the type of calculation used. The difference between the excitonic and conduction bands corresponds to the binding energy electron-hole, E_{ip}:

$$E_{ip} = -\frac{q \cdot q'}{\varkappa r} \qquad\qquad (IV.8)$$

$\varkappa$: dielectric constant of the medium

E_{ip} therefore lies between 2 and 3.5 eV. This implies that the effective dielectric constant is of the order of 2 to 4, which is in fairly good agreement with the values experimentally determined ($\varkappa = 3$–6) [628, 680–683].

It is not clear which model is relevant in the case of polyacetylene: is the absorption band due to an excitonic-type process or due to a transition up to a conduction band? An effective mass m^* close to unity has been found for the charge carriers generated in the 2 eV transition [643]. This supports the second hypothesis. Another clue may be obtained by examining the limiting value of the band gap as a function of the polyene length. The limit value is attained in different ways depending on which model is applicable. The excitonic models lead to a N^{-2} dependence for $C_{2N}H_{2N+2}$ polyenes, whilst calculations based on the one-electron approximation lead to a N^{-1} relationship. It is this latter dependence which is found experimentally [665]. However, no definitive conclusions may be drawn, since complications in the interpretation of the data arise if both electronic correlation and alternance are simultaneously taken into account [684].

Clearer evidence may be obtained by comparing the optical spectra of trans and cis isomers of polyacetylene. Only small differences are observed; the energy gaps are thus close to each other. On the other hand (this will be discussed more fully in Sect. 5.a), for cis-$(CH)_x$, recombination luminescence is detected while trans-$(CH)_x$ is not luminescent at all; cis-$(CH)_x$ is not a photoconductor while trans-$(CH)_x$ is. This provides confirmation that free charges are the photogenerated carriers [685].

c Bond Length Alternation Defects in Polyenes; the "Solitons"

Radicalar defects may be created in polyenic chains. It is easier to visualize the phenomenon by considering "localized" double bonds (Fig. IV.18). The first step is the generation, by some thermal or photochemical process, of a biradical species. It will be seen that the properties of such species have been interpreted within a "soliton" formalism. The correspondence between the usual chemical description and the "soliton" terminology will be given throughout this section. In a second step, the two radicals separate from each other by moving along the polyenic chain. In that way, three distinct domains appear: two phases of type A separated by a phase of type B. A and B phases differ by the placement of the double bond within the polyene (Fig. IV.18).

It is possible, in the Hückel approximation, to calculate the energies corresponding to the radical defects. For a single radical, calculations are made on an odd polyene of general formula $C_{2n+1}H_{2n+3}$, since there is now a odd number of C atoms in the conjugated chain. The following energy levels are then found:

$$E = 0 \tag{IV.9}$$

$$E_j = \pm \left[\gamma_1^2 + \gamma_2^2 + 2 \cdot \gamma_1 \cdot \gamma_2 \cdot \cos \frac{j \cdot \pi}{n+1} \right]^{1/2} \tag{IV.10}$$

$$j = 1, \ldots, n$$

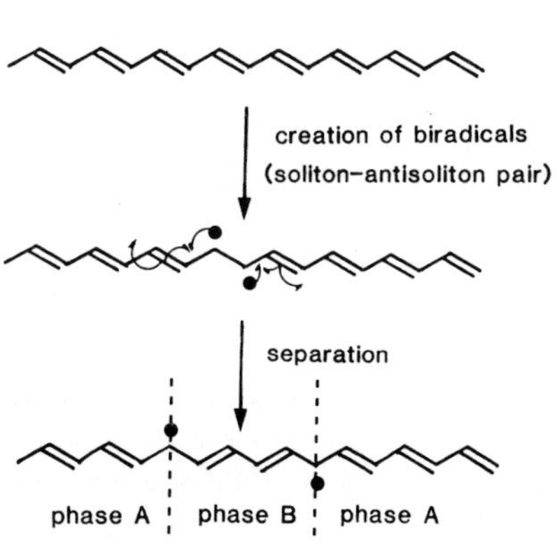

Fig. IV.18. Schematic representation of the creation of radical defects in trans-polyacetylene

Beside the solutions already found for even polyenes, there is in addition a non bonding energy level (E = 0). In the classical band scheme, these levels appear within the band gap, equidistant between the bonding and the antibonding states [540, 637, 686].

It has been calculated [540, 637] that the additional energy required to form a biradical defect is about 0.22 eV (5.1 kcal/mole). At room temperature, the defect states are therefore partly occupied.

The neutral radical defects (spin = 1/2) may be transformed into charged defects (spin = 0) by doping with either acceptors or donors of electrons (Fig. IV.19). Charged defects may contribute to macroscopic electrical transport.

It is noteworthy that similar mechanisms cannot take place in the case of cis-$(CH)_x$ since the two double-bond configurations are not energetically equivalent.

It is possible to quantify the previous observations more precisely. The "soliton model" proposed by Su, Schrieffer, and Heeger will be adopted for this purpose [668, 687, 688]. It has been seen that, for trans-$(CH)_x$, there are two degenerate alternating configurations possible, called phase A and phase B. By going from phase A to phase B double bonds are transformed into single bonds and vice versa. The constitutive atoms of the chain must therefore move slightly when they pass from one form to the other. A parameter Δ_n associated with a displacement u_n of the atom n is defined. In a chain having the configuration A for $n \rightarrow +\infty$ and B for $n \rightarrow -\infty$ there exists a "wall" between the two domains (Fig. IV.20). By analogy with the antiferromagnetic properties of one-dimensional chains, the properties of this wall may be described by a localized "soliton" solution [689, 690]. The "wall" has a certain extension, i.e. the frontier radical defect is more or less delocalized. Within this formalism, the soliton formation energy is calculated to be 0.42 eV; this number must be compared with the value of 0.22 eV obtained in the Hückel approximation. The "soliton" is spread over approximately 14 carbon atoms. The energy required for the translation of the soliton along the polymeric chain is of the order of 0.002 eV.

The effective mass of the soliton can also be evaluated [668, 687, 688]; it is equal to approximately 6 times m_e (the mass of the free electron). This is a remarkably small value considering that bond distortions are involved in the translational process of the soliton migration.

Other calculations within the soliton formalism have been carried out [691]; very similar results are found. The formation energy of the neutral solitons is found to be 0.6 eV, the width of the soliton is 10 C atoms and the corresponding effective mass is 15 times m_e.

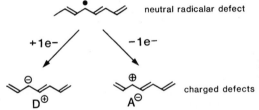

neutral radicalar defect

+1e-   -1e-

$D^{\oplus}$ $A^{\ominus}$

charged defects

Fig. IV.19. Formation of charged defects (charged solitons, spin = 0, charge = ± 1) from radical defects (neutral solitons, spin = 1/2, charge = 0)

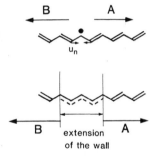

Fig. IV.20. Representation of the frontier occurring between two domains of different alternation configurations (phases A and B)

The soliton model proposed by Su, Schrieffer, and Heeger has been modified to allow a more complete theoretical treatment [692]. The important features of both models are fairly similar and both have successfully accounted for most of the experimental results.

IV.3 Properties of Doped Polyacetylene

Polyacetylene may be "doped" by adding either electron donors or electron acceptors. Two doping mechanisms may be envisaged (Fig. IV.21). In the first, the acceptor attacks the polymeric chain at some neutral defect (neutral soliton); one electron is removed and a positively charged diamagnetic species is formed. In the second mechanism, the dopant reacts with the regular polyenic chain; in this case, a one-electron transfer implies that both a cation (charged soliton) and a neutral radical (neutral soliton) are simultaneously formed. The relative importance of these two doping mechanisms depends on the concentration and on the nature of the dopant.

The electron affinity of $(CH)_x$ is fairly high and its ionization potential is low so efficient doping can be achieved by many different species.

a Dopants and Doping Processes

Both cis and trans-polyacetylene may be doped. However, it has been shown that the doping process induces the isomerization of the cis to the trans form. Is is noteworthy that, in most cases, the quality of trans-$(CH)_x$ isomerized by doping is better than that of thermally isomerized materials [627, 675].

Doping of polyacetylene films may be achieved in several ways [693]. The films may be exposed to a vapor atmosphere of the dopant when this latter is sufficiently

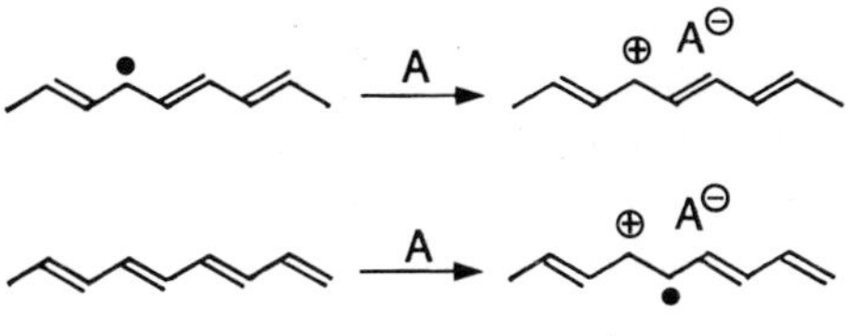

Fig. IV.21. The two main doping mechanisms which can be postulated when an electron acceptor is added to polyacetylene

volatile: I_2, AsF_5, O_2, XeF_2, etc. Films may also be treated by a solution of dopant; I_2 in pentane, sodium naphthalide in tetrahydrofuran, or $FeCl_3$ in nitromethane have been used. Polyacetylene may also be electrochemically doped; if the polymer is used as electrode for electrolyzing a solution of $LiClO_4$ in THF, both n and p doping occurs at the corresponding electrodes [694, 695]. Photoinitiated doping may also be achieved; triphenylsulfonium hexafluoroarsenate in methylene chloride has been used for that purpose [696]. The starting molecule is photodecomposed to yield an active doping agent. Ion implantation [693] has also been used; but, in this case, severe chemical degradation of the organic material has been observed [693, 697].

The detailed doping mechanism depends upon the type of dopant used. With $AgClO_4$ the mechanism is quite simple [698]:

$$(CH)_x + Ag^+,ClO_4^- \rightarrow (CH)_x^+,ClO_4^- + Ag^0 \qquad (IV.11)$$

The cation is quantitatively reduced into the corresponding metal. In the case of IF_5, the doping agent is only partially reduced [699]:

$$3 IF_5 + 2 e^- \rightarrow 2 IF_6^- + IF_3 \qquad (IV.12)$$

$$5 IF_3 \rightarrow I_2 + 3 IF_5 \qquad (IV.13)$$

The decomposition of IF_3 is detected by the appearance of iodine. In some cases, the doping mechanism is rather complicated and ambiguities between several chemical processes still remain. For AsF_5 doping, two slightly different mechanisms have been proposed [698, 700]:

$$(CH)_x + AsF_5 \rightarrow (CH)_x^+,AsF_5^{\cdot -} \rightarrow As_2F_{10}^{2-} \qquad (IV.14)$$

$$2 (CH)_x + 3 AsF_5 \rightarrow 2 (CH)_x^+,AsF_6^- + AsF_3 \qquad (IV.15)$$

Doping by iodine has been studied in great detail. Mössbauer spectroscopy provides evidence that I_2 is transformed into I_3^- and I_5^- [701, 702]; only I_3^- is obtained at low doping levels [644, 648, 681, 702]. The charge transfer is not total between the dopant and the polyenic chain; only approximately 0.8 e^- per I_3 or I_5 subunits is transferred [702, 703].

n-type doping generally has a simpler mechanism. In the case of sodium naphthalide, electron transfer from the anion to the polymeric chain occurs and naphthalene is released:

$$(CH)_x + Naphth^{\cdot -},Na^+ \rightarrow (CH)_x^{\cdot -},Na^+ + Napth. \qquad (IV.16)$$

The monoanion of benzophenone is not a strong-enough reductant to act as an n-type dopant and the corresponding dianion must be used [627, 704].

Care must be taken in studying the physico-chemical properties of n- or p-doped $(CH)_x$ because of the very inhomogeneous dopant distribution within the films [627, 705–710]. Diffusion coefficients of the dopants in the organic material are usually very low; for I_2, in the vapour phase, $D = 10^{-13}–10^{-14}$ cm$^2 \cdot$ s^{-1} [711]. This value is considerably lower than the usual diffusion coefficients of gases in polymer or in solution where $D = 10^{-9}$ cm$^2 \cdot$ s^{-1} [711]. Thus, for example, the relative concentrations of I_5^- and I_3^- can be shown to be different at the surface of the film or in the bulk [712]. Fairly homogeneously doped samples may be obtained at very low vapor pressures [706, 709] or at low dopant concentration [627].

Doped (CH)$_x$ never shows very good stability, though the susceptibility to decomposition does depend on the nature of the dopant. AsF$_5$-doped polymer requires heating to induce degradation with metallic arsenic, F$_2$ being detected amongst the decomposition products [713, 714]. Iodine-doped (CH)$_x$ degrades irreversibly above room temperature. And, whilst all derivatives are air and water sensitive, the n-type polymers are particularly active and produce partially hydrogenated products in the presence of water [586, 605].

b Structural Features

Doping induces fairly drastic modifications of the crystalline structure of (CH)$_x$ as shown by electron microscopy and X-ray diffraction experiments. When cis-(CH)$_x$ is used as starting material for the doping process, the diffraction patterns show a superposition of the lines corresponding to the cis and to the trans isomers. The amount of trans isomer increases with the concentration of the dopant [715]. Additionally, new bands appear in the X-ray pattern. The modifications in the crystallographic structure are reflected by changes in the macroscopic dimensions of the samples [716]. At high doping levels, the main additional feature is the appearance of a broad and diffuse diffraction band at high Bragg angles, corresponding to a periodicity of about 8 Å [601, 657, 715, 717, 718]. X-ray diffraction spacings increase in proportion to the effective Van der Waals thickness of the dopant just as is found for graphite intercalation adducts [601, 717]. The dopants are therefore probably intercalated between the 100 planes of (CH)$_x$ as shown in Fig. IV.22.

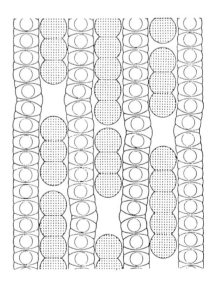

Fig. IV.22. Structural model assumed for iodine-doped polyacetylene. [Reproduced with permission of (601)]

Doping introduces disorder and chemical defects within the polymeric material. These latter may be estimated from the amount of sp^3 carbon atoms determined by solid-state ^{13}C NMR. For iodine-doped $(CH)_x$, the decrease of crystallinity is paralleled by an increase of the sp^3 content [630].

c Optical Properties

Doping induces drastic changes in the optical absorption spectra of polyacetylene (Fig. IV.23). It has been seen that the peak with edge at 1.4 eV and maximum at 1.95 eV is usually attributed to the direct interband transition, i.e. the absorbed photons directly yield free charge carriers in the conduction band. This peak is not significantly modified at low doping levels (Fig. IV.23a); however, an additional peak at 0.9 eV, near the mid band gap, appears [719]. Similar behavior is obtained for various p-type [643, 674, 694, 708, 720] and n-type [721] dopants. At high doping levels, the peak corresponding to the interband transition almost completely disappears, and the broad band around 0.9 eV becomes increasingly more intense [674] (Fig. IV.23b).

It has been seen that doping induces the creation of charged defects, cations or anions, lying in energy near the mid band gap of $(CH)_x$. The absorption which is observed implies transitions between energy levels belonging to the valence or the conduction bands and these newly generated energy states. The soliton model allows a quantitative description of the absorption features of doped $(CH)_x$ [722–724]. However, other interpretations have been given [725, 726].

At high doping levels, in the far infrared region, there is no observable transmission down to 20 cm^{-1} (0.003 eV). It therefore seems that a metallic regime is attained and that the band gap is completely "closed" [680]. The doping removes the bond alternation effect. A regular structure with equalized bond lengths is expected for highly doped $(CH)_x$ [674]. This is in agreement with the soliton model [727, 728]; ab initio electronic structure calculations on lithium doped $(CH)_x$ also predict this behavior [729].

At low doping concentrations, new infrared active modes are detectable in the low energy region [627, 730]. Due to symmetry reasons, phonons corresponding to the polymeric chain with localized double bonds are alternatively Raman or IR active. Whenever a charge is transferred, chain distortion occurs removing some of the symmetry forbiddeness and new IR bands appear. This has been considered as experimental evidence for the soliton picture of doping [731, 732]. However, it has been shown that the predicted IR frequencies are independent of the charge configuration [733].

d Magnetic Properties

As prepared undoped cis-$(CH)_x$, maintained at low temperature (-78 °C), does not exhibit any ESR signal [562, 618]; cis-$(CH)_x$ is therefore relatively free of defects [734]. Isomerization to the trans form takes place when the temperature is increased; simultaneously magnetic defects appear. For thermally isomerized trans-$(CH)_x$, the concentration of paramagnetic centers do not strongly depend on the "thermal history" of the sample; a concentration corresponding to about 1 spin per 3000 carbon atoms is invariably found [538, 618, 627, 641, 735–740]. For cis-$(CH)_x$, slightly isomerized at room temperature, susceptibility values spread over the range of 1 spin per 4000 to 1 spin per 45,000 carbon atoms [627, 737].

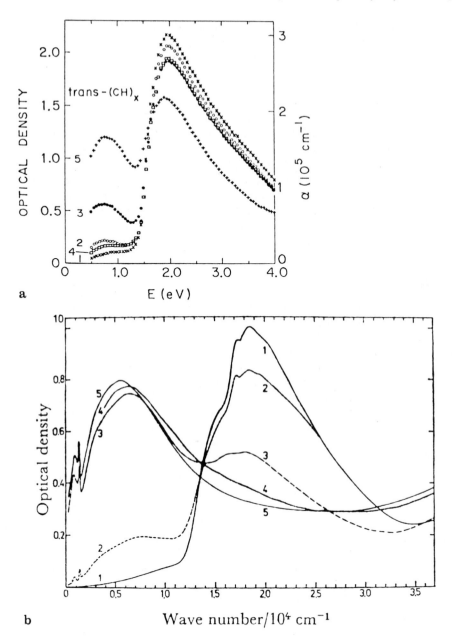

Fig. IV.23. Optical absorption spectra of doped polyacetylene.

a: Effect of doping with AsF_5; 1: undoped; 2: about 0.01% dopant; 3: 0.1%; 4: compensated with NH_3; 5: 0.5% dopant. [Reproduced with permission of (719)].

b: Effect of doping with $ClSO_3H$; 1: undoped; 2 to 5: increasing amount of dopants. The curve 5 corresponds to a conductivity of 250 Ω^{-1} cm^{-1}. [Reproduced with permission of (674)]

The ESR line shape analysis allows further information to be obtained concerning the nature of the paramagnetic centers. The ESR line shape has been shown to depend on the cis to trans ratio [609, 627, 641, 738], the temperature [619, 738, 742], the oxygen content [619, 741], and the homogeneity of the samples [627, 736]. Results obtained from various physico-chemical methods — dynamic nuclear polarization, nuclear relaxation time determinations, ESR lineshape analysis — are in accordance and demonstrate that in cis-$(CH)_x$ the spins are fixed, while in trans-$(CH)_x$ at room temperature the magnetic defects are highly mobile [619, 741, 743]. The diffusion coefficient of the spins along the polymeric chain is $D_\parallel = 5 \cdot 10^{-3}$ cm$^2 \cdot$ s^{-1}; the anisotropy of movement is fairly high since $D_\parallel/D_\perp \sim 10^6$.

It seems that, even for polyacetylene prepared under more or less air-protected conditions, O_2 plays an important role in the magnetic behavior of "undoped" $(CH)_x$. At low temperatures, the spins are trapped by impurities or defects; the concentration of free spins depends on the trapping energy and on the trap concentration. A model has been proposed [619] in which O_2 is bound to the fibrils surface; for vacuum-sealed samples, a molar concentration of O_2 of about 5% is found. Samples handled in air show a far higher O_2 content.

Radicals (neutral solitons) are very mobile within the polymeric chain. Consequently, soliton-soliton annihilation may occur if two of them are present on the same polyacetylene chain (Fig. IV.24). By increasing the concentration of doping agents, neutral solitons are annihilated and give rise to charged defects. Every chain bears either one or zero magnetic centers depending on whether an odd or even number of dopant molecules has reacted. The magnetic susceptibility therefore tends to a value corresponding to one spin for every two conjugated sequences.

Experimentally, a strong decrease of the susceptibility as a function of the dopant concentration has been observed for iodine [734], AsF$_5$ [740], and Na [627]. Since the conductivity increases with the concentration of dopant, it is clear that the paramagnetic centers cannot be identified with the charge carriers [641, 735].

In the highly doped domain, $(CH)_x$ behaves as a metal, and therefore a temperature independent term (Pauli susceptibility) should appear in the overall susceptibility beside the usual Curie component. In the case of iodine [706] and AsF$_5$ [739, 740], the Curie law contribution to the susceptibility decreases to less than 1 ppm in the highly conducting domain [740]; when homogeneous doping is achieved, the Curie contribution may even vanish [709]. The Pauli term is small for concentrations of dopant less than 5%; it abruptly increases above this concentration to reach a value corresponding to a density of states at the Fermi energy of about 0.1 states per eV per C atom [740]. This last point has been disputed however [726]. Attempts to measure the Pauli susceptibility by NMR (Knight shift experiments) do not furnish definite

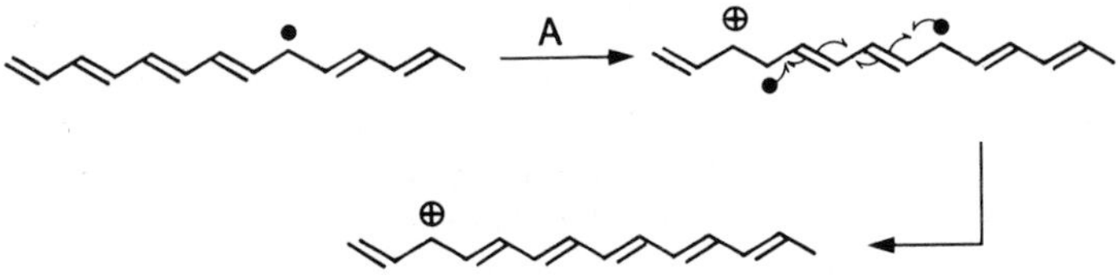

Fig. IV.24. Annihilation of a neutral radical defect by doping with an acceptor

conclusions [744, 745]. Further evidence for the metallic behavior of highly doped $(CH)_x$ is provided by the Dysonian line shape observed in ESR [627, 734, 735, 746].

IV.4 Transport Properties of Polyacetylene

A great variety of dopants has been used to give more or less conducting materials. Adducts corresponding to the formula $[CH(dopant)_y]_x$, where y is the proportion of the dopant and x the degree of polymerization, have been obtained. Depending on the concentration of the dopant, the electrical conductivity of doped polyacetylene may vary over 15 orders of magnitude, from the insulator range ($\sigma \sim 10^{-12} \Omega^{-1} cm^{-1}$) to the metallic one ($\sigma \sim 10^3 \Omega^{-1} cm^{-1}$) (Fig. IV.25) [747]. With increasing amounts of dopant, the electrical conductivity rapidly increases to reach a plateau value, which depends both on the nature of the dopant and on the doping procedure. In the highly doped range, the conductivities are the highest ever obtained for organic polycrystalline samples [552]. In parallel to the conductivity increase, the thermal activation energy estimated by plotting $\log \sigma$ vs $1/T$ decreases. In the case of AsF_5, the activation energy varies from 0.3 eV for undoped $(CH)_x$ down to less than 0.01 eV in the metallic regime. The limiting conductivity is attained for dopant molar ratios, y, lying between 1 and 5 %.

Three different concentration ranges of doping may therefore be distinguished. In the low concentration domain (y < 0.005), the number of charge carriers is proportional to the amount of dopant y; the transport is probably governed by a hopping process. In the highly doped range (y > 0.05), the molecular material behaves more or less like a metal. In the intermediate range of dopant concentration (0.005 < y < 0.05), all the physico-chemical properties vary rapidly and the detailed mechanisms of transport are described with difficulty.

It is almost impossible to obtain the intrinsic conduction properties of polyacetylene. Minute amount of impurities plague most of the experiments carried out. For cis-$(CH)_x$, the conductivities found are spread over a large range of values, the upper limit being $10^{-8} \Omega^{-1} cm^{-1}$ [641, 675, 680–682, 748, 749]; however, the intrinsic conductivity is probably much less than $5 \cdot 10^{-12} \Omega^{-1} cm^{-1}$ [749]. It is even more difficult to determine the intrinsic electrical properties of trans-$(CH)_x$, since the way it is prepared (thermal isomerization) implies the generation of numerous defects. Values around $10^{-5} \Omega^{-1} cm^{-1}$ are commonly found for "undoped" trans-$(CH)_x$; this is clearly far from the intrinsic value [750–752]. Addition of ammonia to trans-polyacetylene leads to a "compensation effect", i.e. n- and and p-type dopants annihilate each other, and conductivities less than $10^{-10} \Omega^{-1} cm^{-1}$ are then obtained [680]. Annealing under argon leads to similar results [753]. The concentration of free charge carriers in "undoped" materials is very low; for cis-$(CH)_x$ it is less than $2 \cdot 10^{16} cm^{-3}$ [749], for trans-$(CH)_x$ the upper limit is $2 \cdot 10^{18} cm^{-3}$ [675, 752]. It is worth noting that these values probably represent the concentration of residual dopants and O_2 is certainly associated with the presence of these dopants [614, 622, 754].

For both cis- and trans-$(CH)_x$, the thermal activation energy for the conduction, E_{act}, depends strongly on the quality of the samples. For cis-$(CH)_x$, E_{act} varies between 0.5 and 1.6 eV [749]; for trans-$(CH)_x$ $E_{act} \sim 0.3$ eV [675, 748]. These values are

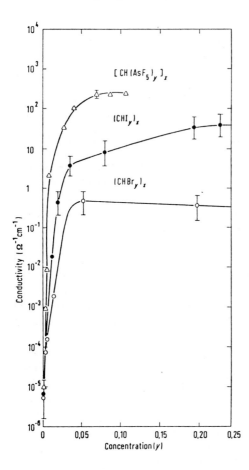

Fig. IV.25. Room temperature electrical conductivity of polyacetylene as a function of dopant concentration $[CH(dopant)_y]_x$. [Reproduced with permission of (747)]

strikingly different from the energy gap determined by optical absorption experiments.

The concentration of magnetic centers in trans-$(CH)_x$ is far higher than the concentration of free charge carriers [627, 735, 755]. The macroscopic low charge carrier mobilities — typically 10^{-5} cm^2/V·s — contrasts with the high spin mobility [619, 741, 743]. *Therefore no doubt subsists that the charge carriers cannot be identified with the paramagnetic centers.*

It has been previously shown how defects may be generated and how they propagate along polyenic chains as trans-$(CH)_x$. Charge transport may take place through two main conduction mechanisms (Fig. IV.26) [756]. In the first one, doping creates a charged defect; an electrostatic interaction E_{ip} between the ionized dopant and the injected charge occurs. This attraction must be overcome in order that the charge may migrate along the polymeric chain. In the second mechanism, there is hopping of an electron between two neighboring chains, one bearing a neutral defect, the other a charged defect. This process ensures a transverse conductivity within the fibrils. These two processes are studied in detail in the next section.

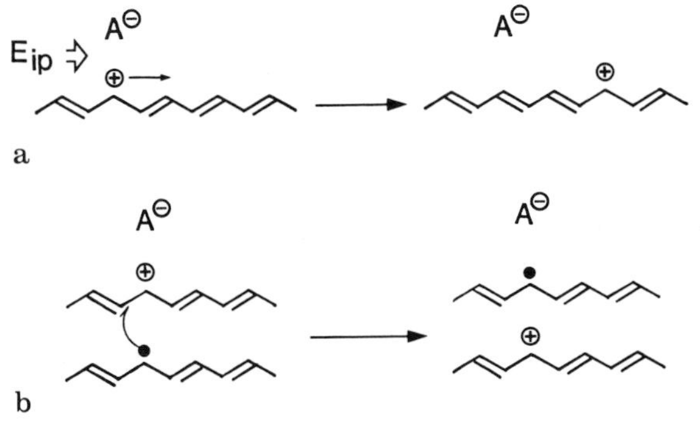

Fig. IV.26. The two main types of conduction mechanisms involved in the transport properties of polyacetylene.
a: Free charged defect diffusion; E_{ip} Coulomb interaction.
b: Electron hopping from neutral to charged defects

a Conduction Mechanisms at Low Doping Levels

At low doping levels, the number of charge carriers generated by doping is proportional to the concentration of dopant. Both charged-defect migration along the polyenic chain and electron exchange between neighboring chains may give rise to this behavior.

In the charged-defect migration process, the population of free charges is mainly governed by the interaction energy E_{ip} within the ionized dopant/injected-charge ion pair. The corresponding conductivity is given by [756]:

$$\sigma_{cm} = \frac{e^2}{kT} \cdot N_{ca} \cdot D_{ch} \cdot \left(\frac{1-n}{n}\right) \cdot x \cdot \exp\left(-E_{ip}/kT\right) \qquad (IV.17)$$

D_{ch}: diffusion coefficient of the charged defects
n: number of charged defects introduced per chain by each dopant ($n < 1$)
x: number of possible states corresponding to E_{ip}
N_{ca}: density of carbon atoms

In the absence of disorder, the diffusion coefficient of the charged defects D_{ch} may be approximated by the diffusion constant of the neutral radicals, $D_{neut} \cdot D_{neut}$ may in turn be estimated from the expression of a one-dimensional Brownian motion of domain walls in contact with thermal phonons. Calculations lead to $D_n \sim 2 \cdot 10^{-2}$ cm^2 $\times$ s^{-1} [675]. If a_c is the lattice constant, this diffusion coefficient corresponds to a diffusion rate D_n/a_c^2 equal to 10^{14} s^{-1}. This value is in good agreement with the NMR relaxation results ($4 \cdot 10^{14}$ radian s^{-1}) [743].

Within the framework of the previous formalism, the thermoelectric power is given by:

$$T_{cm} = \pm \frac{e}{k} \cdot \left[\frac{E_{ip}}{kT} + \ln \frac{(1-n)x}{n}\right] \qquad (IV.18)$$

In the process involving charge transfer between neighboring chains (intersoliton hopping), the energy necessary for electron hopping depends on whether or not there is a dopant molecule near the neutral radical defect. The hopping energy is lowered considerably when this is the case. The corresponding expression of the conductivity is [756]:

$$\sigma_{ih} = 0.45 \cdot \frac{e^2 \cdot W_{ih}(T)}{kT \cdot N} \cdot \left(\frac{\Psi_d}{d_0^2}\right) \cdot \frac{y_n \cdot y_{ch}}{(y_n + y_{ch})^2} \cdot \exp\left(-\frac{2.78 \cdot d_0}{\Psi_d}\right) \quad \text{(IV.19)}$$

y_n, y_{ch}: molar concentrations of neutral and charged defects, respectively
N: number of carbon atoms per conjugated sequences

$\dfrac{W_{ih}(T)}{N}$: frequency proportional to the fraction of time the defects are situated such that the initial and final states are within kT of each other

The exponential factor reflects the electronic overlap between two defects separated by d_0, the mean distance between the dopant molecules:

$$[\text{dopant}] = \left(\frac{4}{3} \cdot \pi \cdot d_0^3\right)^{-1} \quad \text{(IV.20)}$$

Ψ_d is the three-dimensionally-averaged electronic decay length:

$$\Psi_d = (\Psi_{d\parallel} \cdot \Psi_{d\perp}^2)^{1/3} \quad \text{(IV.21)}$$

The values of $\Psi_{d\parallel}$ and $\Psi_{d\perp}$ may be related to the corresponding transfer or resonance integrals.

At every hop, a spin and a charge are transported in opposite directions.

Using the charged-defect model and the electron-hopping model, many properties of polyacetylene may be predicted. At high enough temperatures, the charged-defect migration process along the polyenic chains will dominate because of the exponential temperature dependence of the conductivity. The intersoliton-hopping mechanism is only weakly temperature dependent; it therefore will be predominant at low temperatures. The transition temperature between the two regimes is not easily determined because of the uncertainties attached to the values of the ionized dopant/injected-charge interaction energies and the smallness of the electronic overlap factors entering in the hopping rates [756]. It does seem however, that over the whole temperature range experimentally studied, the intersoliton-hopping mechanism dominates [756].

By reversing the argument, it is possible to evaluate the binding energy E_{ip} by comparing the calculated conductivities (Eq. IV.19) with the experimental results. When AsF_5 is used as dopant, $E_{ip} \geq 0.38$ eV.

It is possible to minimize the ion-pair binding energy E_{ip} by increasing the "local" dielectric constant near the doping molecule and/or the distance between the dopant and the charge carrier. This is achieved by adding solvents which can solvate the ionized dopant; tetrahydrofuran for example, has been shown to drastically modify the conductivity of sodium-doped $(CH)_x$ [757]. A maximum conductivity increase of 10^4 is observed at low sodium-doping levels; an estimation of E_{ip} is then readily made:

$$\exp\left(-\frac{E_{ip}}{kT}\right) = \frac{\sigma_{\text{unsolvated}}}{\sigma_{\text{solvated}}} = 10^{-4} \quad \text{(IV.22)}$$

leading to a value of 0.23 eV, which is a lower limit of the ion-pair binding energy [757].

b Semiconductor-Metal Transition

The semiconductor-metal transition observed when increasing amounts of dopant are added to trans-polyacetylene has been rationalized in two ways. In the first case, over the whole dopant concentration range, $(CH)_x$ is considered to contain highly conducting metallic regions included within a still insulating medium. The transition arises when the metallic domains begin to encounter each other. A "percolation" threshold at which the metallic regions form a continuum should be observed [726]. In the second case, the transition arises when overlapping of conducting domains within every polyenic chain occurs. An oversimplified representation of this model is shown in Fig. IV.27. But this time there in a transition from a true semiconducting regime into a true metallic one [675]. In the first model, the local metallic domains progressively invade the whole material; the corresponding local properties should therefore vary progressively with the doping concentration, and only the macroscopic properties should demonstrate an abrupt transition. In the second model, on the other hand, both macroscopic and microscopic properties should abruptly vary in the vicinity of the transition point.

In the metallic regime, the thermopower is found to be small and linearly dependent with the temperature, but, by contrast, is large and essentially temperature independent at low doping concentrations. The mobility of 10^{-4}–10^{-5} cm^2/V · s observed in the semiconductor domain contrasts with that evaluated in the metallic regime (60 cm^2/V · s) [675].

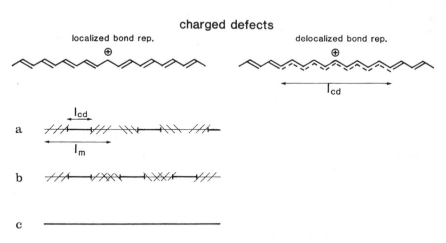

Fig. IV.27. Schematic illustration of the semiconductor to metal transition arising in doped trans-$(CH)_x$. l_{cd}: extent of the delocalization; n_{cd}: number of carbon atoms; l_m: zone over which the charged defect is moving.
a: Low-doping level (semiconductor domain).
b: Semiconductor-metal transition. The zones of width l_m encounter each other.
c: High-doping level (metallic domain)

A more quantitative analysis of the transition phenomena has been obtained from dielectric-constant measurements deduced from the far infrared spectra $(200-700 \text{ cm}^{-1})$ of the doped materials [758]. At a doping concentration corresponding to the transition, the dielectric constant $\varkappa$ diverges according to:

$$\varkappa \sim |y_c - y|^{-s} \qquad \qquad (IV.23)$$

y_c: critical doping concentration
y: concentration of dopant
s: parameter

For doping with AsF_5, the exponent s is equal to 1.1. This value is in agreement with a non-metal to metal transition but not with a percolation transition proposed in the first model, for which s should be equal to 0.7 [758].

The transition (disputed) of the susceptibility, from a Curie behavior to a Pauli type, takes place for much higher concentrations of dopant [709]. This implies the existence of two different transitions, the true metallic regime being attained only at concentrations higher than 5%. In the intermediate domain between the two transitions, the band gap still exists and there is a dense band of localized states [728].

The transition has also been interpreted as being due to the release (depinning) of the free carriers from the trapped charged defects at some critical temperature [759]. In another model [727], the transition arises because of the removal of the bond alternation by doping.

c Metallic Domain

At high doping levels, polyacetylene behaves very much as a classical metal. Several characteristics of the metallic regime appear. A Pauli term becomes an important component of the susceptibility [739, 740]. A term proportional to the temperature becomes apparent in the expression of the heat capacity [760]. The band gap present at low doping levels is closed or is reduced to less than a few tenths of an eV. The far infrared determinations [680] characterizing the band structure of the doped materials have been confirmed by energy-loss experiments [707]. For AsF_5 doping, the thermopower is linearly dependent on the temperature [675, 750]. However, I_2-doping experiments lead to less clear results [761].

Thermoelectric power determinations are of special interest since it is based on a zero-current transport method. The interfibril contacts therefore do not play a significant role. The thermopower may be written in the general case [3]:

$$T = \frac{k^2}{|e|} \cdot \frac{\pi^2}{3} \cdot T \cdot N(E_F) \qquad \qquad (IV.24)$$

$N(E_F)$: density of states at the Fermi level

For $[CH(AsF_5)_{0.15}]_x$, experiments show $N(E_F) \sim 0.2$ states per eV per carbon atom. This value is in agreement with the susceptibility and specific heat measurements [760]. Such features are consistent with a broad energy-band structure [709].

The "intrinsic" conductivity of metallic polyacetylene is much higher than the one measured by dc conductivity. For example, thermal conductivity studies of iodine-doped $(CH)_x$ indicate a ratio $\sigma_{intrinsic}/\sigma_{dc}$ of about 700 [762]. The dc transport is limited by the interfibrillar contacts [643]. At room temperature, heavily doped $(CH)_x$ appears to be constituted of metallic strands; electronic transport from one strand to another occurs through an energy-activated process. Models based on fluctuation-induced tunneling across interfibrillar contacts [763–765] accord well with the electrical conductivity data in the very low temperature range (0.4 K to 1 K) [766]. The inter-fibrillar barrier height has been evaluated to be of the order of $3\text{–}7 \cdot 10^{-3}$ eV, the corresponding barrier thickness is 10–20 Å [675].

d Comparison of the Models with the Experimental Results in the Low Dopant Concentration Domain

In the low dopant concentration domain, it is clear that electron conduction occurs through a hopping process. Two principal hopping mechanisms may be envisaged: the intersoliton-hopping model previously described and a variable range-hopping-type model. In this latter, electrons hop from one energetic well to another through a tunneling mechanism. The energy difference between the initial and final states therefore must be less than kT. The hopping probability is related to the energy difference between the states, but also to the distance separating the wells. The prediction which can be made from the intersoliton-hopping model (IHM) or from the variable-range-hopping-model (VHM) are compared with the experimental results in Table IV.5 [756].

The conductivity is found experimentally to vary rapidly with the dopant concentration. Unfortunately, within the experimental error due to uncertainty of the dopant concentration, it has not been possible to distinguish between the IHM ($\log \sigma \sim$ $\sim [\text{dopant}]^{1/3}$) and the VHM ($\log \sigma \sim [\text{dopant}]^{1/4}$).

The electrical thermopower is found to be high and positive for p-type dopants [709, 750, 755], high and negative for n-type dopants [755]. The magnitude of the thermopower is constant with doping concentrations in the range $10^{-4}\text{–}10^{-3}$. This behavior is in agreement with the IHM and clearly in disaccord with the VHM.

Experimentally, at low doping levels, the variation of conductivity with temperature is lower than that predicted by an exponential dependence [675, 737, 748, 755, 767, 768]. For "undoped" $(CH)_x$, the temperature dependence of σ_{dc} can be fitted equally well with a T^{-n} law (n = 13–14) as with an $\exp(-T)^{1/4}$ relationship. Therefore, both the IHM and VHM are plausible. However, the results do exclude a conduction mechanism occurring through charged-defect migration along the polyenic chain. At low temperature, the temperature dependence of σ changes to that of T^{-1}. Such behavior may be due to a distribution of activation energies caused by some disorder [675].

The ac conductivity is found to be nearly constant in the frequency range $10^1\text{–}10^7$ Hz for lightly doped samples [682] or for "undoped" trans-$(CH)_x$ at room temperature [748]. In this last case, however, σ_{ac} becomes frequency dependent at low temperature. For iodine-doped samples at room temperature, σ_{ac} becomes frequency dependent at 10^{10} Hz [681]. The IHM predicts the onset of a frequency dependence at 10^8 Hz and the VHM at 10^4 Hz, for a dopant concentration of $3 \cdot 10^{-3}$ [756]. The IHM thus is clearly favored.

Table IV.5 Comparison of the predictions of the intersoliton hopping conduction model and of the variable-range hopping conduction model with experimental results obtained for polyacetylene. (After Ref. [227]).

Intersoliton hopping model (IHM)	Variable-range hopping model (VHM)	Experimental results
Conductivity		
1. — depends on the inverse of chain length $\sigma \sim N^{-1}$	— no dependence with chain length	— no experiments
2. — depends on the cis-trans ratio	— small dependence on cis/trans ratio	— depends strongly on the cis/trans ratio
3. — increases with impurity concentration $\ln \sigma \sim [\text{dopant}]^{1/3}$	— rapidly increases with impurity concentration $\ln \sigma \sim [\text{dopant}]^{1/4}$	— increases with impurity concentration [675]
4. — thermal activation energy does not depend on dopant concentration	— thermal activation energy dependent on the dopant concentration	— independent of dopant [767, 768]
5. — varies more slowly than $\exp(-E_{pi}/kT)$ $\sigma \sim T^n \quad n \sim 10$	— varies more slowly than $\exp(-E_{pi}/kT)$ $\sigma \sim \exp\left(-\dfrac{T_o}{T}\right)^{1/4}$	— consistent with the IHM & the VHM [675, 737, 755, 767, 768]
6. — non-ohmic behavior at relatively low electric fields		— ? see [769]
7. — small pressure dependence of σ_{dc} $\dfrac{d \ln \sigma}{dP} \sim -1.39 \dfrac{d_o}{\Psi_d} \varkappa$	— pressure and temperature cannot be separated into different factors	— consistent with IHM [755]
8. — pressure dependence of σ is temperature independent	— pressure and temperature cannot be separated into different factors	— consistent with the IHM [755]
9. — strongly frequency dependent $\sigma(\omega) - \sigma_{dc} \sim \omega \cdot (\ln \omega)^4$ for $\omega > 10^8$ Hz	— frequency dependent $\sigma(\omega) \sim \omega^{0.8}$ for $\omega > 10^4$ Hz	— consistent with the IHM [681, 682, 767]
10. — strongly temperature dependent	— weakly temperature dependent $\sigma_{ac} - \sigma_{dc} \sim T$	— consistent with the IHM $n = 13-14$ [767]

$$\sigma_{ac} - \sigma_{dc} \sim \frac{\omega}{T}\left[\ln \frac{D \cdot \omega}{T^{n+1}}\right]^4$$

Thermopower		
11. — high	— very small	— high [709, 755]
12. — very weakly temperature dependent	— temperature dependent $T_{VHM} \sim T^{1/2}$	— consistent with the IHM [709]

$$T_{IHM} \sim \frac{e}{k} \cdot \left[c^{st} + \ln\left(\frac{kT}{h \cdot \omega_1}\right)\right]$$

Table IV.5   (continued)

Intersoliton hopping model (IHM)	Variable-range hopping model (VHM)	Experimental results
13. — only weakly dependent on impurity concentration	— impurity concentration dependent $T_{VHM} \sim [\text{dopant}]^{1/2}$	— consistent with the IHM $n = 13$–14 [709]
$$T_{IHM} \sim \pm \frac{e}{k} \cdot \left[\frac{n+2}{2} + \ln \frac{y_n}{y_{ch}} \right]$$		
Hall mobility		
14. — small but finite		— no experiments
Spin-diffusion constant		
15. — the in-chain spin diffusion is unrelated to and more efficient than charge transport process	— Einstein relation for in-chain spin diffusion	— consistent with the IHM [743]
16. — the perpendicular spin diffusion is related to the conductivity $D \, [\text{dopant}] \, e^2 = kT \cdot \sigma$	— Einstein relation for perpendicular spin diffusion	— not inconsistent with both models [743]

Following the IHM description of the transport properties, it is possible to evaluate the concentration of residual dopants present in "undoped" trans-$(CH)_x$. The lowest room-temperature conductivity reported so far is about $5 \cdot 10^{-9} \, \Omega^{-1} \, cm^{-1}$ [770]; this corresponds to 10^{-4} charged impurities, i.e. a density of approximately 1–$2 \cdot 10^{18} \, cm^{-3}$.

From NMR relaxation experiments, the spin diffusion constant and its anisotropy have been evaluated ($D_{\parallel} = 4 \cdot 10^{14}$ Hz; $D_{\perp} < 4 \cdot 10^8$ Hz) [743]. A strong anisotropy of the spin-diffusion process is found. The mobility of the charges, on the other hand, is far lower than the values corresponding to the spin-diffusion processes. These facts strongly suggest that in-chain spin transport and charge transport occur via distinct mechanisms. The experiments provide another case of failure for the VHM, which predicts that $D_{\parallel}$ and $D_{\perp}$ should be equal.

In conclusion, most of the experimental results are in accordance with the intersoliton-hopping model and seem to exclude the variable-range-hopping-type mechanisms. The results also exclude another conduction model, which had been mentioned briefly, in which metallic regions are embedded within a poorly conducting medium.

The model predicts that σ_{dc} should vary with temperature as $\exp - \left(\frac{T_o}{T} \right)^{0.5}$ [761], [765]; this cannot be fitted to the experimental results. It also predicts an increasing conductivity at higher frequencies and higher doping levels [682], in disagreement with the experimental evidences [681].

IV.5 Photoelectric Properties and Solar Cells

a Luminescence and Photoconductive Properties of Cis- and Trans-Polyacetylene

Cis- and trans-$(CH)_x$ have markedly different luminescence and photoconductive properties. For cis-$(CH)_x$, a strong luminescence is observed near the interband absorption edge, centered at 1.9 eV (Fig. IV.28). The luminescence turns on sharply for excitation energies greater than 2.05 eV. This emission peaks at 1.9 eV independently of the excitation frequency [685]. This corresponds to a shift of $2.05–1.9 = 0.15$ eV; it presumably arises from a lattice distortion around the photoexcited e^-/h^+ pair. The measured luminescence intensity shows little temperature dependence up to 150 K; at higher temperatures, results are obscured by partial isomerization of the initially cis-$(CH)_x$ film [685]. The luminescence properties of cis-$(CH)_x$ thus seem to arise through a recombination mechanism (Fig. IV.29). In a first step, a pair of free charge carriers is generated by absorption of a photon, some local lattice distortion then occurs to account for the polarity of the e^-/h^+ pair. It is not possible to decide precisely which type of lattice distortion is occurring. A reasonable suggestion is that a rotation around a $C=C$ double bond would give rise to orbitals with reduced overlap, and that this would favor ionized structures. Since the two different cis-configurations have different energies, the positive and negative charges cannot migrate along the polyenic chain. They therefore finally radiatively recombine, giving rise to luminescence. The recombination of the charged carriers occurs very efficiently and only a negligible proportion of them escape to yield to photoconduction. Consequently, cis-$(CH)_x$ does not demonstrate any photoconductivity.

The properties of trans-$(CH)_x$ are strikingly different. Thermal isomerization of cis-$(CH)_x$ into the trans form entirely quenches the luminescence peak. The luminescence intensity in the range 1.4–2.5 eV is less than that of the cis-$(CH)_x$ by at least a factor of 50 [685, 771]. On the other hand, a large photocurrent is observed for incident photon energies corresponding to approximately 1 eV (Fig. IV.30) [685]. The magnitude of the photocurrent in trans-$(CH)_x$ is more than three orders of magnitude greater than the corresponding I_{ph} in cis-$(CH)_x$. Results obtained from different sample preparations show significant variations in the magnitude of the photoresponse below 1 eV (see inset of Fig. IV.30) [677, 685]. For poor quality samples, photocurrents are observed even for photons of very low energies. Addition of ammonia, a donor molecule decreases the low-energy response probably by a "compensation effect": donors neutralize in some way adventitious p-type doping agents [685].

More curiously, the photoconductivity onsets well below the interband absorption even for apparently dopant-free materials (Fig. IV. 31) [685, 771]. Such behavior may arise from a broadening of the absorption edge due to static disorder or to a combination of quantum and thermal fluctuations in $E_g(x)$, the local energy gap [685]. Dynamical fluctuations generate random dynamical distortions in the crystal. According to the Franck-Condon principle, absorption edges are determined by the instantaneous configurations of the lattice (10^{-15} s). Dynamical fluctuations lead to a distribution of short-lived states ($\sim 10^{-13}$ s). Thus, the absorption edge is

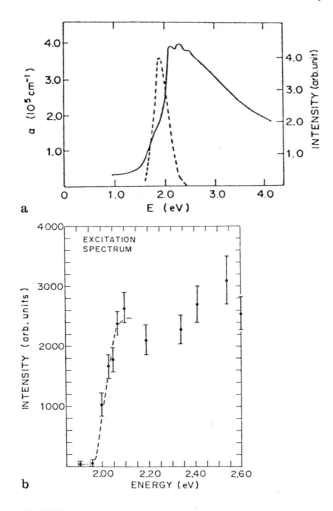

a

b

Fig. IV.28. Luminescence properties of cis-polyacetylene.
a: Comparison of the photoluminescence spectrum (– – –) and the interband absorption spectrum
(solid curve).
b: Excitation spectrum for the luminescence arising from cis-(CH)$_x$. [Reproduced with permission
of (685)]

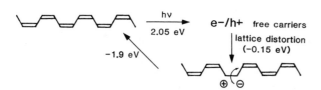

Fig. IV.29. Schematic representation of the luminescence
properties of cis-polyacetylene

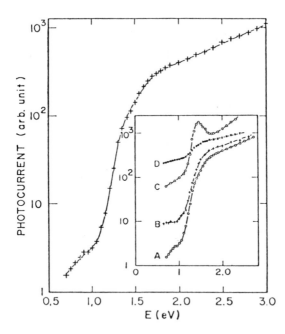

Fig. IV.30. Photoconductive properties of trans-polyacetylene: photocurrent vs incident photon energy. The inset shows the same data with samples of varying quality. Compensation with ammonia converts the I_{ph} spectrum from curve D to curve A. (Reproduced with permission of (685)]

smeared and a tail (Urbach tail) is generated [685]. For photons having energies greater than the band gap E_g, direct e^-/h^+ pairs are produced with a quantum efficiency of the order of unity. However, the minimum energy necessary to generate charged defects is only $\frac{2E_g}{\pi}$ [685, 772, 773]. The fit of the experimental photocurrent below the band edge with the theoretical Urbach absorption tail, leads to a phonon energy of about 0.2 eV; this energy correponds to the mechanism responsible for the random fluctuation in the local energy gap [685].

It is important to distinguish between the photogenerated free charge carriers and the charged soliton-antisoliton pairs (Fig. IV.32). It has been shown that in the presence of an e^-/h^+ pair the lattice is unstable [774]. In a time of the order of the reciprocal of an optical phonon frequency (10^{-13} s), the lattice distorts to form a soliton-antisoliton pair. Thus, whatever the initial process generating the charge carriers, the formation of charged defects occurs in $(CH)_x$ in a time of the order of 10^{-13} s [685]. The energy which is necessary to create a soliton-antisoliton pair (E_s) is less than the energy gap. E_s has been found to be equal to $\frac{2E_g}{\pi}$ in the standard models [668, 687], this leads to $E_s \sim 0.9$ eV for a band gap $E_g \sim 1.4$ eV. For a single charge carrier, the soliton-like state is stabilized by approximately 0.3 eV over the free electron or holes. ESR studies recently have confirmed that light illumination at photon energies greater than E_g generated spinless charge carriers [775].

A more illustrative way of rationalizing the photoconductive properties of trans-$(CH)_x$ may be found by considering the chemical mechanisms involved. The sudden polarization concept has been evoked to explain the formation of ionized species from photoexcited polyenes [776] (Fig. IV.33). For zwitterionic species to exist,

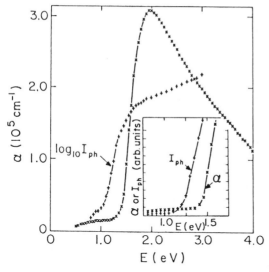

Fig. IV.31. Comparison of the photocurrent (log I_{ph}) and optical absorption coefficient (α_a) as a function of incident photon energy. The inset shows a linear representation of I_{ph} and α_a versus photon energy; a threshold of I_{ph} well below the interband absorption transition is observed. [Reproduced with permission of (685)]

twisting of the C=C double bond must occur; if not, the overlap between the two orbitals is such that the species cannot be described by an ionic state. The degree of charge separation as a function of the twist angle has been calculated in the case of s-cis, s-trans-hexatriene (Fig. IV.33 b) [776]. The charge separation peaks very sharply at $\Theta = 90°$, it remains significant between 89° and 91°; this represents a 2° window for the polarization to occur. When $\Theta = 90°$, the overlap integral between the two C atoms is zero, charge may be exchanged only through the exchange integral γ_{ab}. It has been demonstrated that a slight dissymmetry is a prerequisite for polarization. A pyramidalization of the C atom bearing the negative charge strongly stabilizes the charge separation [776].

In the light of the "sudden polarization concept", it is possible to propose a chemical mechanism for the photoconductive behavior of trans-$(CH)_x$. In a first step, at some point a twisting of the polyenic chain occurs (Fig. IV.34). Simultaneously or not, a photon is absorbed giving rise to two charged-defects of opposite sign (charged soliton-antisoliton pair). The charge-separation process is highly facilitated by the absence of orbital overlap in the twisted C-C bond. It is tempting to assign the value

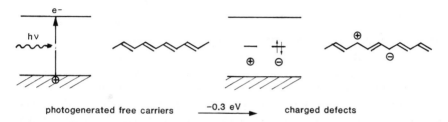

Fig. IV.32. Schematic representation of the photogeneration of free charge carriers and pairs of charged defects (soliton-antisoliton pair) with the corresponding band diagrams. (After Ref. [685])

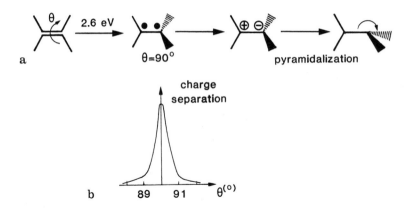

Fig. IV.33. Illustration of the sudden polarization effect.
a: Formation of a zwitterionic state.
b: Charge separation as a function of twist angle in the lowest excited state of hexatriene.
The difference in energy between the biradical ground state and the zwitterion is 3.5 eV. (After Ref.
[776])

of 0.2 eV, found in the Urbach mechanism for the phonon contribution, to the energy
necessary to twist the polyenic chain. However, no experimental evidence is yet
available. It is worth remarking that no free charge carriers are generated as intermediates during this process. The energy threshold for this mechanism is thus approximately $E_s \sim 0.9$ eV, i.e. the energy needed to create a pair of charged solitons. This
is in agreement with the photoconductive threshold previously noticed. Calculations
have been carried out considering the generation of soliton-antisoliton pairs at
phonon energies less than the band gap as a tunneling problem [777].

b Junctions Properties and Molecular Solar Cells

It has been previously seen that the photocurrent I_{ph} onsets well below the energy
gap, in a region where the optical absorbance is negligible. These results were obtained
with undoped trans-$(CH)_x$/Al contacts [685, 771]. $(CH)_x$ lightly doped with AsF_5
[677, 778] or with HCl [779] has also been studied. Different sample configurations
have been adopted for determining its photovoltaic properties; sandwich-type and
surface-electrode-type devices lead to very similar results [677, 778]. Sample handling
was carried out in an inert atmosphere; however, short exposures to air (1–2 min)
were unavoidable during the mounting of the sample in the vacuum metallization
chamber [677, 778]. Although the photocurrent starts to increase above approximately
1 eV, well below the optical band edge, by contrast, the corresponding photovoltage
is in close coincidence with the absorption peak (Fig. IV.35). A plot of the square
of the photovoltage vs photon energy demonstrates more clearly the correlation
between the photovoltage threshold and the optical absorption edge. By extrapolation,
a band gap equal to 1.48 eV is found, in good agreement with the previous deter-

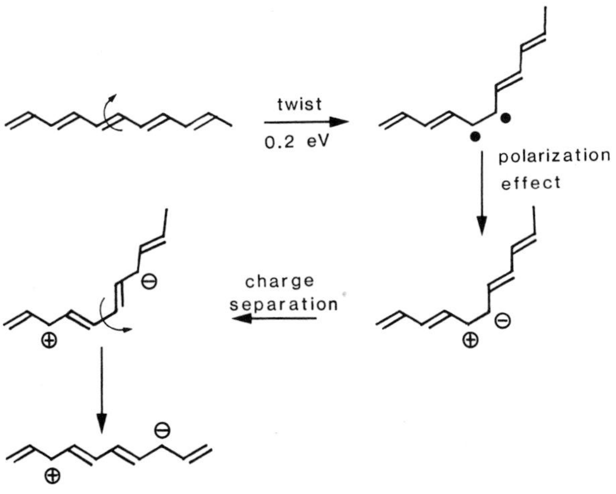

Fig. IV.34. Schematic representation of the generation of charged defects through a sudden polarization-type mechanism

minations [677, 778]. The straight line from this plot indicates a relationship of the type:

$$\alpha_a \sim (E - E_g)^{1/2} \tag{IV.25}$$

α_a: absorption coefficient
E_g: threshold energy (energy gap)

as would be expected for transitions in a three-dimensional-like continuum [677]. This could indicate that strong interchain interactions occur; in a one-dimensional system, a dependence on $E^{-1/2}$ is expected [677]. Alternatively, exciton states may be involved [677, 780]. This last explanation is reenforced by the fact that the magnitude of the photocurrent is dependent on the reverse-bias voltage [779]. For low energy irradiation (2 eV, 6300 Å), the photocurrent is strongly dependent on the biasing voltage. At shorter wavelengths (2.75 eV, 4500 Å), the photocurrent depends less on reverse voltage and it saturates at a lower voltage. An Onsager model may be applicable to the electric field and wavelength dependences of the photocurrent [779]. While the interband nature of the main optical absorption peak can hardly be questioned, it does seem, however, that the interaction between charge carriers of opposite sign is far from being negligible.

The energy threshold for the photovoltage seems to differ from the energy threshold for the photocurrent. Are these two results incompatible [685, 771]? The photocurrent is, in the general case, determined by the quantum efficiency for generating charge carriers. If this latter is only moderately electric-field dependent, the magnitude of the photocurrent generated in the bulk semiconductor is close to that arising from photons absorbed near the interface in the space-charge region. There is no obligatory relationship between the photocurrent and the junction characteristics. But such a correlation obviously must exist for the photovoltage. *It consequently seems that the direct generation of charged soliton-antisoliton pairs arising below the optical edge is a quite inefficient process for modifying charge equilibria at the Schottky contact.*

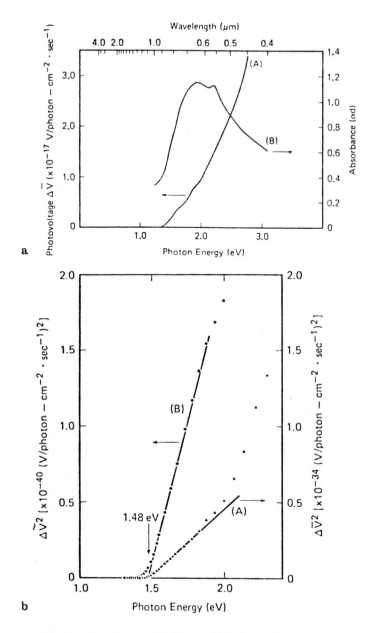

Fig. IV.35. Photovoltaic properties of trans-polyacetylene lightly doped with AsF_5 (doping level $y = 0.006$).

a: A: Open-circuit photovoltage for an indium Schottky contact.

   B: Absorption spectrum of undoped $(CH)_x$.

b: Plot of the square of the photovoltage vs photon energy.

   A: Curve normalized to the incident photon flux.

   B: Curve normalized also for the indium absorption.

[Reproduced with permission of (677, 778)]

The dark electrical properties of $(CH)_x$/metal [754, 778, 779, 781–788] or $(CH)_x$/semiconductor [752, 754] junctions have been thoroughly studied. For thermally isomerized undoped trans-polyacetylene, contacts are ohmic with high work-function metals and rectifying for the others (Fig. IV.36). Metals having work functions less than 4.5 eV (In, Bi, Al, Pb, Sn) lead to rectifying contacts, while those having a work function greater than this value (Ag, Sb, Cu, Au) are ohmic [754]. Only in one case [786] has a diode effect been found with Au and Pt contacts. At high forward voltages, the I–V curves become linear [778], showing that the bulk resistance then determines the conduction properties. For lower forward voltages, a classical Schockley equation is found in most cases; a plot of log I against V yields a straight line [754, 784, 786–788]. The simplest form of this Schockley equation is:

$$J = J_{sat} \cdot \left[\exp\left(\frac{e \cdot V_a}{A_o \cdot kT} \right) - 1 \right] \tag{IV.26}$$

J_{sat}: saturation current
V_a: applied voltage
A_o: perfection factor

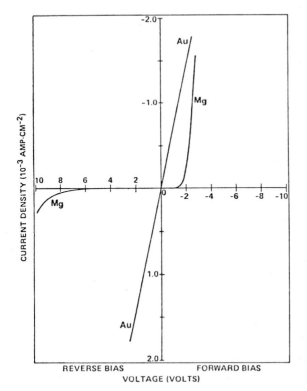

CURRENT DENSITY $(10^{-3}$ AMP-CM$^{-2})$

REVERSE BIAS   FORWARD BIAS
VOLTAGE (VOLTS)

Fig. IV.36. Current-voltage characteristics of trans-$(CH)_x$/Au and trans-$(CH)_x$/Mg contacts. The polyacetylene is thermally isomerized and is not intentionally doped $(\sigma \sim 10^{-7} \, \Omega^{-1} \, cm^{-1})$. [Reproduced with permission of (782)]

with

$$J_{sat} = A^* \cdot T^2 \cdot \exp\left(-\frac{\Delta E_{MS}}{kT}\right) \qquad\qquad (IV.27)$$

A^*:   Richardson constant
ΔE_{MS}: barrier metal-semiconductor

It is derived by assuming a thermoionic emission of charge carriers across the interfacial barrier.

It is worth noting that the plot of log J vs V is linear only over a small part of the whole curve [754]. The perfection factor (A_o) greatly differs from unity ($A_o = 2$–4). Moreover, $\dfrac{d \log J}{dV}$, which should be inversely proportional to the temperature, is found to be virtually temperature independent [754]. All these facts demonstrate that the thermoionic emission model provides only a very partial description of the experimental results. Tunneling effects through the barrier [754] or image-forces effects have been evoked to rationalize the experimental findings [779, 786].

In all the I–V characterizations previously described, the samples were handled, as far as possible, in an inert atmosphere. However, short air exposures (1–5 min) were unavoidable. A question arises on the effect of O_2 on the electrical properties of $(CH)_x$-based junctions? Trans-$(CH)_x$/In contacts have been the object of detailed studies [785] (Fig. IV.37). By O_2 doping, the resistivity of the sandwich cell diminishes, the rectification increases from 75 for "undoped" $(CH)_x$ to 100 in the presence of O_2. Lightly AsF_5-doped samples show the same behavior [785]. It is not possible to decide whether $(CH)_x$ in strict absence of dopant (O_2 or others) would show the same I–V behavior. The chemical nature of the ionized species responsible of the space-charge region cannot be assessed.

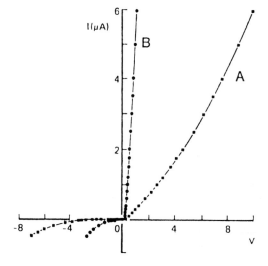

Fig. IV.37. Current-voltage characteristics of "undoped" and O_2-doped $(CH)_x$ forming a Schottky contact with In.
Curve A: undoped $(CH)_x$
$\sigma = 3 \cdot 10^{-8} \, \Omega^{-1} \, cm^{-1}$
Curve B: O_2 doped $(CH)_x$
$\sigma = 2.5 \cdot 10^{-7} \, \Omega^{-1} \, cm^{-1}$
[Reproduced with permission of (785)]

The electrical properties of p-n junctions with alternatively p- or n-type-doped $(CH)_x$ have been described [789] as well as the characteristics of photoelectrochemical cells [754, 790, 791].

Numerous publications discuss the ac electrical properties of $(CH)_x$ contacts [682, 749, 752, 754, 778, 779, 781, 785, 788]. It is important to determine first the parameters R_p and C_p, the equivalent parallel resistance and capacitance (see Chap. II), as a function of the frequency of the applied current under zero-bias condition. It is then possible to determine in which frequency domain the barrier parameters may be obtained. Experiments with trans-polyacetylene have been carried out with ohmic (Au) or blocking (In) electrodes (Fig. IV.38) [682]. For Au contacts, the resistivity is independent of the frequency in the range $10-10^6$ Hz indicating that it is bulk properties which are being measured. The magnitude of the resistivity varies with doping. For In contacts, the resistivity is frequency dependent. Only at high frequencies are the bulk values for the undoped and doped states obtained. It may be assumed that, at low frequencies, the intrinsic barrier characteristics are observed. The break frequency between the two domains is 11.5 kHz for AsF_5-doped and 75 Hz for O_2-doped samples [682]. For "undoped" $(CH)_x$, the plateau value of the resistivity is only obtained around 10 Hz [682]. It is worth remarking that low concentrations of O_2 seem to reversibly dope trans-$(CH)_x$, as electrical properties return to their original values on overnight pumping [682]. The frequency dependence of C_p gives very similar results [682]. Experiments at very low frequencies (10^{-3} Hz) have been carried out on cis- and trans-$(CH)_x$ [749]. Ohmic Au or rectifying Al contacts were employed. As previously mentioned, samples were unavoidably exposed to air over short periods of time. In the case of Al contacts this may induce the formation of a significant aluminum oxide layer which can invalidate the ac measurements. For the rectifying contacts t-$(CH)_x$/Al, the plateau values of C_p and the conductance G_p are obtained for frequencies lower than approximately 10^{-1} Hz. However, high temperatures (150–190 °C) must be used to observe the transition [749].

It is now clear that the characteristics of the space charge region are obtained only at fairly low frequencies. This precaution being taken, the voltage dependence of the capacitance may be studied to determine the number of ionized dopants in the depletion region and the corresponding space charge region depth. For lightly AsF_5-doped t-$(CH)_x$/In junctions, a plot of $1/C^2$ vs V gives a straight line [785] (Fig. IV.39). The modulation frequency was in the range 100–1000 Hz, apparently sufficiently low to obtain the true junction capacitance. The slope and intercept of the $1/C^2$ vs V plot permits calculation of the depth of the depletion layer (190 Å), the built-in potential (0.9 V) and the density of ionized dopants ($2.2 \cdot 10^{18}$ cm^{-3}) [785]. Assuming a complete thermalization of the charge carriers with the traps, the hole mobility may be calculated from the conductivity value ($\sigma = 10^{-4}$ Ω^{-1} cm^{-1} for AsF_5-doped sample) and is found to be $3 \cdot 10^{-4}$ cm^2/V $\cdot$ s [785]; this value is in the expected range for polycrystalline materials. In t-$(CH)_x$/n-CdS junctions, the number of ionized dopants is found to be very comparable [752]. However, fairly erratic values may be obtained depending upon the conditions of sample handling. It has been previously seen that the $(CH)_x$ films prepared by the Ziegler-Natta method are highly inhomogeneous. The surface in contact with the glass is shiny, while the other is dull. This difference of aspect is reflected in the crystallinity of the sample determined by X-ray diffraction experiments. The crystallinity is 80% for the dull surface and less than

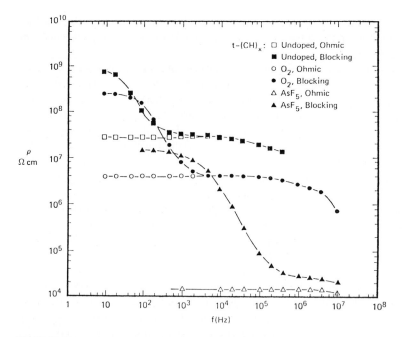

Fig. IV.38. Frequency dependence of the resistivity of trans-polyacetylene for ohmic (Au) and blocking (In) contacts.

Undoped trans-$(CH)_x$: from thermally isomerized cis-$(CH)_x$.

O_2-doped: exposed to air.

AsF_5-doped: exposed to AsF_5.

For both dopants, the doping level is less than 0.1 % as estimated from the resistivities. [Reproduced with permission of (682)]

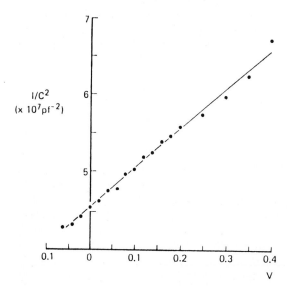

Fig. IV.39. Capacitance-voltage plot for lightly AsF_5-doped trans-$(CH)_x$/In junction. The modulation frequency is in the range 100–1000 Hz. [Reproduced with permission of (785)]

10% for the shiny one [779]. This is also apparent from the C–V measurements. A trap density of $1.4 \cdot 10^{18}$ cm^{-3} is found for the shiny surface while $7.4 \cdot 10^{16}$ cm^{-3} for the dull one [779].

As is the case for all junctions formed with molecular semiconductors, deep levels coexist in the forbidden gap with shallow donor or acceptor levels. This may give rise to a frequency dependence of the junction parameters due to the long relaxation times associated with these deep centers. A model has been proposed in the case of t-(CH)$_x$/In contacts to take into account this phenomenon [778]. The deep levels are calculated to be 0.16 eV above the Fermi level; the associated relaxation time being of the order of 10^{-2} s [778].

From both I–V and C–V measurements, it is certain that the simplest form of the thermoionic emission model for analyzing the junction properties of polyacetylene is inadequate. In particular, the values determined for the barrier heights cannot be related reliably to any physico-chemical properties of the interface [785]. Moreover, due to the thickness of the Schottky barriers, image forces and tunneling effects may be effective in modifying the electrical behavior.

Polyacetylene may be used as the semiconducting part of an organic solar cell [754, 779, 783, 784, 787, 788]. The open-circuit voltage under irradiation, V_{oc}, the short-circuit photocurrent I_{sc}, and the fill factor FF are the main parameters describing the solar cell characteristics, as discussed in Chap. II (Table IV.6). The devices described have all been exposed to air over more or less extended periods of time; measurements are usually made under vacuum or in a flow of nitrogen. The absolute quantum yield efficiency, i.e. the ratio of the short-circuit current to the photon flux, reaches 10% for photon energies of the order of 1.95 eV for t-(CH)$_x$/Al contacts (0.2 mW/cm^2) [754]. This value includes a correction for aluminum opacity. The quantum efficiency drops abruptly for photons of lower energies. The quantum efficiency saturates above 2.5 eV [779]. The semitransparent blocking electrode always becomes negative under illumination, electrons are transferred from the organic material to the blocking electrode. Doping with I_2 and AsF$_5$ simultaneously decreases V_{oc} and I_{sc} [788]. HCl-MeOH doping, on the other hand, seems to increase V_{oc} but decrease I_{sc}. However, these comparisons must be viewed with care due to the difficulties of obtaining reproducible results from cell to cell. The variations of I_{sc} and V_{oc} with the incident photon flux Φ_o have been found to follow the relationship:

$$V_{oc} \sim \log (\Phi_o) \qquad\qquad\qquad (IV.28)$$

$$I_{sc} \sim \Phi_o^a \qquad\qquad\qquad (IV.29)$$

the parameter a being in the range 0.5–0.8 [779, 788]. This characteristic is found for most of the molecular semiconductors; no satisfying explanation has yet been given, though recombination of the charge carriers within the space-charge region has been assumed [779]. This dependence of the photocurrent upon the incident photon flux severely limits the conversion efficiencies of the solar cells under high illumination conditions. The "uncorrected" energy conversion efficiencies vary from 0.02 to $6 \cdot 10^{-8}\%$ depending on the dopant used and on the nature of the blocking electrode (Table IV.6). Far higher yields are sometimes reported in the literature. These high

Table IV.6 Characteristic parameters and power-conversion efficiencies of polyacetylene-based solar cells.

System	White light Irradiation[a] (mW/cm^2)	V_{oc} (mV)	I_{sc} (μA/cm^2)	FF	Stand.[b] Yield (%)	Ref.
Electrodag/t-(CH)$_x$/M						
Undoped (CH)$_x$						
M = In	24	87.5	—	0.25	0.017	[787]
In	100	88	190	0.25	0.0041	[788]
Al	50	250	67	0.15	0.005	[788]
	70	310	63	0.21	0.006	[783, 754]
Pb	50	103	1.23	0.25	$6 \cdot 10^{-5}$	[788]
Doped (CH)$_x$						
I$_2$ M = Al	50	25	0.34	—	—	[788]
AsF$_5$ = Al	50	6.4	1.26	—	—	[788]
Pb	50	0.4	0.3	0.25	$6 \cdot 10^{-8}$	[788]
Au/t-(CH)$_x$/Al						
Doped (CH)$_x$						
HCl-MeOH						
$(10^{-3}-10^{-4} \, \Omega^{-1} \, cm^{-1})$						
M = Al	40	400	40	0.25	0.01	[784]
	7	320	35	0.26	0.04	[779]

V_{oc}: open-circuit photovoltage; I_{sc}: short-circuit photocurrent; FF: fill factor.
[a]: white light, uncorrected for transmission through the semitransparent metallic electrode; yields are multiplied by 10–30 if this correction is taken into account. [b]: for the definition of the conditions of the "standardized yield" see Sect. III.3 f.

values (0.1–1 %) are obtained by including a correction for the opacity of the semi-transparent metallic electrode or/and by working under very low illumination conditions (1 μW/cm^2 or less). The exposure of the solar cells to air leads to a significant decrease in the magnitude of the short-circuit photocurrent [754, 779]. This is probably due to the formation of highly insulating oxide layers at the interface between the molecular semiconductor and the metal.

V The Main Other Molecular Semiconductors

Two well representative examples of molecular semiconductors — phthalocyanines and polyacetylene — have been previously described in detail. Most of the basic concepts which have been introduced are valid for all organic systems. An overview of the main other molecular materials will be given in this section in the light of the previous conclusions.

The first system, aromatic derivatives, is a typical example of molecular crystals. An accurate prediction of their molecular properties may be given knowing the photochemical and chemical properties of the very first elements of the aromatic series. Graphite can be considered as the ultimate element of the aromatic series.

Charge-transfer complexes, while being constituted of aromatic-type compounds, demonstrate very different material properties. In charge-transfer complexes, cooperative properties become predominant and the characteristics of the material are hardly predicted from a knowledge of the properties of the isolated subunits, but are strongly dependent, among other factors, on the details of the stacking within the crystals.

Hence properties of the main aromatic systems have been explored, the more complex polymeric systems may be studied. Two polymers are distinguished: poly-sulfurnitride and polydiacetylene. They are both prepared through solid-state syntheses. Their structural and chemical homogeneities are orders of magnitude better than most of the other polymeric materials. Reliable charge-transport properties may be deduced from the various experimental determinations. It will be seen that this is not always the case for the polymers prepared in a more conventional manner. In these latter, as polyphenylacetylene, polyphenylene, pyrolyzed polyacrylonitrile, and polypyrrole, for example, the modelization of the transport properties is far more difficult, especially in the semiconducting range.

Finally, a general description of the photovoltaic devices whose photoactive part is constituted by molecular materials is developed.

V.1 Aromatic Hydrocarbons and Graphite

Aromatic hydrocarbons have been extensively studied in the past as candidates for forming organic semiconductors. Good correlations have been observed between their physico-chemical properties and their structure (Fig. V.1). Most of the various

202

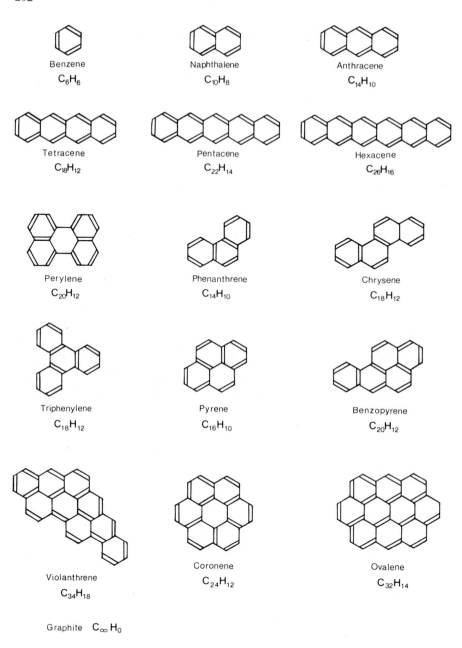

Benzene
C_6H_6

Naphthalene
$C_{10}H_8$

Anthracene
$C_{14}H_{10}$

Tetracene
$C_{18}H_{12}$

Pentacene
$C_{22}H_{14}$

Hexacene
$C_{26}H_{16}$

Perylene
$C_{20}H_{12}$

Phenanthrene
$C_{14}H_{10}$

Chrysene
$C_{18}H_{12}$

Triphenylene
$C_{18}H_{12}$

Pyrene
$C_{16}H_{10}$

Benzopyrene
$C_{20}H_{12}$

Violanthrene
$C_{34}H_{18}$

Coronene
$C_{24}H_{12}$

Ovalene
$C_{32}H_{14}$

Graphite $C_\infty H_0$

Fig. V.1. The main aromatic hydrocarbons, from benzene to graphite

energy parameters (EP) may be empirically evaluated through a correlation function of the following type [792]:

$$EP = a + b \cdot k^{-n} \tag{V.1}$$

a, b, k: constants
n: number of π electrons

The energy parameter EP can be either the gas-phase ionization potential I_g, the crystalline ionization potential I_c, the gas-phase electron affinity A_g, or the band gap E_g. The previous relationship is obeyed only for structurally closely related aromatic derivatives. In the polyacene series for example (Fig. V.2 and Table V.1), all the energy parameters vary regularly with the number of π electrons. The constant k remains unchanged whatever the energy parameter considered. The ionization energy I_g decreases from 9.2 eV for benzene to 6.55 eV for pentacene while the corresponding band gaps go from 6.0 eV to 2.2 eV. At the same time, the corresponding polarization energies are nearly constant.

The relationship between the energy parameters and the number of π electrons is also dependent upon the molecular symmetry of the aromatic compounds (Fig. V.3).

The lowest singlet and triplet excited-state levels follow similar relationships. For highly condensed benzene ring systems, the first singlet 0-0 transition as well as the first charge-transfer state tend towards the conduction band edge [6]. A fairly accurate description of the energy level diagrams may therefore be given for simple aromatic compounds (Fig. V.4). The extrapolation to higher-molecular-weight compounds is valid.

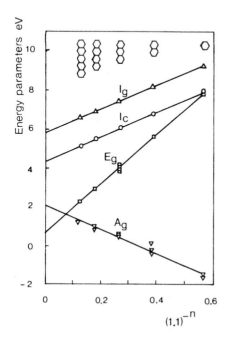

Fig. V.2. Dependence of the energy parameters of polyacenes as a function of the number of π electrons. (After Ref. [792])

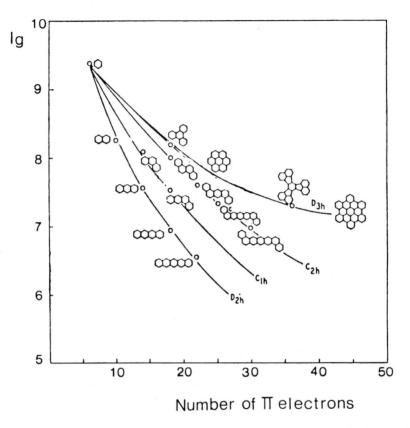

Fig. V.3. Gas-phase ionization energies (in eV) as a function of the number of π electrons for different molecular symmetries. [After Ref. [793]]

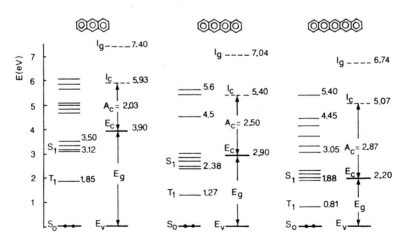

Fig. V.4. Energy-level diagrams for anthracene, tetracene, and pentacene. (After Ref. [6])

Table V.1 Experimental energy parameters in eV for polyacene derivatives. (After Ref. [6] and [792]).

	I_g	I_c	A_g	E_g	E_{pe} [a]
Benzene	9.2		−1.4	6.0	1.34
Naphthalene	8.1	6.8	−0.2	5.0	1.39
Anthracene	7.4	5.75	0.61	3.9	1.42
Tetracene	6.9	5.40	0.8	2.8	1.44
Pentacene	6.55	5.04	1.2	2.2	1.45

[a] Crystal polarization energy.

The Davydov splitting, which is an indication of the interactions arising within the materials, increases from 0.03 eV for anthracene up to 0.09 eV for pentacene [6]. The magnitude of the intermolecular forces is therefore not dramatically dependent upon the size of the π-electron system.

The mechanism of charge-carrier motion in aromatic crystals remains largely unknown despite the enormous amount of works devoted to elucidate it. The domains of validity of the band or hopping models are rarely clearly defined, A transition from band to hopping motion has been experimentally observed in naphthalene single crystals by studying the temperature dependence of electron mobility (Fig. V.5). Three regions may be distinguished. From 150 to 324 K the mobility is nearly temperature independent. This probably corresponds to a hopping motion region [22, 27, 794, 795]. A transition domain is observed between 150 and 100 K. In the third range, below 100 K, the electron mobility rapidly increases when the temperature is decreased. The mobility is then exponentially dependent on the temperature. The corresponding activation energy is 0.0058 eV [6].

Graphite may be considered as the limiting case of the aromatic polycyclic series [796]. Carbon atoms are arranged in layers having a hexagonal structure (Fig. V.6) [797]. Within each layer, the C-C bond distance is uniform and equal to 1.421 Å.

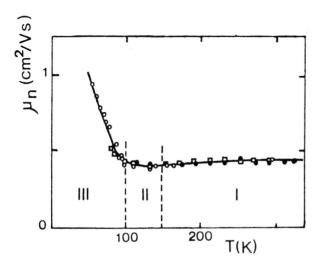

Fig. V.5. The temperature dependence of the electron mobility μ_n in the **c′** direction of naphthalene single crystals. (After Ref. [6])

No bond alternation occurs in this two-dimensional system. The planes are stacked such that half of the carbon atoms are directly over carbon atoms in the adjacent layer in the ABAB structure (Fig. V.6). The ideal interlayer distance is 3.353 Å [798]. This value is however rarely reached for most samples, since stacking disorder leads to significantly higher interplanar distances.

Synthetic graphite may be prepared by decomposing a hydrocarbon gas and hot pressing [799]. The product is called "highly oriented pyrolytic graphite" (HOPG). It is characterized by a good c-axis orientation and a random orientation of the a axis of the different layers (Fig. V.6). This method furnishes single crystals of quality approaching that of natural, but rarely encountered, single crystals. Most of the experimental results further discussed in this section have been obtained with HOPG.

The band structure of graphite presents several peculiarities. A high anisotropy of properties is naturally expected because of the huge difference in bond strength in directions perpendicular or parallel to the layers. Calculation of the electronic structure may be carried out, to a first approximation, in a two-dimensional model. In this case, the π and π^* bands are degenerate for reasons of symmetry at some points of the Brillouin zone through which the Fermi level passes. The graphite is therefore a zerogap semiconductor. The density of states at the Fermi level and the number of free charge carriers are low.

In a three-dimensional model, the interlayer interactions are taken into account. The overlap between adjacent planes is 0.39 eV compared to 3.16 eV in a single layer plane. The conduction anisotropy is therefore important (Table V.2). In pristine graphite the in-plane conductivity is of the order of $2 \cdot 10^4 \, \Omega^{-1} \, cm^{-1}$. The conduction anisotropy is higher than 10^3. The high conductivity observed in the a direction is due to an extremely large mobility of the charge carriers $(1.0–1.2 \cdot 10^4 \, cm^2/V \cdot s)$ [36]. On the contrary, the carrier concentration is relatively low $(10^{19} \, cm^{-3})$. Graphite is therefore a typical semimetal. The density of states at the Fermi level is approximately 0.01 states per eV and per carbon atom.

The loose coupling of the layers in graphite allows the formation of intercalation compounds by insertion of acceptors or donors. Graphite intercalation derivatives therefore consist of two-dimensional carbon layers separated by intercalated atoms or molecules [797]. The in-plane electrical conductivity is significantly increased by the introduction of donors or acceptors within the lattice (Table V.2) due to a con-

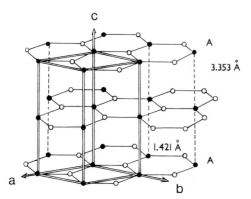

Fig. V.6. The ideal crystalline structure of graphite. The calculated corresponding density is 2.268. (After Ref. [798])

Table V.2 Room temperature in plane (σ_a) and perpendicular to the plane (σ_c) conductivities of graphite (HOPG) and various intercalates. (After Ref. [36]).

	σ_a (Ω^{-1} cm^{-1})	σ_c (Ω^{-1} cm^{-1})	σ_a/σ_c
Pristine HOPG	$2.5 \cdot 10^4$	8.31	$3.0 \cdot 10^3$
Intercalates			
Li	$2.4 \cdot 10^5$	$8 \cdot 10^4$	14
K	$1.1 \cdot 10^5$	$1.94 \cdot 10^3$	56
HNO$_3$	$1.6 \cdot 10^5$	1.8	$1.7 \cdot 10^5$
Br$_2$	$2.2 \cdot 10^5$	1.6	$1.4 \cdot 10^5$
AlCl$_3$	$1.6 \cdot 10^5$	6.1	$2.6 \cdot 10^4$

siderable increase in the charge-carrier density — 0.8 to $1.2 \cdot 10^{21}$ cm^{-3} for doping with acceptors — to be compared with 10^{19} cm^{-3} for pristine graphite. Simultaneously the mobility is decreased by a factor of three to ten (0.9–$3.2 \cdot 10^3$ cm^2/V · s). The conduction anisotropy is either increased or decreased depending on the dopant considered. This reflects:

(i) the modification of the interlayer spacing which is necessary to accommodate the dopant;
(ii) the orbital mixing between the carbon π electrons of graphite and the valence electrons of the intercalant. These two factors have opposite effects on the conduction anisotropy [800].

Aromatic hydrocarbons and graphite allow us to draw several conclusions concerning the transport properties of molecular materials:

(i) the properties of both the ground- and excited-state levels may be accurately predicted knowing the properties of the very few first elements of the series;
(ii) molecular materials may show extremely high charge-carrier mobilities of the order of 10^4 cm^2/V · s;
(iii) the mobility corresponding to the intermolecular transport of charge carriers perpendicular to the plane does not significantly vary from low-molecular-weight aromatic hydrocarbons up to graphite. The Van der Waals interactions are not strong enough to cause sufficient orbital overlap between the subunits whatever the size of the conjugated double-bond system, and even in high-quality crystals, at room temperature, the transport properties proceed through various hopping mechanisms.

V.2 Metallo-Organic Derivatives

It has been seen in the last section that π-π overlaps only lead to small mobilities of the charge carriers — at least when completely full or empty π orbitals interact. Metallo-organic derivatives permit us to consider also d-d or d-n orbital overlaps (Fig. V.7). It is, however, fairly difficult to favor the formation of infinite metallic chains with metallo-organic complexes. The platinum-platinum single-metal bond is among the longest known; the M-M distance is 2.77 Å in the metal. The thickness of aromatic systems is 3.4 Å. The packing of the aromatic cores therefore seems to prevent the formation of strong metal-to-metal bonds. For steric reasons, small non-bulky ligands such as CN^-, CO, $C_2O_4^{2-}$ must be used to complex the metal ion. The use of connecting subunits such as pyrazine, imidazole, or μ-oxo groups, allows to envisage the coupling of the electronic cores of the aromatic systems through d-n overlaps and to avoid the previous steric limitations.

The electrical properties of a fairly large number of metallo-organic complexes have been studied (Fig. V.8.). Doped single crystals are obtained through co-crystallization of the metallo-organic complexes with an acceptor: bromine, iodine, trinitrobenzene, trinitrofluorenone, or tetracyanoquinodimethane. By doping, the single-crystal conductivity is increased by several orders of magnitude up to the range 10^{-4}–10^2 Ω^{-1} cm^{-1}. However, in no case does important metal-to-metal overlap occur. In the case of octamethyltetrabenzoporphyrinato nickel(II) [344], the iodine-doped crystals show a structure where the metal ions are superposed. The Ni-Ni distance (3.778 Å) is even higher than for standard molecular crystals due to the ruffled conformation of the macrocyclic ring. In nickel diphenylglyoxamate derivatives, the metal-to-metal distance is 0.5 Å shorter [806, 807], but it is still too long to permit strong d-d overlaps.

The electrical properties of doped metallo-organic materials are generally very similar to those found for metallophthalocyanines. For tetramethylporphyrinato nickel(II) iodide, (TMPNi)I, the electrical conductivity along the needle axis ranges with crystal from 40 to 270 Ω^{-1} cm^{-1} [802]. This is comparable with the value found for tetrabenzoporphyrinato nickel(II) iodide (150–330 Ω^{-1} cm^{-1}) and is slightly lower than for PcNiI (260–750 Ω^{-1} cm^{-1}) [802]. Octamethyltetrabenzoporphyrinato nickel(II) iodide whose Ni-Ni distance is significantly larger (3.778 Å) [344] demonstrates a conductivity one order of magnitude lower (4–16 Ω^{-1} cm^{-1}).

The temperature dependence of the conductivity of (TMPNi)I is similar to many other molecular systems. The conductivity reaches a maximum at a temperature $T_m \sim 115$ K; it corresponds to a carrier mean free path of l in units of intermolecular spacing [802]. Below T_m the conductivity rapidly drops.

The properties of the molecular materials are dramatically different when d-d orbital overlaps become a predominant factor in the conduction processes. An enormous amount of work has been devoted to the design, the synthesis, and the study of one-dimensional metallic chains [812–815]. Tetracyanoplatinate derivatives are of course the archetypes of such systems.

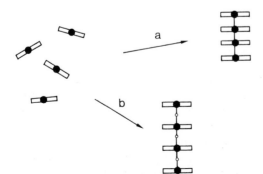

Fig. V.7. The formation of molecular materials through formation of metal-metal bonds (path a) or through the intervention of a connecting subunit (path b)

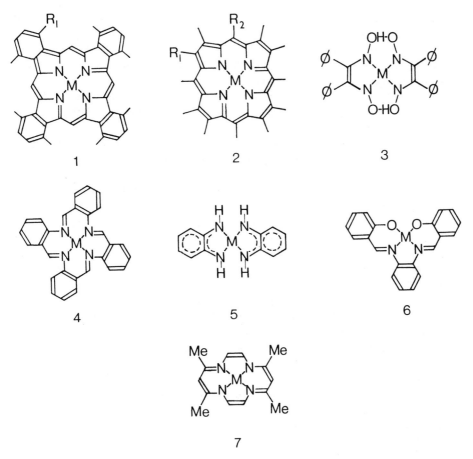

Fig. V.8. The main metallo-organic complexes whose electrical properties have been studied. (1) R_1 = H tetrabenzoporphyrin [378]. R_1 = Me octamethyltetrabenzoporphyrin [344]. (2) R_2 = Et; R_1 = H octaethylporphyrin (OEP) [801]. R_1 = Me; R_1' = H tetramethylporphyrin [802]. $R_2 = C_6H_5$; R_1 = H tetraphenylporphyrin (TPP) [803–805]. (3) bis(diphenylglyoxime) [806, 807]. (4) [808]. (5) bis(orthophenylenediimine) [809]. (6) [810]. (7) tetraaza-tetramethylannulene [811]

Tetracyanoplatinate ions $Pt(CN)_4^{2-}$ are known since Knop's discovery in 1842 of "anomalous" platinum-cyano complexes having "a gold bronze luster". The corresponding structure of these compounds was not understood for more than a century [813]. Platinum complexes form one-dimensional metallic chains through overlaps of the metal d orbitals (Fig. V.9). The higher energy bands are formed from the $e_g(d_{xz}, d_{yz})$, $a_{1g}(d_{z^2})$, $b_{2g}(d_{xy})$, and $b_{1g}(d_{x^2-y^2})$ metal-ion orbitals. The electronic configuration of Pt(II) is d^8. For isolated complexes in solution, the configuration is therefore $(e_g)^4(a_{1g})^2(b_{2g})^2$. When the isolated tetracyanoplatinate ion is approached along the z axis by two other ions, a tetragonal distortion occurs changing the relative energies of the orbitals: the d_{z^2} orbital is strongly destabilized as it points at the incoming ions while the d_{xy} orbital energy decreases (Fig. V.10). In an extended Hückel model, the bottom of the d_{z^2} band is composed of the symmetrical combination of the d_{z^2} orbitals, while the top of the d_{z^2} band is formed via the antisymmetrical combination [816].

The d_{z^2} is normally fully occupied. By oxidation, electrons are removed from the top of the d_{z^2} band. The antibonding nature of this band is therefore lowered; a strengthening of the platinum-platinum bond and a concomitant shortening of the metal-metal bond are therefore expected. This is indeed observed (Table V.3). The metal-to-metal distance in undoped complexes is of the order of 3.50 Å; this value corresponds to a fairly low conductivity ($5 \cdot 10^{-7} \Omega^{-1}$ cm^{-1}). Upon oxidation, the M-M distance decreases up to 2.80 Å, only 0.03 Å longer than the value found in platinum metal [817]. The conductivity simultaneously increases to $2.3 \cdot 10^3 \Omega^{-1}$ cm^{-1}. This last value is, however, hardly reproducible from crystal to crystal. Minute amounts of impurities are sufficient to highly perturb the conduction experiments. For $K_2Pt(CN)_4Br_{0.3}$ (KCPBr) for example, values ranging from 10^{-4} to 830 Ω^{-1} cm^{-1} have been reported.

In tetracyanoplatinate derivatives, an overall conductivity increase of approximately 9 orders of magnitude is therefore possible by doping. This is due both to the augmentation of the density of charge carriers and to higher mobilities. The conduction anisotropy is of the order of 10^5 [813].

The temperature dependences of the conductivity are very similar for all the tetracyanoplatinate complexes studied. As the temperature is lowered, the conductivity

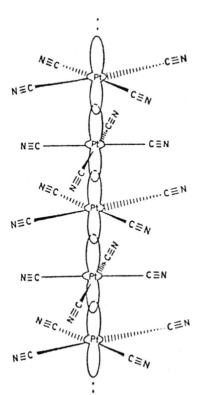

Fig. V.9. Schematic representation of the one-dimensional metallic chains formed from tetracyanoplatinate ions. [Reproduced with permission of (813)]

BANDS   LEVELS

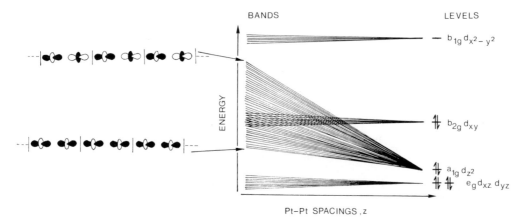

— $b_{1g} d_{x^2 - y^2}$

⇅ $b_{2g} d_{xy}$

⇅ $a_{1g} d_{z^2}$
⇅ ⇅ $e_g d_{xz} d_{yz}$

ENERGY

Pt–Pt SPACINGS , z

Fig. V.10. Electronic energy levels of isolated $Pt(CN)_4^{2-}$ and the corresponding band diagram as the ions approach each other along the **z** axis. A schematic representation of the molecular orbitals of the top and of the bottom of the d_{z^2} band has been shown. (After Ref. [813] and [816])

Table V.3 Platinum-to-platinum distances for various tetracyanoplatinate derivatives and the corresponding conductivities. (After Ref. [813] and [817]).

	Pt-Pt (Å)	σ_a ($\Omega^{-1} cm^{-1}$)
Pt(metal)	2.775	$9.4 \cdot 10^4$
$K_2[Pt(CN)_4](H_2O)_3$	3.50	$5 \cdot 10^{-7}$
$K_2[Pt(CN)_4]Cl_{0.32}(H_2O)_3$	2.87–2.88	$2 \cdot 10^2$
$K_2[Pt(CN)_4]Br_{0.3}(H_2O)_3$	2.88–2.90	4–830
$Rb_2[Pt(CN)_4](FHF)_{0.40}$	2.80	$50–2.3 \cdot 10^3$

increases slightly to reach a maximum (around 270 K for KCPBr) and then decreases more or less abruptly. All the materials become insulators at low temperatures. However, between room temperature and 30 K, the conductivity fall is 10^{12} for KCPBr, whereas for the bifluoride compound the factor is only 10^4 [817].

Although by far the most extensively studied, the tetracyanoplatinate derivatives are not the only compounds which can form one-dimensional metallic chains. For steric reasons, the d-d overlaps is enhanced for square planar complexes. The d^8 configuration favors this geometry [818]. All the metals belonging to the group VIII B, from Fe to Pt, have an accessible d^8 electron configuration (Fe^0, Co^I, Ni^{II} for the first row) [814]. However, only Rh^I, Ir^I, and Pt^{II} complexes have been shown to form well-defined, one-dimensional metallic chains. This reflects the propensity of M-M bond formation decreasing from the third to the first row [814].

Poorly conducting complexes may be obtained with platinum ions within different coordination spheres. Magnus' Green salt $Pt^{II}Cl_4 \cdot Pt^{II}(NH_3)_4$ has been shown to be constituted of one-dimensional platinum chains with a 3.245 Å spacing [813]. The different subunits are staggered by 28°. The platinum can be substituted by

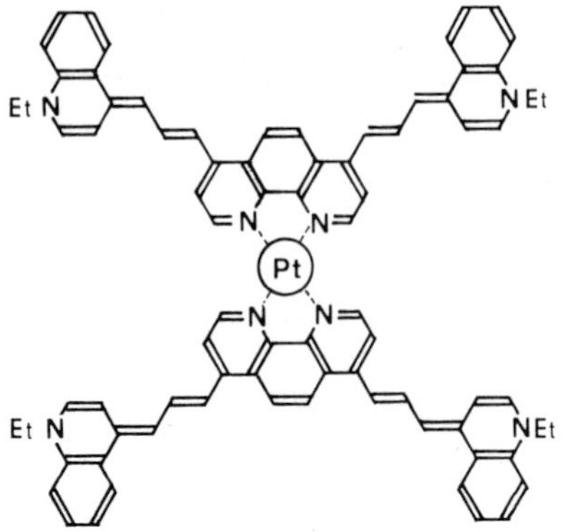

Fig. V.11. Structure of the "excitonic superconductor" proposed by Little. (After Ref. [820])

palladium(II) ions. All these salts have low conductivities (10^{-8}–10^{-9} Ω^{-1} cm^{-1}). Substitution of the amino groups on the cation hinders close approach and prevents the formation of isomorphous materials.

One-dimensional stacks of divalent haloamine complexes of platinum have been reported. Among them Pt en X$_2$ (en: ethylenediamine, X = Cl, Br) complexes have been the best characterized [813]. In the solid state the platinum-platinum separations are 3.39 Å and 3.50 Å for the chloro- and bromo-derivatives, respectively. The crystal structure of bipy Pt^{II}Cl$_2$ (bipy: bipyridyl) with a Pt-Pt spacing of 3.45 Å has also been reported [813, 819].

Metallic chains with the metals in different oxidation states may also be formed. Palladium, platinum, and gold complexes form one-dimensional mixed-valence compounds in the solid state [813]. $M^{II}L_4 \cdot M^{IV}L_4X_2$ derivatives with M = Pt, Pd, L = NH$_3$, en; X = Cl, Br, I have been shown to form poorly conducting complexes, where the one-dimensional chain is formed of M^{II} ... X-M^{IV} repeat units. These materials behave more as isolated M^{II} and M^{IV} compounds rather than M^{III} derivatives as confirmed by the longer M^{II} ... X distance found by X-ray diffraction studies [813].

Finally, it is worth mentioning the structure of the "excitonic superconductor" proposed by Little [820] (Fig. V.11). This complex is expected to form a linear metal chain through overlapping of the platinum d orbitals. The platinum ions are complexed by a highly polarizable substituted phenanthroline in such a way that the conduction electrons travelling in the platinum spine may be coupled by the electronic polarization of the ligands. This should lead to a "room-temperature superconductor". As it has been stated [821]: "the proposed compound is extremely complicated and would be difficult to prepare. Even if this could be done, it is by no means obvious that the monomers would stack in the manner needed". Several attempts have been made to synthesize and crystallize such derivatives [822].

V.3 Charge Transfer Systems

Charge-transfer (CT) systems [8, 797, 815, 823, 824, 828–855] are defined as materials where one or several electrons are exchanged between two constituents: an electron donor D and an electron acceptor A:

$$A + D \rightleftarrows A^- + D^+ \tag{V.2}$$

A huge variety of acceptors and donors have been used (Table V.4). The extent of the charge transfer within the solid material may be estimated from the redox potential in solution of the individual substituents [823, 824]. The most successful couple has been tetrathiofulvalene (TTF) allied with the electron acceptor tetracyanoquinodimethane (TCNQ). Based on that system, several hundred different donor-acceptor complexes have been synthesized and studied. After some general considerations on CT complexes, the two examples, TTF-TCNQ and (tetramethyltetraselenofulvalene)$_2$ X, will be treated in detail.

Table V.4 The main electron donors and electron acceptors used for forming charge transfer complexes.

Donors	Structure	$E_{1/2}^a$	Ref.
N-Ethylcarbazole		+1.25	[823]
		+0.68	[823]
Hexamethylbenzene (HMB)		+1.57	[865]
Tetramethyl-p-phenylene diamine, (TMPD)			
Tetrathiotetracene (TTT)		+0.08	[826]
Tetrathiofulvalene (TTF)	X = S, R = H	+0.33	[825]
Tetraselenofulvalene (TSeF)	X = Se, R = H	+0.48	[825]
Tetramethylthiofulvalene (TMTTF)	X = S, R = Me	+0.27	[825]

Table V.4 (continued)

Donors or Acceptors	Structure	$E_{1/2}{}^a$	Ref.
Alkali	Li$^+$, K$^+$, Cs$^+$, Rb$^+$	−3.0	[823]
Triethylammonium (TEA)	Et−$\overset{+}{N}$H−Et Et	−2.8	[823]
N-Methylpyridinium (NMPy)		−1.28	[823]
N-Methylquinolinium (NMQn)		−0.86	[823]
N-Methylacridinium (NMAd)		−0.41	[823]
Trinitrofluorenone (TNF)		−0.45	[823]
Tetracyanoquinodimethane (TCNQ)		+0.13	[823]
11,11,12,12-Tetracyano-naphtho-2,6-quinodimethane (TNAP)		+0.21	[824]
Tetracyanoethylene (TCNE)		+0.15	[824]
Tetracyanobenzene		−0.71	[826]
p-Chloranil		+0.01	[827]
2,3-Dichloro-5,6-dicyano benzoquinone (DDQ)		+0.51	[827]

^{a}: versus saturated calomel electrode.

Charge-transfer systems present several pecularities. Consider a material in which the intermolecular interactions are weak, before the occurrence of the charge transfer the constituting molecules are neutral and diamagnetic. By transferring one electron from the donor to the acceptor, two radical ions are formed. The density of charge and the density of spin within the material are strictly correlated and they both have the periodicity of the lattice. When electron-electron or electron-phonon interactions become important, new cooperative phenomena occur: charge density waves (CDW) and spin density waves (SDW) may form. The periodicity of CDW, λ_s, is related to the lattice constant a_c and to the amount of charge transfer ϱ, $\varrho = 2a_c/\lambda_s$. CDW form "superstructures" commensurate or not with the underlying lattice depending on the charge transfer. When electron-electron interaction dominates, SDW may form; SDW are commensurate or not with the lattice, but are not accompanied with a lattice distortion. SDW do not contribute to the conductivity.

Charge transfer within the donor-acceptor complex populates previously empty orbitals and partially empties fully occupied levels. Non-bonding or antibonding interactions are therefore transformed into bonding interactions between partially occupied orbitals. The intersubunit interaction energy is strengthened, and inter-molecular distances should decrease. This is indeed observed in most CT complexes.

The knowledge of the charge-transfer properties in solution is a good starting point, though only a first approximation, for the understanding of the solid-state behavior of CT systems [856]. The manifestation of the amount of charge transfer is detected, when acceptor and donor molecules are dissolved in a solvent, by the appearance of a new optical band which corresponds to an intermolecular CT transition:

$$A, D \xrightarrow{\Delta E_{CT}} A^-, D^+ \tag{V.3}$$

The corresponding energy, ΔE_{CT}, is related to the gas-phase ionization potential I_g and to the electron affinity A_g through the relationship:

$$\Delta E_{CT} = I_g - A_g - E_{coul} \tag{V.4}$$

The coulombic term E_{coul} includes various contributions arising from the environment.

a Different Types of Charge-Transfer Systems

When donors and acceptors are co-crystallized from a solvent, radical ion salts (RIS) or charge-transfer complexes (CTC) may be formed depending on the nature of the counterion. RIS involve two charged species, one with one unpaired π electron and the other, the counterion, with a closed shell configuration. Counterions can be inorganic atoms or molecules, organic derivatives, or metallo-organic compounds. In CTC both charged species possess one unpaired electron. In the solid state, various stoichiometries are possible for both CTC and RIS complexes. Pyrene-tetracyano-ethylene, TTF-TCNQ, are typical 1:1 complexes. Tetramethyltetraselenofulvalene-perchlorate crystallizes in a 2:1 ratio. Various other stoichiometries are possible such as Cs_2-$(TCNQ)_3$, dibenzylbipyridilium-$(TCNQ)_4$, or even TTF-$Br_{0.73}$.

A great variety of molecular packings is possible depending upon the structure of the components (Fig. V.12). Alternated packing is often encountered for CTC complexes where both acceptor and donor have similar sizes, every column is constituted alternatively of donors and acceptors. Pyrene-tetracyanoethylene (Structure 1 in Fig. V.12) and bis 8-hydroxyquinolinatopalladium (II)-chloranil (Structure 2) crystallize in an alternated packing. In the former case, the mean interplanar distance is 3.323 Å, close to the Van der Waals value [857].

Segregated packing shows different columns for donors and acceptors. In TTF-TCNQ (Structure 3), the interplanar distances are constant in both columns, 3.17 Å between TCNQ molecules and 3.47 Å between TTF molecules [858]; these values are significantly smaller than the Van der Waals distances: 3.40 Å and 3.70 Å, respectively. The molecular overlap within the columns is shown in Fig. V.13 for the system tetramethylthiofulvalene-TCNQ [864]. The molecules are not superposed to each other, σ-σ overlap seems to be favored in a staggered conformation.

In segregated packing, zigzag structure is possible within the columns (Structure 4). Tetramethyl-para-phenylenediamine-$(TCNQ)_2$ [859] and (morpholinium)$_2$-$(TCNQ)_3$ [860] are examples of this type. In these cases both regular or alternated interplanar distances are possible. Isolated dimers may also be formed as in trimethylbenzimidazolium-TCNQ (Structure 5) [861].

When inorganic closed shell counterions are implied (Structures 6 and 7) either dimerization or regular packing may be observed. In tetramethyl-para-phenylenediamine-ClO_4, interplanar cation distances are strongly alternated (3.10 Å, 3.62 Å) [862]. In (tetramethyltetraselenofulvalene)$_2$-ClO_4, the cations are regularly spaced (3.63 Å) but a zigzag structure is observed (Fig. V.14) [863].

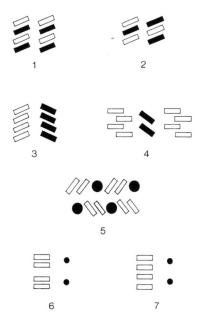

Fig. V.12. Schematic representation of the different packings in charge-transfer complexes and ion-radical salts.
1, 2: Alternated packing; 3, 4: Segregated packing; 5: Isolated dimers; 6, 7: Packing encountered with inorganic counterions

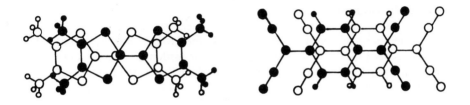

Fig. V.13. View direction normal to the mean molecular planes showing the molecular overlaps in columns of tetramethyltetrathiofulvalene cations and TCNQ anions. (After Ref. [864])

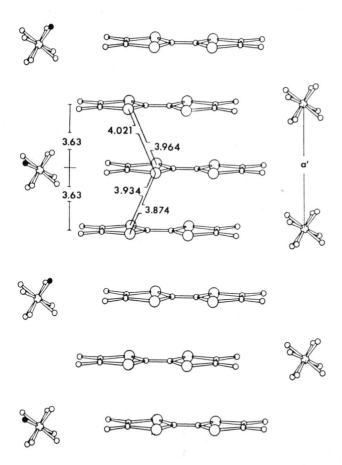

Fig. V.14. Side-view of the stack in (tetramethyltetraselenofulvalene)$_2$-ClO$_4$. Because of disorder, the eight possible positions of oxygen atoms are indicated. [Reproduced with permission of (863)]

b Charge Distribution

In the vast majority of donor-acceptor crystals, the ground state is ionized, while no transfer between the donor and the acceptor occurs within the complex in solution; this is the case for TTF-TCNQ [871]. In the gas phase the energy needed to transfer one electron is simply given by:

$$I_g - A_g \tag{V.5}$$

If ionized states are stable within the crystal, it requires that the Madelung energy E_M — the sum over all the sites of the repulsive and attractive coulombic interactions — must be greater than $I_g - A_g$. The net electrostatic binding energy E_B will depend on the amount of charge transfer ϱ, following the equation [872]:

$$E_B(\varrho) = -\varrho[\varrho|E_M| - (I_g - A_g)] \tag{V.6}$$

Within this model, both charge delocalization within the stacks and electronic correlation are neglected. Various calculations of the binding energy have been carried out on segregated systems (Fig. V.15). For TTF-Br$_\varrho$, the charge-transfer amount is varied by changing the TTF/Br ratio with the assumption of a complete transfer from TTF to Br. When the concentration of Br$^-$ is small, the attractive anion-cation interactions dominate the repulsive terms. Strong Coulomb repulsion between charged TTF$^+$ molecules will keep the charges apart along the stack. Coulombic attraction TTF$^+$/Br$^-$ forces the charge to be localized close to each other. When the concentration of Br$^-$ is increased, the repulsive anion-anion and cation-cation interactions become progressively predominant and the binding energy increases [872]. An optimum donor-acceptor ratio is therefore theoretically predicted: this has been indeed experimentally observed (Fig. V.15). In the case of TTF-TCNQ, the binding energy as a function of the charge transfer is found to be monotonically increasing. Experimentally, it is observed that the amount of charge transfer is of the order of 0.6 in the ground state (Table V.5). Supplementary contributions must

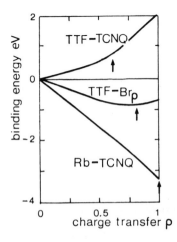

Fig. V.15. Binding energy as a function of composition or degree of oxidation calculated for TTF-TCNQ, TTF-Br$_\varrho$, and TCNQ-Rb. Arrows indicate the values experimentally observed. [Reproduced with permission of (872)]

Table V.5 Experimental charge-transfer amounts for various charge-transfer systems.

System	Amount of Charge Transfer	Conditions	Ref.
TTF-TCNQ	0.59	T > 250 K	[866]
	0.55	T < 180 K	[866]
TSeF[a]-TCNQ	0.63	T = 95 K	[867]
HMTTF[b]-TCNQ	0.72	RT	[868]
TTT[c]-$I_{1.5-1.6}$	0.68	RT	[869]
TTT-$I_{2.7}$	0.80–0.85	RT	[869]
TTF-Chloranil[d]	0.2	T = 300 K	[870]
	0.6	T = 15 K	[870]

[a] Tetraselenofulvalene; [b] hexamethylenetetrathiofulvalene; [c] Tetrathiotetracene; [d] mixed stack phase.

therefore be invoked. It has been calculated that the energy required to transfer one electron from TTF to TCNQ is 4.2 eV. The electrostatic energy of a crystalline array of $TTF^+/TCNQ^-$ is -2.3 eV, the contribution of the polarization of the atoms composing the molecules only add a stabilization energy of -0.09 eV. The energy of delocalization of the electrons along the stack further stabilizes the system by only -0.25 eV [873]. In consequence, polarization and delocalization energies are not sufficiently large to account for the existence of an ionic ground state in TTF-TCNQ [872–877].

In segregated stacks, Coulomb repulsion within the columns can be reduced in two ways:

(i) by incomplete charge transfer from donor to the acceptor as in TTF-TCNQ,

(ii) by deviation from 1:1 stoichiometry (Fig. V.16).

On-site electron-electron repulsion is another important electrostatic contribution within the material:

$$A^-, A^- \rightleftarrows A, A^{2-} \tag{V.7}$$

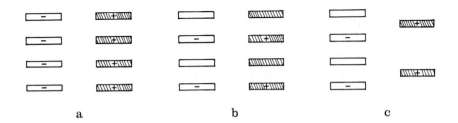

a　　　　　　　　　　b　　　　　　　　　　c

Fig. V.16. The main three types of ionic charge-transfer systems:
a — Simple salt with complete charge transfer as in TCNQ-Rb.
b — Simple salt with partial charge transfer as in TTF-TCNQ.
c — Complex salt with a 2:1 stoichiometry as in (tetramethyltetraselenofulvalene)$_2$-PF$_6$. (After Ref. [872])

The electron-electron repulsion is highly dependent upon the size of the π-electron system involved in the diionized species [878]. The energy of formation of diionized species is particularly important for the charge-transport properties of the materials.

c Charge-Transport Properties

The charge-transport processes occurring in CTC and RIS systems are extremely various and a general rationalization of their electrical properties is fairly difficult. An attempt has been reported for segregated systems based on the conductivity dependences on temperature [879]. The various systems are classified into three categories:

(i) Class I. The conductivity is strongly temperature activated, $\log \sigma$ vs $\dfrac{1}{T}$ is linear, the room temperature conductivity (σ_{RT}) is generally of the order of 10^{-6} to $10^0 \, \Omega^{-1} \, cm^{-1}$.

(ii) Class II. The conductivity shows a broad weak maximum (σ_m) at a temperature T_m, σ_{RT} is of the order of $100 \, \Omega^{-1} \, cm^{-1}$.

(iii) Class III. The conductivity demonstrates a sharp maximum at a temperature T_m, $\sigma_{RT} \sim 500–1000 \, \Omega^{-1} \, cm^{-1}$.

The different temperature behaviors of the conductivity have been parametrized assuming that the usual factorization into charge-carrier densities and mobilities is valid:

$$\sigma(T) = e \cdot n(T) \cdot \mu(T) \tag{V.8}$$

The following general expression has been proposed for the temperature dependence of the conductivity:

$$\sigma(T) = a \cdot T^{-\alpha} \cdot \exp\left(-\frac{E_{act}}{T}\right) \tag{V.9}$$

a: constant

The carrier concentration is proportional to $\exp\left(-\dfrac{E_{act}}{T}\right)$; $T^{-\alpha}$ represents the temperature dependence of the mobility [879]; α is a parameter depending upon the sample quality. A schematic representation of the temperature dependences of these three factors is shown in Fig. V.17.

The behavior of Class-II compounds follows Eq. V.9. Reduced expressions are found for Classes I and III. For Class I, $E_{act} > 2000$ K and the mobility is temperature independent, whereas in Class III the mobility is temperature dependent and the density of charge carriers remains constant. A microscopic model of the temperature dependence of the mobility has been proposed based upon the electron-phonon interactions [879].

In TTF-TCNQ — a typical Class-III system (Fig. V.18) — the room-temperature mobility is approximately 2 cm^2/V $\cdot$ s with a density of conduction electrons equal to the amount of charge transfer (0.6 per couple of molecules) [880]. The maximum of

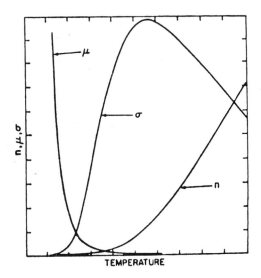

Fig. V.17. Schematic temperature dependences of charge carrier concentration $n \sim \exp\left(-\dfrac{E_{ac}}{T}\right)$ mobility $\mu \sim T^{-\alpha}$ and conductivity $\sigma = n \cdot e \cdot \mu$. [Reproduced with permission of (879)]

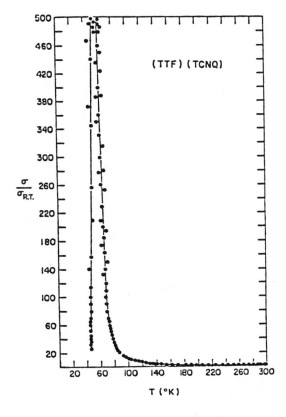

Fig. V.18. Temperature dependence of the normalized conductivity of TTF-TCNQ single crystals. [Reproduced with permission of (881)]

Table V.6 Conductivity data for some highly conducting charge-transfer systems belonging to Class III.

System	$\sigma_{\parallel}{}^a$	$\sigma_{\parallel}/\sigma_{\perp}{}^a$	T_m (K)	$\dfrac{\sigma_m}{\sigma_{RT}}$	T_c (K)	Ref.
	$(\Omega^{-1}\,cm^{-1})$					
TTF-TCNQ	600	500	59	20	38, 49, 54	[882–885]
TTF-Br$_{0.71-0.76}$	300–550				200	[887–888]
TMTTF-TCNQ	350		60	15	34	[883]
HMTTF-TCNQ	500	80–170	75–80	3.5	43, 49	[883, 889–891]
TSeF-TCNQ	800	300	40	12	28	[883, 892]
TMTSeF-TCNQ	1200	100–250	65	7	57	[883, 893]
(TMTSeF)$_2$-PF$_6$	540	50–30,000	18	200		[847, 894, 895]
HMTSeF-TCNQ	2000	30	No	3.5	No	[883]
HMTSeF-TNAP	2400	80–100	47	6.0		[896, 897]
TTT-(TCNQ)$_2$	20–160		90	2–3.5		[898]
TTT-I$_{1.5-1.6}$	600–1200	160	40–80	2–3	100	[869, 899, 900]

TMTTF: tetramethyltetrathiofulvalene; HMTTF: hexamethylenetetrathiofulvalene; TSeF: tetraselenofulvalene; TMTSeF: tetramethyltetraselenofulvalene; HMTSeF: hexamethylene tetraselenofulvalene; TTT: tetrathiotetracene; TNAP: tetracyanonaphthaquinodimethane; $\sigma_{\parallel}$ and $\sigma_{\perp}$: conductivities parallel and perpendicular to the chains, respectively; σ_m: maximum conductivity occurring at T_m; T_c: transition temperatures; ^{a}: data at 300 K.

conductivity occurs at $T_m = 59$ K; σ_m/σ_{RT} strongly varies from sample to sample with typical values around 20 [882–885]. At the maximum conductivity, the mobility increases up to 50–300 cm^2/V · s depending on the quality of the sample [880]. The main examples belonging to Class III are described in Table V.6. The seleno-derivatives of TTF seem to yield significantly higher room-temperature conductivities ($\sim 2000\ \Omega^{-1}\,cm^{-1}$). This is very probably due to the better Se-Se overlaps within the stacks. The substitution of TCNQ by the more delocalized tetracyanonaphthoquinodimethane does not dramatically improve the conduction properties of the materials.

Non-stoichiometric radical ion salts behave very similarly to charge-transfer complexes. In the case of (tetramethyltetraselenofulvalene)$_2$-PF$_6$, the maximum of conductivity occurs at 18 K with exceptionally high corresponding conductivities ($10^5\ \Omega^{-1}\,cm^{-1}$) [847, 894, 895].

The pressure strongly affects the transport properties of the charge-transfer systems. In hexamethylenetetraselenofulvalene-TCNQ at liquid helium temperature the mobility is 1400–3700 cm^2/V · s at 7 kbars [880, 901]. Under these conditions, the system is semimetallic.

Most of the alkali derivatives of TCNQ belong to the Class I. The room-temperature conductivity of Cs$_2$-(TCNQ)$_3$ along the b axis is approximately $10^{-3}\ \Omega^{-1}\,cm^{-1}$; it decreases exponentially on cooling down to $10^{-16}\ \Omega^{-1}\,cm^{-1}$ at 90 K [903]. This has been interpreted as indicative of a 0.6 eV intrinsic band gap. The electron mobility is 0.65 cm^2/V · s at 300 K. This value is very close to the one obtained for typical neutral molecular crystals. In the systems M-TCNQ where M is an alkali metal it has been shown that the physico-chemical properties and crystal structures depend on the size of the cation considered [904]. Smaller ions (Li, Na, K) yield room-temperature conductivities of the order of $10^{-3} \, \Omega^{-1} \, cm^{-1}$ and activation energies equal to 0.3–0.4 eV. The biggest ions (Rb, Cs) lead to $\sigma_{RT} \sim 10^{-2}$ with $E_{act} \sim 0.2$ eV [904]. In Na-TCNQ, the room-temperature single-crystal mobility is 0.08 $cm^2/V \cdot s$ [905].

Most of the ammonium derivatives of TCNQ belong to Class II [906–908]. In single crystals of N-methylphenazinium-TCNQ, acridinium-TCNQ and quinolinium-TCNQ, the room-temperature conductivities are in the 70–400 $\Omega^{-1} \, cm^{-1}$ domain. The corresponding mobilities are in the range 0.2–0.5 $cm^2/V \cdot s$. As for the previous M-TCNQ systems, the conduction is ensured by the TCNQ columns. The substitution of large bulky cations to alkali ions rises the conductivity by several orders of magnitude. The decreasing of the coulombic interaction between the charge carriers and the cations very probably favors the delocalization within the stacks.

Because of their structure, it is expected that both CTC and RIS should show highly anisotropic conducting properties. Conduction anisotropies at room temperature vary from 30 to 500 for most Class III systems (Table V.6). The anisotropy is governed by the type of packing, the relative arrangement of the stacks, and the amount of overlap between adjacent columns. One-dimensionality gives rise to several original phenomena:

   (i) localization of charge carriers by disorder becomes of a crucial importance in 1-D systems;
   (ii) Peierls and Kohn instabilities can take place;
   (iii) charge density waves may form;
   (iv) long-range order may be destroyed by large fluctuation effects.

In real 1-D substances, there is always a finite coupling between adjacent conducting chains via electron-phonon and/or electron-electron interactions. Both 1-D and 3-D behaviors may be observed depending on the experimental conditions and on the physical properties studied.

d Tetrathiofulvalene-tetracyanoquinodimethane (TTF-TCNQ) and Related Complexes

The first publication on TTF-TCNQ appeared in 1973 [909]; since then thousands of papers have been devoted to the characterization and the study of this system. Sample quality has been definitely demonstrated to be a crucial factor in determining the room-temperature conductivity σ_{RT}, the maximum value of σ, and the corresponding temperature T_m [882]. The following values are presently retained: $\sigma_{RT} = 660 \pm 130 \, \Omega^{-1} \, cm^{-1}$, $\sigma_m/\sigma_{RT} > 20$, $T_m = 58$ K. However, even in this case, the role of disorder and impurities is probably not negligible [910].

At room temperature, the charge carriers — the electrons — are localized and the transport may be considered as taking place by a diffusion process [893]. Hall measurements indicate mobilities of the order of 2 $cm^2/V \cdot s$ [880, 911]; at 60 K mobility reaches 50 $cm^2/V \cdot s$ [880]. The mean free path in the direction perpendicular to the chain axis is approximately 10^{-3} lattice constant [912], TTF-TCNQ at room temperature may therefore be regarded as a 1-D system.

Careful examination of the conductivity data reveals two peaks at 53 K and 38 K [922–924]. The former corresponds to a second-order phase transition [922] while the latter is assigned to a first-order phase transition [925, 926].

Below 60 K, TTF-TCNQ may be regarded as a semiconductor. In this temperature range, $\sigma_{\parallel}$ and $\sigma_{\perp}$ decrease strongly with temperature, the apparent activation energy being temperature dependent [927]. At liquid helium temperature $\sigma_{\parallel} \sim 1\,\Omega^{-1}\,cm^{-1}$ [927].

Several phase transitions have been detected by X-ray diffuse scattering [866, 913–917] and inelastic neutron scattering [918–921] measurements. At low temperatures, 1-D distortions within the columns (b-direction) are observed. The period of this "superlattice" is related to the amount of charge transfer. Such phenomena have been attributed to the existence of charge-density waves, i.e. modulations of the electron density along the columns. This modulation may propagate or not depending on the commensurability between their period and the lattice period. At 53 K, corresponding to a peak in conductivity data, and 49 K, CDW on both TTF and TCNQ chains order. At 38 K, CDW are locked with the modulation wave-vector corresponding to a*/4 [928]. A fourth phase transition has been detected at 46 K from small anomalies in heat-capacity measurements and electrical data [925, 929]; no interpretation has been proposed for this transition.

Attempts have been made to estimate the respective contributions of the TTF and TCNQ chains in the charge-transport processes. At room temperature, conductivity is mainly ensured by the TCNQ chains, $\sigma_{TCNQ} \sim 5\sigma_{TTF}$ [930]. The metal-insulator transition at 53 K affects primarily the TCNQ chains [925, 931–937], the density of states at the Fermi level suddenly lowering. The transition at 38 K results from a long-range ordering within the donor stacks [935]. Between 38 K and 53 K, the TTF chains dominate the conductivity processes [924, 932, 937].

Detailed studies of the electrical properties of TTF-TCNQ as a function of pressure have also been carried out. The primary effect of pressure is to enhance the molecular overlap and correlatively the bandwidth (Fig. V.19) [938]. Experimentally, the conductivity along the stacks attains $5 \cdot 10^3\,\Omega^{-1}\,cm^{-1}$ at 30 kbar; in the same time the b parameter decreases from 3.82 Å down to 3.6 Å. The linear compressibility along the b direction is about 0.18 % per kbar [939]. The decrease of the b parameter modifies the mobility of the charge carriers but it also affects the amount of charge transfer and the transverse conductivity. The temperature-pressure phase diagram of TTF-TCNQ has been established from the transport properties (Fig. V.20) [940].

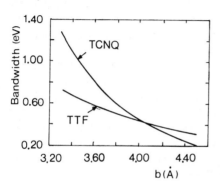

Fig. V.19. Effect of pressure on the electrical properties of TTF-TCNQ: calculated bandwidth versus the **b**-axis parameter. (After Ref. [938])

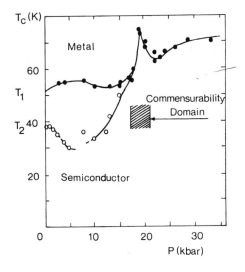

Fig. V.20. Temperature-pressure phase diagram of TTF-TCNQ derived from transport properties. (After Ref. [940])

Above 15 kbar, the two well-defined transitions at ambient pressure $-T_1 = 53$ K, $T_2 = 38$ K — collapse. At 19 kbar ($T_c = 74$ K) an anomaly appears in the phase diagram; the periodicity of the charge-density waves becomes commensurable with the crystalline lattice. This is also reflected on the conductivity: the pinning to the lattice of the commensurate CDW induces a significant conduction decrease. This indicates that beside the usual single-particle mechanisms, the conduction in TTF-TCNQ may be due to collective modes such as CDW [941–944]. Defects within the material have been created by neutron irradiation [945]. Small concentrations of defects ($2 \cdot 10^{-3}$) lead to a pinning of the CDW. The single-particle conductivity is in the same time much less affected [945].

TTF-TCNQ and its selenium analogue TSeF-TCNQ are isostructural. Moreover, they both show very similar room-temperature conductivities ($\sigma_{\parallel} = 800\ \Omega^{-1}\ \text{cm}^{-1}$) [883, 892, 946] and conductivity temperature dependences. The maximum of conductivity appears at 38–42 K for TSeF-TCNQ; a metal-semiconductor transition is observed at 29 K [924, 946, 947]. This latter transition is, as for TTF-TCNQ, strongly pressure sensitive ($T_c = 44$ K at 9 kbar) [948]. The charge transfer between TSeF and TCNQ is 0.63 for temperatures above T_c [867, 949]. The conductivity in the metallic domain is equally balanced between TSeF and TCNQ chains [930]. Contrary to TTF-TCNQ, the susceptibility behavior of metals is not observed for TSeF-TCNQ; in the whole temperature range, susceptibility seems to arise mainly from the TCNQ chains [950, 951]. CDW also participate in the conductivity processes [952]. Both TSeF and TCNQ stacks order at the same temperature T_c. In the semiconductor domain, the energy gap has been evaluated to be of the order of 0.04 eV (250 K) [924, 953, 954]. The replacement of sulfur atoms by selenium atoms should favor molecular overlap between the subunits; increased longitudinal overlap could explain the lower transition temperature T_c [955]. Simultaneously, transversal overlap favors the coupling between TSeF and TCNQ chains [956], the transport remaining however diffusive [847].

Hexamethylenetetraselenofulvalene-tetracyanoquinodimethane, HMTSeF-TCNQ, is the most highly conducting complex of the TTF-TCNQ family. Room-temperature parallel-conductivity is of the order of 1800 Ω^{-1} cm^{-1}; the electrical anisotropy ($\sigma_\parallel/\sigma_\perp \sim 30$) is much less than for previously related systems [948]. HMTSeF-TCNQ and TTF-TCNQ have different stack arrangement (Fig. V.21). The salient feature is a very short selenium-nitrogen distance (3.10 Å), far smaller than the value predicted from the respective Van der Waals distances (3.50 Å). This suggests a significant interchain π-bonding [847]. The temperature dependence of conductivity shows no anomaly down to 6 mK [883]. However, thermopower measurements indicate a metallic behavior only down to 40 K [957]. Moreover, the system is diamagnetic below 100 K [958]. HMTSeF-TCNQ therefore seems to be a semimetal at low temperature [901, 959, 960]. More recent conductivity experiments indicate a smooth transition at $T_c = 24$ K [902]. From 1.4 to 24 K, the average mobilities are temperature independent, $\bar{\mu} \sim 1400$–3700 cm^2/V $\cdot$ s [902]. Conductivity changes are mostly associated with modification of the number of charge carriers, about 1 carrier per 500–1000 molecular subunits [901]. T_c decreases under hydrostatic pressure but never vanishes [847]. The mobility under pressure at low temperatures rises to $4 \cdot 10^4$ cm^2/V $\cdot$ s [901]. CDW corresponding to 1-D incommensurate distortions are observed in the whole temperature range. The amount of charge transfer is 0.74 [867, 917]. Due to the large interchain integrals, the Peierls instability cannot induce a metal-insulator transition [961]. However, the 3-D character is not sufficient to maintain the metallic property of the system in the entire temperature range; even under high pressure, HMTSeF-TCNQ becomes semimetallic at low temperature.

e (Tetramethyltetraselenofulvalene)$_2$-X and Related Radical-Ion Salts

Tetraseleno derivatives of TTF form among the most interesting radical-ion salts. Tetramethyltetraselenofulvalene, TMTSeF, yields 2:1 salts which exhibit a variety of physical properties: metallic behavior, supraconductivity, spin density waves (SDW), etc.

Single crystals of (TMTSeF)$_2$X are obtained by electrochemical oxidation [895, 962]:

$$\text{TMTSeF} \xrightarrow{-e^-} (\text{TMTSeF})^+ \tag{V.10}$$

$$\text{TMTSeF} + (\text{TMTSeF})^+ + X^- \rightarrow (\text{TMTSeF})_2\text{-X} \tag{V.11}$$

X is an inorganic anion (PF$_6$, AsF$_6$, ClO$_4$, ReO$_4$) coming from the salt, insuring a suitable ionic strength within the medium. (The negative charge of the anions will be omitted for simplicity reasons) (TMTSeF)$_2$-X salts form an isomorphous family with a triclinic unit cell (Fig. V.22) [963]. Planar TMTSeF molecules are arranged in zigzag columns (a direction). Selenium-selenium distances both in the stacks and between neighboring columns are significantly lower than the Van der Waals values [964, 965]. Interchain coupling is therefore an important factor in determining the physical behavior of the material. Upon cooling, the largest structural changes involve the interstack Se-Se separation [966].

Charge transfer between the inorganic anions and TMTSeF is complete. The stoichiometry being 2:1, the system possesses a half-filled band. Although stacking along the **a** axis shows a slight dimerization, there is no evidence for a gap at high

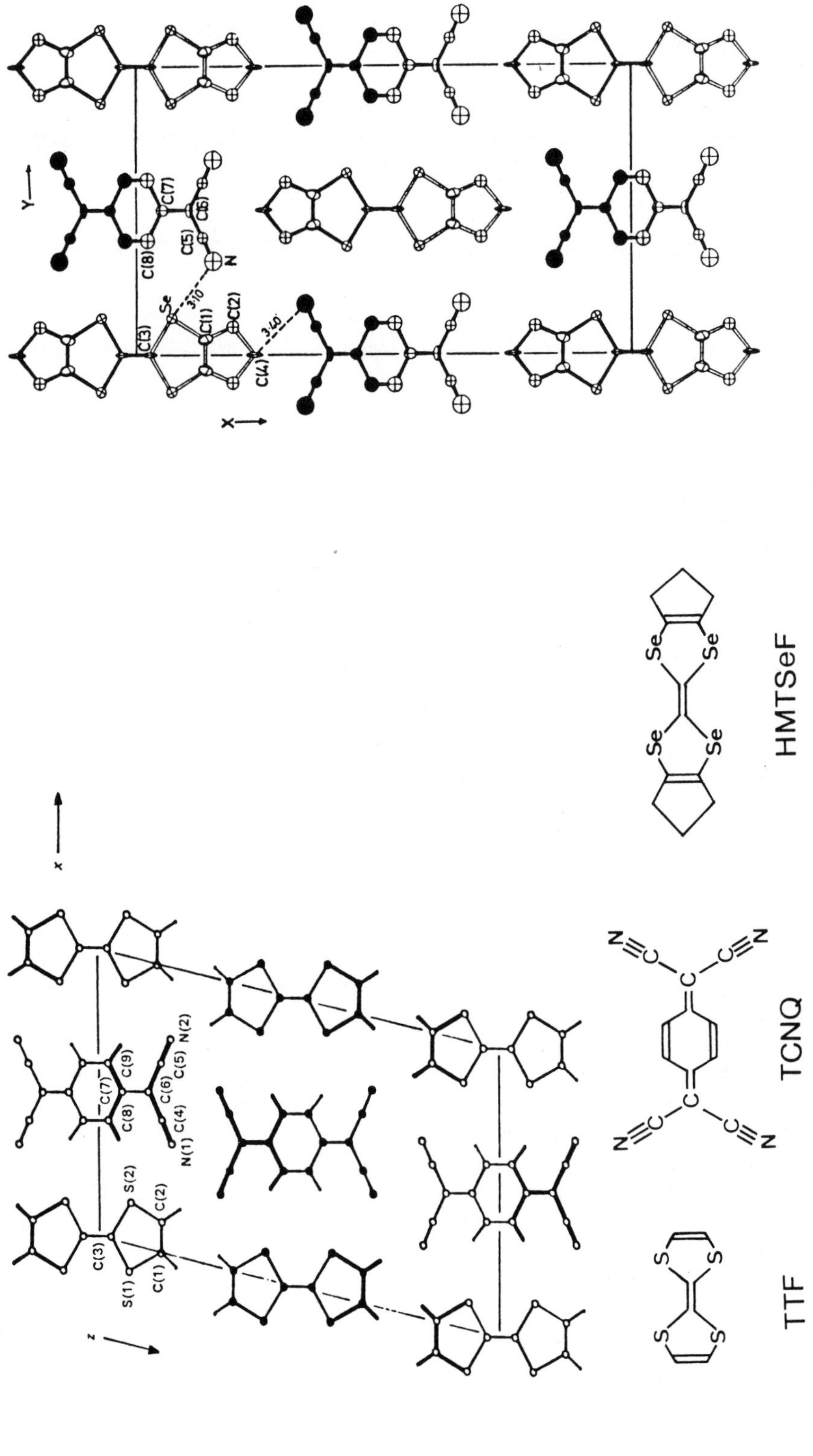

Fig. V.21. Comparison of the crystalline packing of TTF-TCNQ and HMTSeF-TCNQ. (After Refs. [858, 1190])

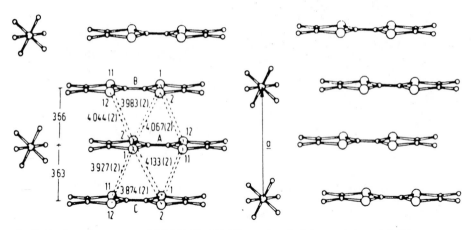

Fig. V.22. Crystal structure of $(TMTSeF)_2$-PF_6. [Reproduced with permission of (963)]

temperature and the materials are metallic. The symmetry of anions plays an important role in determining the physical properties. The anions reside in a site with an inversion center, non-centrosymmetric anions (ClO_4, ReO_4) consequently occupy two equivalent positions; this influences the periodicity of the lattice.

The conductivity of $(TMTSeF)_2$-PF_6 is maximum at 20 K; it also corresponds to the maximum of mobility of the charge carriers, 10^5 cm^2/V $\cdot$ s to be compared with the value at room temperature $\mu_{RT} = 10^3$ cm^2/V $\cdot$ s [895]. At lower temperatures, spin-density waves develop. Simultaneously, electronic spectra show the opening of a gap [967, 968]. This is also evidenced by the magnetic properties of the material; while the susceptibility is definitely not zero, there is no corresponding ESR signal [969]. Spins are strongly coupled into an antiferromagnetic state and the antiferromagnetic resonance is shifted out of the accessible field range [967]. NMR studies confirm the presence of strong local magnetic field below the transition temperature and the antiferromagnetic ordering [970, 971]. In the low temperature domain, the conductivity of $(TMTSeF)_2$-PF_6 remains high (10^2–10^3 Ω^{-1} cm^{-1}) and not thermally activated [972]. The corresponding mobility is extremely high, $\mu = 10^5$–10^6 cm^2/V $\cdot$ s at 4 K [894, 973, 974]. The density of states is correspondingly very small and the system may be considered as a small bandgap semiconductor or a semimetal.

Other radical ion salts based on TMTSeF with centrosymmetric anions behave very similarly to $(TMTSeF)_2$-PF_6 [895, 971, 975–979]. In particular, the electrical properties as a function of temperature are quite comparable. All the systems, while anisotropic, cannot be considered as 1-D materials [979, 980, 981]. In the $(TMTSeF)_2$-X series, the computation of the electronic structure gives transfer integrals of the order of 0.374 eV in the stack direction (**a** direction) and 0.023 eV in a perpendicular direction (**b** direction) [982]. The transfer integral along the third axis is probably of the order of 1 meV [983].

The metal-semiconductor transition is very sensitive to the application of a hydrostatic pressure. For $(TMTSeF)_2$-PF_6, the transition temperature decreases by increasing the pressure up to 8 kbar. Above 10 kbar, the resistivity abruptly drops around 0.9 K (Fig. V.23) [847]. This last transition has been demonstrated to yield a super-

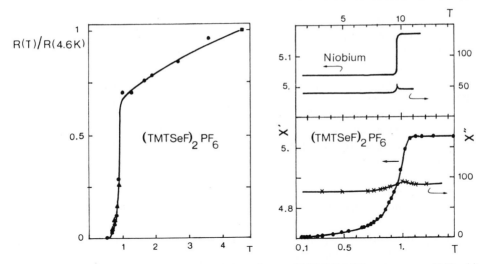

Fig. V.23. The temperature dependences of the resistance (R(T)/R(4.6 K)) and ac-susceptibility (χ)) of (TMTSeF)$_2$-PF$_6$ under a hydrostatic pressure of 12 kbar. A superconducting transition is observed at 0.9 K. A niobium marker is used for calibration. [Reproduced with permission of (847)]

conducting state [847, 984, 985]. Similar observations have been reported with other centrosymmetric anions: AsF$_6$ [978], SbF$_6$ [986], TaF$_6$ [986], or non-centrosymmetric anions: ReO$_4$ [987], ClO$_4$ [988–990]. In particular (TMTSeF)$_2$-ClO$_4$ is a super-conductor under ambient pressure at 1.22 K [964, 986, 991, 992]. The superconducting state is extremely sensitive to irradiation-induced defects [993, 994].

Pressure-temperature phase diagrams have been established in some cases (Fig. V.24). The superconduction domain is in competition with a state where spin-density waves are formed. SDW do not imply a lattice distortion and, under certain conditions, both SDW and supraconducting state may probably coexist with no defined transition

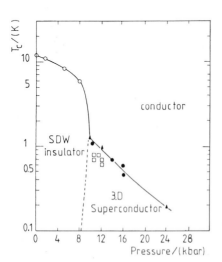

Fig. V.24. Pressure-temperature phase diagram for (TMTSeF)$_2$-PF$_6$. [Reproduced with permission of (995)]

between them [996, 997]. The application of a magnetic field suppresses the conductivity and SDW are observed [979, 998–1000].

In the $(TMTSeF)_2$-X systems, the effect of anions on the metal-insulator transition (T_{MI}) and the critical superconduction temperature (T_{SC}) is of the utmost importance (Table V.7). Non-centrosymmetric anions — tetrahedral (ClO_4, ReO_4 ...), planar (NO_3), or non-symmetrical (SO_3F) — generally lead to transition temperatures T_{MI} largely superior to centrosymmetric ones. In two cases, however, this transition is not observed at all; $(TMTSeF)_2$-NO_3 for example, remains metallic down to 50 mK [995]. Above T_{MI} (200 K), the iodate salt has conductivities significantly lower than its homologues and it demonstrates a temperature-activated mechanism [863].

The metal-insulator transition may be due either to the formation of spin-density waves or to anions ordering [1010, 1011]. In this last case, a distortion of the organic stacks occurs [1012]. At room temperature, the anions are statistically disordered; below T_{MI}, the anions order and the correlated potential induces the opening of a gap in the band diagram (at the Fermi level) generating a nonmagnetic semiconducting ground state [1009]. This state is in competition with the superconducting state.

Table V.7 Characteristics of various (tetramethyltetraselenofulvalene)$_2$-X systems.

Anion	T_{MI}^a (K)	Origin^c	T_{SC}^b (K)	P_{SC}^b (kbar)	Ref.
Octahedral					
PF_6	12–17	SDW^d	0.9	6.5	[895, 985, 1001–1004]
AsF_6	12–16	SDW^d	1.1	12	[895, 978, 1005]
SbF_6	17	—	0.4	11	[895, 986]
TaF_6	—	—	1.4	12	[986]
NbF_6	12	—	No		[863]
Tetrahedral					
ClO_4	No	SDW$^{d, e}$	1.4	1 bar	[964, 986, 991]
ReO_4	180	AO^f	1.3	9.5	[1006–1008]
BF_4	40	AO^f	—		[895]
BrO_4	200	AO^f	—		[1009]
IO_4		semiconductor			[863]
Planar					
NO_3	No		No		[895, 1007]
Non-symmetrical					
SO_3F	86		2–3	6	[966]

^{a}: Metal-insulator transition temperature; ^{b}: critical temperature and pressure of superconduction; ^{c}: origin of the metal-insulator transition; ^{d}: spin density wave; ^{e}: under magnetic field; ^{f}: anion ordering.

V.4 Polysulfurnitride and Polydiacetylene

At first sight, polysulfurnitride, $(SN)_x$, and polydiacetylene have nothing in common. However, they are both synthesized by polymerizing single crystals of the monomer. They therefore possess two qualities very rarely encountered simultaneously: a polymeric covalent backbone which allows good mobilities of the charge carriers, and an original way of elaboration which minimizes the quantity of structural and chemical defects present in the material. Are these systems the "voie royale" towards molecular semiconductors? It will be seen that a few important problems have still to be solved.

Polymerization of single crystal monomers is known for more than half a century. It was observed in 1932 that trioxane crystals, exposed to formaldehyde vapour, are polymerized to polyoxymethylene [1013]. Very importantly, the polymer molecules lie parallel to the **c** axis of the original monomer crystal. There is therefore a strict correlation between the structures of the monomer crystal and the polymeric material. The term "topochemical polymerization" has been proposed to designate such class of reactions where the orientation of the monomer molecules within the crystal is the predominant factor in the polymerization course [1014, 1015] (for reviews see [39, 40, 1016–1021]). The following conditions are required to realize topochemical polymerizations:

  (i) little energy must be released or absorbed during the polymerization process;
 (ii) the changes in density on going from the monomer to the polymer must be as low as possible;
(iii) the crystalline structure of the polymeric crystal must not be too different from the initial starting material;
(iv) the molecules must be mobile enough within the crystal to approach each other to distances less than approximately 3 Å to allow the chemical reaction to proceed. Diffraction studies of molecular motion have shown that displacements of the order of 0.3 Å and rotation of about 3–4° are possible in typical molecular materials [1022, 1023].

These conditions are naturally fairly difficult to fulfill and only a few monomers, such as S_2N_2 and diacetylene derivatives, may give rise to topochemically polymerized materials having potentially interesting electrical properties.

Several reviews have been devoted to the synthesis and the properties of polysulfurnitride, $(SN)_x$ [1024–1026]. This polymer was first obtained at the turn of the century by vaporizing S_4N_4 through a gauze of silver wool [1027]. At that time, it had been already suggested that the material could present some metallic character. A partial elucidation of the polymerization mechanism was achieved only a long time after this first discovery [1028]. In a first stage, the starting material S_4N_4 is converted into S_2N_2:

$$S_4N_4 + 8 Ag \rightarrow 4 Ag_2S + 2 N_2 \tag{V.12}$$

$$S_4N_4 \xrightarrow{Ag_2S} 2S_2N_2 \tag{V.13}$$

S_2N_2 is a colorless diamagnetic crystalline solid, stable only at low temperatures. It is cyclic with a square planar D_{2h} symmetry (Fig. V.25). Its electronic structure is constituted of six π electrons and four unshared electron pairs, superimposed on the π bonding. The polymerization of S_2N_2 occurs even at fairly low temperatures; it takes approximately 8–9 weeks for the polymerization to occur at liquid nitrogen temperature [1029].

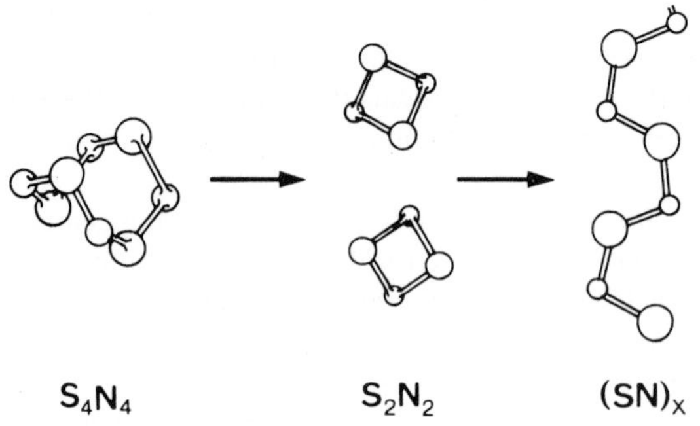

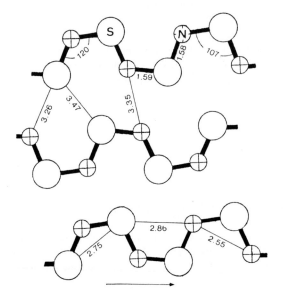

S_4N_4 S_2N_2 $(SN)_x$

Fig. V.25. Schematic illustration of the preparation of $(SN)_x$ from S_4N_4. [Reproduced with permission of (1025)]

In $(SN)_x$, the sulfur and nitrogen atoms form approximately planar infinite chains. The deviation from planarity is only 0.17 Å for both S and N atoms [1024]. All the interchain contacts are at normal Van der Waals distances, except those between sulfur atoms. Several structure determinations indicate significantly different closest S to S distances, from 3.10 Å to 3.48 Å [1030–1032]. These variations obviously affect the calculated band structure of the polymer. In all cases a strong interchain coupling may be expected (Fig. V.26).

The electrical and optical properties of $(SN)_x$ are highly peculiar. Crystals of $(SN)_x$ exhibit a metallic luster. The conductivity along the chains at room temperature is of the order of 1000 Ω^{-1} cm^{-1} for most samples. There is a small conductivity maximum — i.e. 3–5 times the room-temperature conductivity — at 33 K [1024]. However, it has been further shown that this maximum moves to lower temperatures when the quality of the crystal sample is improved. The room-temperature conductivity of high quality $(SN)_x$ is 4000 Ω^{-1} cm^{-1}; the conductivities increase with no maximum

Fig. V.26. Projection of the polysulfurnitride structure onto the 102 plane. The bond distances shown are only indicative. [Reproduced with permission of (1033)]

up to a superconduction transition which occurs at 0.26 K [1024]. Superconduction in $(SN)_x$ is original in two respects. Firstly, it is observed in a material containing no metallic element. Secondly, a polymeric material is shown to be able to become superconducting. The anisotropy of the conductivity is approximately 50 at room temperature; it increases to 10^3-10^4 in the low-temperature region for high-quality samples [822].

In conclusion, $(SN)_x$ may be regarded as an anisotropic three-dimensional semi-metal. The strong interchain interactions resulting from the S-S contacts prevent the occurence of a Peierls distortion. $(SN)_x$ can therefore undergo a transition to a super-conducting state at low temperatures.

Polydiacetylene is the second typical electroactive polymer which can be obtained through topochemical reactions [39–41, 1019, 1033–1036]. The crystallinity and the structural perfection of polydiacetylene single crystals are generally higher than for $(SN)_x$. It has been first demonstrated at the end of the sixties that certain diacetylene derivatives may give rise to highly crystalline polymeric materials [1037]. The polymerization may be initiated by either X-ray or UV irradiations or by thermal treatment (Fig. V.27). The polymerization of the diacetylene subunits can be achieved via a simple rotation around the gravity center of the molecules. The reaction can therefore proceed homogeneously without disturbing the parent crystal lattice. During the polymerization process, the structural units are translationally invariant; in some cases, the side groups of the diacetylene do not have to move at all during the polymerization. More than 100 diacetylene derivatives are known to give rise to topochemical polymerizations.

The polymeric crystals obtained from diacetylene derivatives are of exceptional qualities. Nearly defect-free macroscopic crystals may be elaborated; macroscopic chain lengths may be obtained. Chains consisting of 4,000–10,000 repeat units with equivalent lengths of 2–5 μm have been observed [43, 1038, 1039].

All polydiacetylenes investigated so far may be classified as insulators. Room-temperature conductivities of the order of 10^{-6} to $10^{-12}\ \Omega^{-1}\ cm^{-1}$ have been measured [1040–1042]. The determination of electron mobilities by dc photoconduction [1043], space-charge-limited currents [1044] and electron injection from NaK, Na, or Ca electrodes [46] has been performed. Mobilities of the order of 10^3 to $10^5\ cm^2/V \cdot s$ have been found. The carriers can therefore travel over distances of the order of 1 mm before being trapped [1045]. The low value of the conductivity

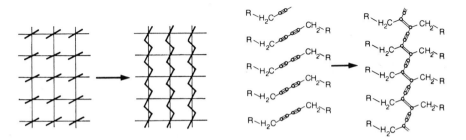

Fig. V.27. Schematic representation of the topochemical polymerization of diacetylenes. A rotational diffusion is sufficient to transform the monomer single crystal into the polymeric material. [Reproduced with permission of (1019)]

therefore reflects an extremely low density of free charge carriers. Moreover, poly-diacetylene single crystals cannot be doped so far. The electrical properties are highly anisotropic and the ratio of the charge-carrier mobilities parallel and perpendicular to the chain axis is of the order of 10^3 [1046, 1047].

A large number of theoretical works has been devoted to the calculation of the band structure of polydiacetylene derivatives [669, 1048–1050]. Simple Hückel calculations [1048] furnish the correct order of magnitude for the electron mobility ($\mu \sim 10^3$ cm^2/V · s). Ab initio calculations further predict a high asymmetry for electron and hole motions [1049]; at 300 K the mobility of holes, $\mu_p \sim 1700$ cm^2/V · s, are calculated to be 65 times higher than the electron mobility $\mu_n \sim 26$ cm^2/V · s.

The correlation of optical and charge-transport properties of polydiacetylenes have been thoroughly studied [1042, 1051]. By comparing the absorption spectrum of poly-2,4-hexadiyn-1,6-diol-bis(p-toluene sulfonate) (PTS) (R = —CH$_2$OTs) with the yield of charge-carriers production, it is clear that the strong optical absorption band near 620 nm cannot be assigned to a valence-to-conduction-band transition (Fig. V.28). The action spectra for charge-carrier production are distinctly different from the absorption spectrum of PTS. The outsets of optical and photoconductivity peaks are separated by approximately 0.5 eV. The peak photocurrent-to-dark-current ratio is approximately 300 at 300 K and superior to 600 at 120 K. A significant number of charge carriers is produced beyond the optical absorption band, peaking at 760 nm. This phenomenon is probably associated with impurities contained in the material [1042]. Similar effects have been noticed in the polyacetylene case. The defect extra-band is more likely due to a defective polymer chain than to a chemical impurity. As a matter of fact, chemical impurities not incorporated within the

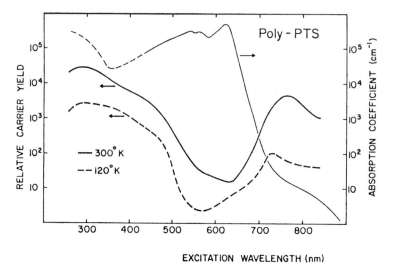

Fig. V.28. Optical absorption spectrum of PTS (R = —CH$_2$OTs) compared with the relative carrier yields at 300 K and 120 K under irradiation. The electric field is ∼ 1300 V/cm, oriented approximately in the chain direction. [Reproduced with permission of (1042)]

conjugated backbone would probably form antitraps rather than traps, their HOMO's and LUMO's remaining outside the forbidden band-gap energy.

However, at higher energies than the maximum of the absorption band, a valence-to-conduction-band transition may be responsible for the photoconduction observed [41, 1051]. This last transition is "buried" in the excitonic band which possesses a higher oscillator strength. The corresponding band gap for the generation of free charge carriers varies from 2.1 eV to 2.6 eV, depending on the nature of the side groups of diacetylene, $R = -CH_2OTs$ or $-(CH_2)_4OCONHC_6H_5$, respectively [41, 1051]. Polydiacetylene is thus one of the rare cases where a fairly complete and accurate band diagram may be given (Fig. V.29). The strongly allowed optical band corresponds to the transition at 2.0 eV. This state may produce free charge carriers with an efficiency Φ_{pe}, Φ_{pe} is constant or varies slowly with the wavelength of the irradiation. At higher energies, the valence-to-conduction-band transition occurs (2.1–2.6 eV).

Polydiacetylene derivatives and polyacetylene have very similar ionization potentials [669]. However, they behave extremely differently toward doping. Polyacetylene may be doped up to the metallic state. In the polydiacetylene case, the intrinsic dark conductivity is of the order of $10^{-17}\,\Omega^{-1}\,cm^{-1}$ at 300 K. No complexes having conductivities higher than $10^{-5}\,\Omega^{-1}\,cm^{-1}$ have ever been found.

Polymerization of diacetylene derivatives do not proceed in solution melt or in the liquid crystalline state [42]. However, diacetylene substituted with long paraffinic chains may be polymerized in mono- or multilayers [42, 1052–1055]. Multilayers of defined thicknesses are built up on quartz substrates using the Langmuir-Blodgett technique, and then polymerized by exposure to UV light. Multilayered membranes and vesicles have also been polymerized [1056–1059].

Solid-state polymerization of diacetylene derivatives has also been initiated by high pressures (5–30 kbar) [1060]. The simplest diacetylene C_4H_2 has, on the other hand, been polymerized from the vapour state by deposition onto an organic polymer substrate, such as polyethylene or teflon [1061]. However, the polymer formed in this last case contains pendant acetylenic groups and thus does not possess the characteristic polydiacetylene conjugated backbone.

Soluble polydiacetylene compounds have been synthesized [1062–1064]. These derivatives, unlike all other polydiacetylenes, are soluble in chloroform, tetrahydrofuran or dimethylformamide. The color of the solutions changes from yellow to blue or red when the nature of the solvent is varied. It has been shown that this corresponds to a conformational change from a fully planar conformation in the red and blue forms to a non-planar weakly conjugated conformation in the yellow ones [1062–1064].

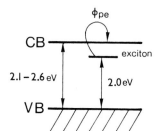

Fig. V.29. Schematic representation of the excitonic and conduction levels in polydiacetylene derivatives. (After Ref. [1051])

Polydiacetylene derivatives therefore furnish molecular materials having extremely high degrees of purity, high crystallinities, while being made of almost infinite polymeric chains. The mobility of the charge carriers is extremely high (10^3 or more), the same order of magnitude as for the in-plane mobility in graphite. However the density of free carriers is so low that polydiacetylene materials are insulators rather than molecular semiconductors.

V.5 Polymethines

Polymethine dyes represent a class of historically important colorants [505, 1065]. Polymethines are composed of electron-acceptor and electron-donor groups, connected via a conjugated chain (Fig. V.30). Three main classes of dyes may be distinguished. Two of them are symmetrical: the cationic cyanines and anionic oxonols. Delocalization of the positive or negative charge may readily occur; the two forms are energetically equivalent. The merocyanines are unsymmetrical, the electro-accepting and donating groups are different. The amount of the ground-state charge transfer may be varied by changing the oxidation or reduction potentials of the substituents. At the origin, the term merocyanine was only employed when the donor and acceptor moieties were part of heterocyclic systems. Linear chromophores have been more recently included under this denomination.

The electronic properties of cyanine dyes have been extensively studied. They present in the visible region two strong absorption bands. The transitions are polarized along the longitudinal molecular axis. The lowest energy band of non-aggregated dyes in solution is fairly narrow, indicating low changes in bond lengths upon excitation (Fig. V.31). The maximum absorption wavelength increases linearly with the number of conjugated double bonds (Table V.8). A shift of roughly 100 nm per double bond (vinylene shift) is noticed. This is in contrast with polyenic systems where bond alternation occurs. The energy gap at infinite chain length should thus be zero for symmetrical polymethines. It seems, however, that this is not the case and that polymethine-cyanines are not stable if the energy gap is lower than about 0.9 eV (1380 nm) [505, 1067]. This corresponds to approximately 11 double bonds for the connecting conjugated backbone.

In the case of unsymmetrical dyes, the vinylene shift is no longer constant with increasing chain length (Table V.8). As for polyenes, the vinylene shift regularly

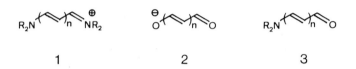

| 1 | 2 | 3 |

Fig. V.30. The three main classes of polymethine dyes
1: cationic polymethine-cyanines; 2: anionic polymethine-oxonols; 3: neutral polymethine-merocyanines

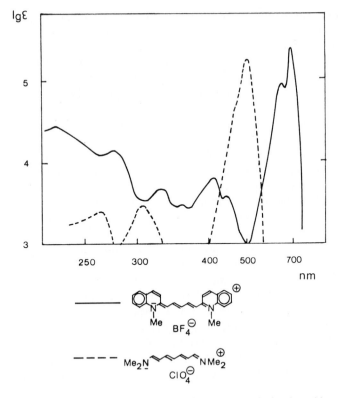

Fig. V.31. Optical absorption curves of two symmetrical polymethine-cyanines. (After Ref. [505])

Table V.8   Maximum absorption wavelengths (in nm) as a function
of the number of double bonds (n) for the three main classes of polymethine dyes
    1: polymethine cyanine
    2: polymethine oxonol
    3: polymethine merocyanine
In parentheses are indicated the logarithm of the absorption coefficient. (After Ref. [505] and [1066]).

n	1[a]	2[b]	3[a]
0	224 (4.16)		
1	312.5 (4.81)	267.5 (4.43)	283 (4.57)
2	416 (5.08)	362.5 (4.75)	361.5 (471)
3	519 (5.32)	455 (4.88)	421.5 (4.75)
4	625 (5.47)	547.5 (4.80)	462.5 (4.81)
5	734.5 (5.55)	(644)	491.5 (4.83)
6	848 (5.34)		512.5 (4.86)

[a] In methylene chloride; [b] in dimethylformamide

decreases when the conjugated chain is lengthened. Intermediate cases between polyenic and cyanine behavior may be found depending on the nature of the electro-attracting or electrodonating moieties.

The excited-state properties of polymethines are dramatically dependent upon the structure of the dyes considered (Table V.9). Unlike the absorption maxima, the maximum of the fluorescence emission is not directly related to the number n of double bonds in the polyenic connecting chain. However, the fluorescence quantum yields seem to increase with n. Additional increase of the fluorescence yield may be obtained by rigidizing the dye structure. Cross-linking between the 1,1' positions of the dye 1 via a $-CH_2-CH_2-$ group produces a 160-fold increase in the fluorescence quantum yield. Cross-linking with a $-CH_2-$ group results in a 1000-fold increase in fluorescence quantum yield Φ_F. It seems therefore that the excited single-state properties are strongly dependent upon the degree of steric constraints within the molecule. It has been proposed [1069] that the twisting of the heterocyclic rings around the polyenic chain enhances the rate of internal conversion to the ground state. The excited-state energy is then dissipated by intramolecular vibrations and rotations. The inclusion of a $-CH_2-$ linkage forms a six membered ring which increases the planarity of the molecule. Both the fluorescence quantum yield and the fluorescence lifetime are then enormously increased. The $-CH_2-CH_2-$ linkage allows some distortion of the polyenic chains and therefore leads to less spectacular increases. The singlet excited-state lifetime can vary from less than 5 ps to more than 6000 ps, depending on the "rigidity" and the molecular structure of the polymethine dye considered.

PPP calculations have been carried out on $Me_2N-(CH)_{2n+1}=O$ to elucidate the distribution of the charges in the ground and excited states [1065]. The ground state

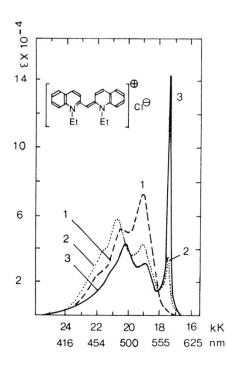

Fig. V.32. Effect of aggregation on the absorption spectrum of pseudoisocyanine. The aggregation process is favored at the higher concentrations. 1: 10^{-5} M; 2: $5 \cdot 10^{-3}$ M; 3: $3 \cdot 10^{-2}$ M; t = 20 °C in water. (After Ref. [1072])

Table V.9   The main optical properties of polymethine dyes. (From Ref. [1068]).

	λ_{max}^{abs} (nm)	ε_{max} ($\times 10^4$ cm^2 M^{-1})	λ_{max}^{fluo} (nm)	τ_g (ps)	Φ_F
1	525 (491)	5.4	635	<5	0.001
2	575	0.9	—	<5	—
3	523	2.0	576 (622)	6100	1.00
4	516	9.0	534 (563)	800	0.16
5	607 (563)	12.4	625	13	—
6	562	4.3	578	450	—
7	559	12.6	574	243	0.05
8				165	0.05
9				22	—
10				1250	0.49
11				54	—

λ_{max}^{abs} : maximum absorption wavelength with the corresponding absorption coefficient (ε_{max})
λ_{max}^{fluo} : maximum fluorescence wavelength
τ_g : fluorescence lifetime
Φ_F : fluorescence quantum yield
Determinations in methanol.

	$R_1=R_2$	A	X
1	C$_2$H$_5$	H	I
2	CH$_3$	CH$_3$	ClO$_4$
3	–CH$_2$–	H	I
4	–C$_2$H$_4$–	H	I

	$R_1=R_2$	A	X	Z
5	C$_2$H$_5$	H	Cl	–CH=CH–
6	C$_{18}$H$_{37}$	H	I	–O–
7	C$_2$H$_5$	H	Cl	–S–

8   R=C$_2$H$_5$

9   R=C$_2$H$_5$

10   R=C$_2$H$_5$

11   R=C$_2$H$_5$

of the merocyanine is relatively polar, due to a drift of electrons from the donor nitrogen atom to the acceptor carbonyl group. The amount of charge transfer in the ground state is of course related to the respective oxidation and reduction potentials of the substituents. In the singlet excited state, the charge separation is even more important. The negative charge on the oxygen atom increases at the expense of the amino group. In the same time, the bond orders show a much greater uniformity in the excited state than in the ground state. This is generally true for most of the polymethines.

It is well known that the absorption and emission characteristics of dyes are highly dependent upon their concentration in solution. The formation of dimers or higher-molecular-weight aggregates is responsible of this phenomenon [1070, 1071]. The effects of the formation of aggregates is particularly spectacular for some dyes such as 1,1'diethyl-2,2'cyanine(pseudoisocyanine) (Fig. V.32) [1072–1075]. The aggregated form is characterized by a narrower absorption band — called the J band — and by an almost coinciding fluorescence peak [1072, 1073]. This reflects a very important interaction between the individual transition moment of the molecules within the aggregates.

The photoconductive properties of polymethine cyanines have been determined [1076]. Monocrystals of polymethine dyes can hardly be grown, and photoconduction determinations have been carried out principally on thin films. The results found are extremely sensitive to the details of preparation of the films, and definitive conclusions can be hardly drawn. It seems, however, that cyanine dyes constitute a distinct class of photoconductive dyes, differing from the others by the lack of a high density of energetic traps for the charge carriers [1076]. A few studies on single crystals have however been reported (see [1191] page 646).

Cyanine dyes substituted with long paraffinic tails have also been synthesized [1077–1079]. These dyes may be organized into mono- or multilayers deposited on glass substrates. These assemblies of "specifically designed architecture" have been shown to demonstrate cooperative properties in energy-transfer or charge-migration phenomena [1077–1079].

Polymethines therefore represent an original class of molecular materials with regard to two aspects. First, for short enough polymethine chains, no bond alternation occurs for the symmetrical dyes. This is of the utmost importance as to the applicability of the soliton model to such systems. Second, a high degree of interaction is possible between the molecular subunits in the solid state and even within aggregated species in solution. In consequence, both the intramolecular and the intermolecular characteristics of these dyes significantly differ from all the other chromophores. Unfortunately, accurate determinations of charge-carrier mobilities and other related electrical and photoelectrical properties of these materials have not been possible because of the difficulty of growing single crystals.

V.6 Polymeric Conjugated Systems

Beside polydiacetylene and polymethine, a huge number of other polymeric conjugated systems have been elaborated. Polyphenylene, polyphenylacetylene, polypyrrole, and pyrolyzed polyacrylonitrile are among the best characterized molecular materials. They will therefore be discussed in some detail.

a Polyphenylene and Related Materials

Poly(p-phenylene) is among the very few polyenic systems where the degree of π delocalization along the polymeric chain may be extensively varied by small conformational changes (Fig. V.33). In model compounds, the angle between the phenyl rings is about 22°. The width of the highest occupied π band is strongly affected by this rotation angle. For a 22° angle between the phenyl rings, the calculated bandwith is 3.5 eV; it increases to 3.9 eV for coplanar phenyls and decreases to 0.2 eV for perpendicular phenyls [1080].

Poly(p-phenylene) may be formed by treating benzene with catalysts such as $AlCl_3$-$CuCl_2$, containing an oxidant beside the classical Friedel-Craft catalyst [552, 573, 1081]. Electrochemical methods may also be employed [1082]; in a first step, benzene is oxidized to the corresponding cation radical which, in turn, yields a radical by loosing a proton, and dimerization then gives a biphenyl molecule. The product of reaction — biphenyl — is more easily oxidized than the starting benzene molecule (1.85 V against 2.4 V). In consequence, further reaction may readily take place. Polymerization occurs up to the precipitation of high-molecular-weight compounds from the reaction medium. The gas-phase ionization potential (I_g) of the first oligomers of benzene have been determined by mass spectrometry measurements (Table V.10) [1083]. From the dimer to the hexamer, I_g varies from 8.95 eV to 7.67 eV. The polymerization process is therefore strongly driven to high-molecular-weight compounds. However, a huge variety of side reactions may interfere and highly branched, cross-linked, infusible and insoluble materials are usually obtained. To avoid these difficulties, a solid-state synthesis of polyphenylene has been reported [1084]. Powders, films or single crystal plates of polyphenylene oligomers (biphenyl, p-terphenyl, p-quaterphenyl, p-quinquephenyl, p-sexiphenyl) are treated with a strong Lewis acid such as AsF_5 (400 torrs) at room temperature for periods up to 24 hours. Doped polymeric materials are then obtained with average chainlength of the order of 9 monomer units.

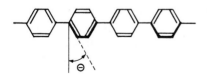

Fig. V.33. Schematic representation of poly-(p-phenylene) in a twisted conformation. Θ is the angle between the phenyl rings

Table V.10 Gas-phase ionization potentials obtained from mass spectrometry measurements for the first elements of the poly(p-phenylene) series. Column 1 indicates the number of phenyl rings. (After Ref. [1083]).

	I_g (eV)
2	8.95
3	8.78
4	8.08
5	8.18
6	7.67

The intrinsic properties of poly(p-phenylene) can be hardly determined because of the low purity of the materials available. The conductivity of undoped polyphenylene is inferior to $10^{-11} \, \Omega^{-1} \, cm^{-1}$ [573]. ESR studies show spin concentrations varying from 10^{18} to 10^{21} spin per gram — 1 electron for 8 phenylene units for "undoped" materials [573]. NMR studies show that the paramagnetic species are localized [1085]. The concentration of spins largely depends on the work-up procedure. The paramagnetic species are therefore associated with uncontrolled chemical impurities.

Heavy doping may transform poly(p-phenylene) into a poorly conducting metal. Conductivities of the order of $10^{2} \, \Omega^{-1} \, cm^{-1}$ are obtained by treating the polymer with AsF_5 (450 torrs) at room temperature [1086, 1087]. A hole mobility of 10 cm²/ V · s has been reported for the AsF_5 doped polymer [1084]. Reductive dopants yield significantly lower apparent mobilities [1084].

Structurally related polymers with an heteroatom connecting the phenyl rings have been elaborated. Poly(p-phenylene sulfide) (PPS) is the most widely studied derivative [1088–1093]. PPS is a melt- and solution-processible polymer prior to doping. PPS is soluble in solvents such as diphenyl ether at temperatures above 200 °C; PPS melts at approximately 285 °C when it is prepared from p-dichlorobenzene and sodium sulfide [1093]. Films from solution or melt can therefore be easily obtained. Doping is achieved as previously by exposure to AsF_5; after treatment, a blue-black material with a conductivity in the range 1–10 $\Omega^{-1} \, cm^{-1}$ is obtained. During doping there is evidence that polymeric chains undergo cross-linking [1093].

b Substituted Polyacetylenes, Phenylacetylene

Substituted acetylenes are potential precursors of a large family of conjugated double bond polymers. Among them, phenylacetylene has been by far the most studied; however, a few other substituents have been considered: R = CH_3 [606, 1094, 1095], R = F [1096], R = CN [1097, 1098], R = CH_2Cl [1098], R = CF_3 [1098], R = $CH_2-C_6H_4-NH_2$ [1099], R = Naphthyl- [1100]. The cyano group for example, has a very high dipole moment (3.9 debye) [1101] which should induce important perturbations of the conjugated chain. The substituent effects are, however, very often obscured by other phenomena, due to the noncontrol of the chemical and structural homogeneities of the material. Polyphenylacetylene deserves more attention, however.

Polyphenylacetylene (PPA) is synthesized from the corresponding substituted acetylene; the catalysts used are the same as those reviewed for polyacetylene (see Table IV.1) [1098, 1100, 1102–1107]. PPA may be obtained under three forms, differing by the geometry around the double bond and the crystalline state: cis and trans amorphous materials, and cis crystalline form [1103]. Amorphous cis-PPA is an orange powder soluble in nonpolar solvents. It is prepared using $MoCl_5$ as a catalyst [1102]. Amorphous trans-PPA is a reddish-brown powder obtained with WCl_6 [1102]. Crystalline cis-PPA is the only insoluble fraction; it is an orange compound obtained by using ferric acetylacetonate or rhodium trichloride catalysts [1104]. An irreversible cis-trans isomerization takes place near 120 °C [1103, 1108, 1109]; simultaneously, the solubility of the sample is highly increased. Irreversible degradation of the polymer becomes significant above 150 °C [543, 1110], i.e. intramolecular cyclizations [1109] or chain scissions [1110] occur.

The structure of the crystalline cis form has been studied by X-ray diffraction [1111] and solid-state ^{13}C NMR [1112]. Both methods agree to indicate a cis-cisoid helix structure. The solubility is strictly related to the amount of disorder within the material [1113].

The conductivity of pellets of PPA is very low (10^{-14}–10^{-18} Ω^{-1} cm^{-1}). Susceptibility measurements indicate a concentration of spins of the order of 10^{16}–10^{17} per gram [1102, 1114]. Reflectance spectra indicate a fairly large band gap (2.3 eV) for the undoped polymer [1115, 1116]. All the determinations on undoped PPA must, as for the other systems, be considered with care due to the poor control of the material.

PPA may be doped with usual electron acceptors such as iodine, AsF$_5$, dicyano-dichloroquinone (DDQ), or tetracyanoquinodimethane (TCNQ). Doping is generally achieved in solution. PPA demonstrates conductivities of the order of 10^{-4}–10^{-5} Ω^{-1} cm^{-1} after doping with AsF$_5$ or iodine [1117]. DDQ or TCNQ induce much lower conductivity increases ($\sigma \sim 10^{-8}$ Ω^{-1} cm^{-1}). In these last two cases, an n-type conduction is found from the measurement of the Seebeck coefficients [1114]: the charge carriers are therefore localized on the ionized dopants rather than on the polymeric chain. PPA may also be doped with AsF$_5$ in the vapor phase inducing considerably higher conductivities ($2 \cdot 10^{-2}$ Ω^{-1} cm^{-1}) [1115]. Nevertheless, a metallic conductivity is never obtained. The concentration of paramagnetic species remains much lower than the dopant concentration; the acceptor dopants therefore generate spinless PPA cations [1117]. For iodine-doped PPA, it has been convincingly

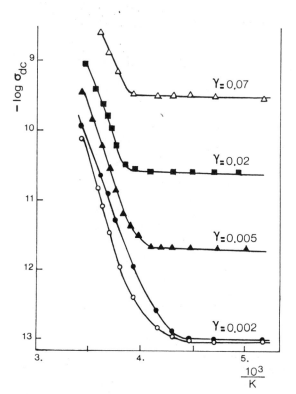

Fig. V.34. Temperature dependence of the conductivity of AsF$_5$-doped polyphenylacetylene pellets (cis-cisoid form). (After Ref. [1113])

demonstrated that the conductivity is of an almost purely ionic nature [1115]. The various charge-transport mechanisms must therefore be preferably determined on AsF_5-doped samples.

In the range 110–300 K, the conductivity of AsF_5-doped PPA shows two different regimes with a transition temperature $T_c = 257$ K (Fig. V.34). At low temperatures, the conductivity is temperature independent. For $T > T_c$ the dc conductivity becomes thermally activated [1113], and the corresponding activation energy decreases when the concentration of dopant increases. It is difficult to assign a precise transport mechanism to these temperature dependences of the conductivity. The low-temperature region has been thought to be dominated by a tunneling or hopping mechanism through low-energy barriers. The high temperature range could correspond to a hopping process between localized states in the band tails [1113]. This last assumption seems to be supported by the frequency dependence of the conductivity [1118].

The photoconduction properties of AsF_5-doped PPA have been determined [1117]. The photoconductivity peaks at about 500 nm; this wavelength approximately corresponds to the band-gap energy determined from the reflectance spectra on undoped polymers.

c Pyrolyzed Polyacrylonitrile

Pyrolysis of saturated polymers may also produce conjugated backbones. The most demonstrative example is furnished by polyacrylonitrile (PAN) (Fig. V.35). High-quality samples of PAN may be obtained electrochemically; this method of polymerization furnishes a partially oriented material, the strong dipolar cyano group being oriented in the electric field used for the polymerization [1101, 1119].

Polyacrylonitrile has, as expected, a very low conductivity ($\sigma < 10^{-9}\,\Omega^{-1}\,cm^{-1}$) [1120]. By heating at 200 °C, the conductivity sharply increases to $10^{-3}\,\Omega^{-1}\,cm^{-1}$,

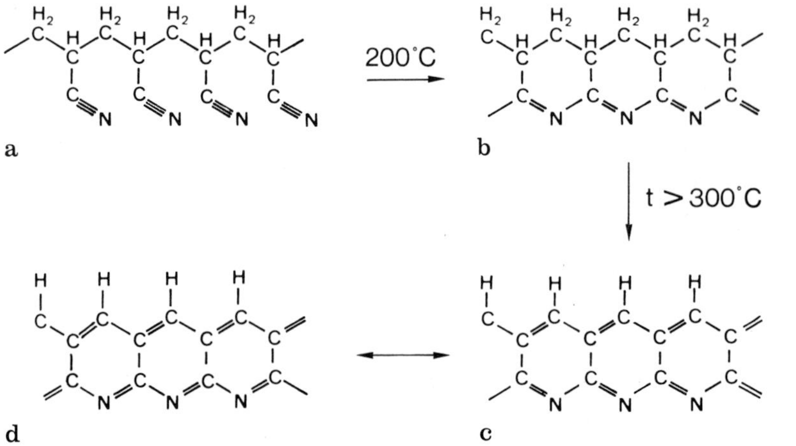

Fig. V.35. The different polymers obtained by pyrolysis of polyacrylonitrile.
a: before pyrolysis; b: singly conjugated polymer; c: doubly conjugated polymer (cis conformation); d: doubly conjugated polymer (trans conformation). (After Ref. 1121]

while the conduction activation energy decreases down to 1.5 eV [15, 937, 1121–1123]. It could be shown by infrared spectroscopy that at this temperature the pendant cyano groups react with one another to form a singly conjugated polymer via cyclization (Fig. V.35) [1119, 1121, 1124–1126]. Under further heating (T > 300 °C) the conductivity still increases up to $5 \, \Omega^{-1} \, cm^{-1}$. This corresponds to the formation of a doubly conjugated polymer by dehydrogenation of the saturated polymeric backbone. Graphitization seems to appear only at about 600 °C [15].

Pyrolysis is, however, not a very selective way to prepare high-quality conjugated polymers. Various side reactions such as reticulation processes occur during the cyclization step [1119, 1127]. Consequently, the charge-transport mechanisms may be hardly explored in detail. For the polymer formed around 200 °C, the charges must travel through $(-C=N-)_x$ chains [1128]; the presence of nitrogen atoms within the conjugated chain induces a mixing of the σ and π bands. Electron energy-loss spectroscopy has shown that the charge carriers are strongly delocalized [1122]; as the temperature of pyrolysis is increased, the spectra become more and more similar to those of graphite.

Theoretical calculations have been carried out on pyrolyzed PAN [1129, 1130], the relative energy of the σ and π bands is controversed. An estimate of the width of the highest occupied π band is 4.3 eV [1130].

A few other pyropolymers such as pyrolytic polyimide [1131], polyvinylmethylketone [1132], polydivinylbenzene [394]; and polyvinylidene chloride [394] have been described in the literature. Very comparable results have been obtained.

d Polypyrroles

A convenient way of synthesizing polypyrrole (PP) films consists in the oxidative electrolysis of pyrrole [1133]. During the electrolysis, the polymer is simultaneously doped, the counterion coming from the electrolyte dissolved in the solution. When tetrabutylammonium perchlorate, for example, is used as supporting electrolyte, a doped polymer with 3 pyrrole units per perchlorate anion is obtained. The "undoped" polymer may be prepared by neutralizing the dopant by reduction, in absence of air or water [1134].

No detailed structure of polypyrrole has been reported so far. Diffraction studies indicate that PP is a highly disordered material [1135]. A reasonable structure may be postulated from comparison with model compounds (Fig. V.36). ^{13}C NMR data are consistent with pyrrole units linked through the α-carbon atoms; a planar structure is generally postulated, but no definitive proof has been yet given. The calculated ionization potential of PP (3.9 eV) is approximately 0.8 eV lower than for cis- or trans-polyacetylene [1130]. The experimental value is in accordance with the calculated I_g [1137]. The band gap of PP is fairly large (calculated: 3.6 eV [1130], experimental: 3.2 eV [1134]). Consequently, the electron affinity is small and n-type doping should be difficult, whereas p-type doping should readily occur even with weak electron acceptors. The calculated width of the highest occupied π band is 3.8 eV, much smaller than the corresponding one in polyacetylene and of the same order of magnitude than that of poly(p-phenylene). Experimentally, it is indeed found that the highest conductivity obtained for doped PP is only $600 \, \Omega^{-1} \, cm^{-1}$ (Table V.11).

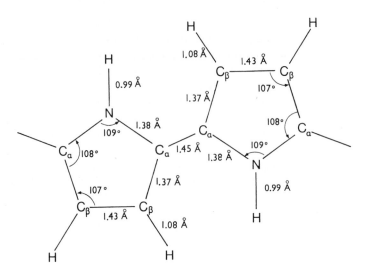

Fig. V.36. Reasonable structure which can be postulated for polypyrrole from comparison with model compounds. Such geometry has been used in theoretical calculations (1136). (After Ref. [1135])

Table V.11   Room temperature conductivities of doped polypyrrole films.

System	σ_{RT} $(\Omega^{-1}\,cm^{-1})$	Ref.
PP	$< 10^{-5}$	[948]
O_2	10^{-2}	[1138]
I_2	600	[1134]
BF_4^- or PF_6^-	30–100	[1139]
ClO_4^-	60–200	[1139]
HSO_4^-	0.3	[1139]
$CF_3CO_2^-$	12	[1139]

The charge-transport mechanism may hardly be determined with certainty from the temperature dependence of the conductivity. The relationship found,

$$\sigma \sim \exp\left(-\frac{T_o}{T}\right)^{1/4}$$
, agrees with a three dimensional variable range-hopping

mechanism [1139]. On the other hand, the paramagnetic species determined by ESR seem to be highly mobile.

The optical spectrum of PP doped with BF_4^- presents two peaks near 1 eV and 3 eV (Fig. V.37). The peak near 1 eV is bound to the doping of PP and disappears by applying a reductive potential to the film. It thus probably corresponds to some charge-transfer excitation [1121]. The 3 eV peak can be due to an interband transition; however, no definite evidence has been yet furnished.

As for most other molecular materials, the conductivity of polypyrrole is O_2 sensitive. Doped films are less sensitive to the presence of oxygen; but even in that case, the corresponding magnetic properties are modified.

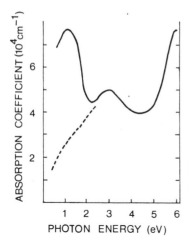

Fig. V.37. Optical spectrum of BF$_4$-doped polypyrrole. The 1 eV peak can be removed electrochemically. (After Ref. [1140])

A few substituted derivatives of polypyrrole have been described. All of them show lower conductivities than PP in the doped state. The N-methyl derivative, for example, has a conductivity of $10^{-3}\ \Omega^{-1}\ cm^{-1}$ for optimum doping [1139]. This is very probably related to steric hindrances which induce a twist of the polymeric chain [1130].

V.7 Molecular Solar Cells

a Squaric Acid and Merocyanine

Interest in molecular solar cells — solar cells which use a molecular semiconductor — was renewed at the end of the seventies because of the work on merocyanine-type derivatives [92–93, 1141]. Such compounds had been previously used to make solar cells [1142, 1143], but they demonstrated yields as low as 0.02% for monochromatic light [1143]. New devices, as M_1/merocyanine/M_2, show energy-conversion efficiencies of 0.7% ($V_{oc} = 1.2$ V, $I_{sc} = 1.8$ mA/cm^2) under white light illumination (78 mW/cm^2) with no correction for opacity of the metallic electrodes [93]. The reason of such spectacular increase is not yet clearly established but a few clues can be found in the light of the previous chapters.

In merocyanine-based devices as in other systems, the magnitude of the photo-current is extremely sensitive to the presence of added gases (ambient air, iodine vapour, etc.); the photovoltaic response is multiplied by 10 in presence of I_2 [1141]. The photocurrent action spectrum closely matches the absorption curve (Fig. V.38) [94]. On the contrary, the quantum efficiency of production of charge carriers per absorbed photon does not follow the absorption of the merocyanine film [92–94]. The quantum efficiency monotonically increases when the irradiation wavelength decreases. If this is verified, it demonstrates that

(i) the relaxation of the excited state up to the vibrationally equilibrated S_1 does not occur.

(ii) the e^-/h^+ pair, once formed, is strongly bound and needs some "extra-energy" to yield free charge carriers.

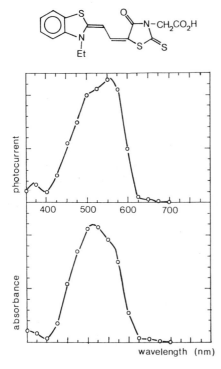

Fig. V.38. Short-circuit photocurrent of glass/Al/ merocyanine/Ag devices (illumination through aluminum electrode) and absorption spectrum of the merocyanine film (thickness 2600 Å). (After Ref. [94])

In M_1/merocyanine/M_2 solar cells, it also seems that "active regions" situated near the interfaces are formed. They extend as for PcM devices over a few hundred angstroms within the molecular material. It has been stated that the yields of the solar cells is limited by the amount of photons absorbed in the "active region" [92–94]. Merocyanines very commonly have narrow absorption profiles (width ~ 100 nm) (Fig. V.39). In this case, only 15% of the total solar output may be absorbed. The absorption curves of squaric-acid thin films are broader; for the same thickness the amount of absorbed photons may be multiplied by 2 or 3. Squaric-acid-based devices are, however, slightly less efficient than the merocyanine ones [1148, 1149]; the factors limiting the energy conversion efficiencies therefore are not only related to absorption profiles of the dyes.

A correlation between the ease of oxidation of the merocyanine and the photo-voltaic quantum yield has been postulated [1141]. However, such relationships are very difficult to establish definitely because of the poor reproducibility of the photo-voltaic characteristics of the cells. The crystallinity, the nature and the distribution of the dopant, and the stability of the dye under irradiation are, among others, parameters which are rarely thoroughly mastered.

The influence of chlorine in glass/Al/merocyanine/Au cells has been studied [1144]. Doping is achieved either

(i) by exposing the completed cell to dopant vapour,

(ii) by exposing only the aluminum prior to the deposition of the dye, and

(iii) by evaporating the merocyanine in presence of the dopant. The best yields have been obtained by using the second method [1144]. The nature of the dopant influences

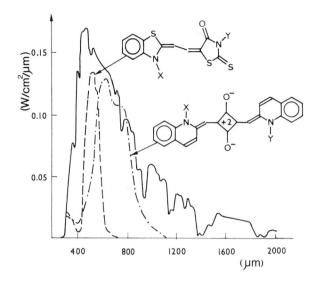

Fig. V.39. Spectral distribution of sunlight irradiation (AM1) compared to the absorption spectra of merocyanine and squaric acid thin films. (After Ref. [92])

the electrical characteristics of the cells (Table V.12). For undoped cells, the power-conversion efficiency based on incident light is inferior to $5 \cdot 10^{-6}$. These cells are prepared and tested without breaking the vacuum at any stage; the influence of O_2 is therefore minimized. Air or other donor molecules (NO_2, I_2, Cl_2) increase the efficiencies by many orders of magnitude. In the same time, the dark conductivity remains unchanged, indicating a minor role of dopant in generating free charge carriers in the dark.

Devices involving two different dyes have also been prepared [1145, 1146]. These devices have been abusively called p-n junction, while neither n- nor the p-type doping process were identified or mastered in any way. "n-Type" material used is malachite green, rhodamine B, or triphenylmethane; the "p-type" layer is constituted

Table V.12 Photovoltaic characteristics of Al/merocyanine/Au cells in presence of dopants. The dopant is introduced by exposing the aluminum layer prior deposition of the organic layer. Irradiation is carried out through the aluminum electrode. (After Ref. [1144]).

Dopant	Al % trans mittance	Φ_0 (mW/cm^2)	I_{sc} (mA/cm^2)	V_{oc} (mV)	FF	$\varkappa_{pc}$	$\varkappa'_{pc}$	Φ
Undoped	~25	50	10^{-5}	<1	—	$<5 \cdot 10^{-6}$	—	—
Cl$_2$	43	90	1.0	730	0.39	0.31	1.3	11.2
I$_2$	29	39	0.28	701	0.39	0.20	1.2	9.6
NO$_2$	27	90	0.50	625	0.41	0.16	1.0	10.6
air	~25	57	0.02	557	0.25	0.005	—	0.3

Φ_0: incident photon flux
$\varkappa_{pc}$: sunlight (AM2) power conversion, efficiency based on incident light
$\varkappa'_{pc}$: monochromatic (633 nm) power conversion efficiency
Φ: quantum yield of generation of carriers at 633 nm based on light absorbed by the dye

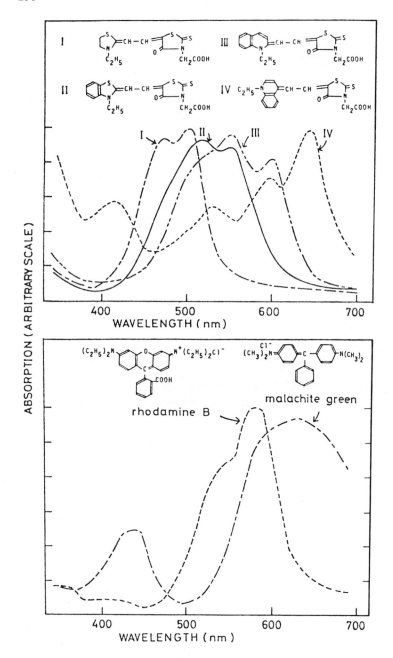

Fig. V.40. Molecular structure and absorption spectra of merocyanine dyes (left: p-type dyes; right: n-type dyes). [Reproduced with permission of (1145)]

of merocyanine. NESA/malachite green/merocyanine/Ag cells have been more particularly studied [1145] (Fig. V.40). The current-voltage curves show a fairly high asymmetry. The dark current seems to follow the Schockley equation with, however, high perfection factors (~ 3). The photovoltaic characteristics of such cells are very similar to the previous ones: $V_{oc} = 0.75$ V, $I_{sc} = 0.13$ mA/cm^2, conversion efficiency $= 0.05\%$ [1145–1146]. It has been shown that exposure of the cells to chlorine dramatically improves the sunlight-conversion efficiencies [1146]. As for the standard Schottky devices, the "active region" able to generate photocarriers is very thin, of the order of 200–300 Å [1146].

Mixture of dyes have also been used for making solar cells [1147]. The photo-current action spectra of Al/mixed dyes/Ag cells closely follow the absorption spectra of the mixture.

Single crystals of 1,1'-diethyl-2,2'-quinocyanine-(TCNQ)$_2$ have been shown to give rise to photovoltages under illumination [1150]. This has been attributed to a Dember effect where the mobility of electrons is superior to the mobility of holes. Merocyanine thin films have been deposited on single crystals of TiO$_2$ [1151]; the rectification ratio of TiO$_2$/merocyanine/Au devices attains 6 at ± 0.4 V, indicating a fairly low assymetry of the interfaces.

b Aromatic Derivatives

Many studies have been devoted to solar cells constituted of aromatic derivatives. Naphthalene [1152], plasma polymerized naphthalene [1153], perylene [1154], coronene [1154], anthracene [1155], tetracene [480, 1156–1158], pentacene [1159], tetrathio-tetracene [1160], p-quaterphenyl [1161], dibenzothiophene [1162], crystal violet [1163], and various polyacenes [1164] have all been used in molecular solar cells. Very similar results have been obtained in all cases. As previously, the photocurrent action spectra follow the absorption curves; only a thin layer near the interface is active in generating photocarriers; photovoltaic characteristics are highly dependent upon the fabrication procedure and the presence or not of donor dopants, in particular, ambient air. In the following sections only the peculiarities of the aromatic systems will be mentioned.

The photoconduction in dibenzothiophene crystals is suppressed by addition of tetracene [1162]. Tetracene is a fluorescence quencher of the excited state of dibenzo-thiophene; quenching therefore indicates that the formation of free charge carriers involves the first singlet excited state.

The photovoltage induced in aromatic devices is often constituted of two compo-nents [1158, 1162, 1165]; a fast photovoltage response is followed, and generally rapidly dominated, by a more slowly rising photovoltage of opposite sign [1158]. The slow response process seems to be more effectively induced by photons corres-ponding to the spectral region where the crystals absorb weakly [1158]. When the temperature is raised, the slow photovoltage component decreases in importance [1158]. Photoinduced detrapping of charges, large surface recombination rates, and exciton-annihilation mechanisms have been evoked to rationalize these obser-vations.

The energy conversion efficiencies of aromatic-based devices are, by at least one order of magnitude, lower than for merocyanine. The best results have been obtained for Al/dye 1/dye 2/Au devices [1157] with dye 1 = aceanthraquinoxaline or tetracene and dye 2 = $HNEt_3(TCNQ)_2$ or trinitrofluorenone. Open circuit voltages ranging from 0.3 to 0.9 V and short circuit currents equal or inferior to 1 $\mu A/cm^2$ have been reported [1157].

Doping experiments have also been carried out [1160]. The charge-transfer complex tetrathiotetracene-o-chloranil may be transformed into an n-type material on introduction of group I and IV metals; p-type conduction may be obtained with group II and III metals [1160]. The additives are introduced by dissolving the CT complex in non polar solvents in presence of the corresponding metals or metal nitrides; these later are decomposed in the presence of ethyl alcohol before reprecipitating the doped mixture. Because of the way of preparation and the noncontrol of the ambient during the elaboration of the devices, these results must, however, be considered with some suspicion.

Aromatics are basically very poor electron acceptors; the use of highly reducing metallic electrodes nevertheless has been proposed to form Schottky-type devices [1154]. Lithium, sodium or potassium thin films are deposited upon the aromatic (phenylene, violanthrene, etc.); further measurements are carried out under high vacuum. The photovoltage of the junctions thus formed indicates, as expected, a negative sign for the metallic electrodes. In two cases, however, the metallic electrode is positive; it seems surprising that sodium or lithium could act preferably as electron acceptors.

c Aromatic Liquid Crystals

Only two publications relate the use of aromatic liquid crystals for making junctions and solar cells [1166, 1167]. The existence of "liquid crystalline semiconductors", however, poses several fundamental problems.

Liquid inorganic semiconductors have been studied [1168]. Silicium melts at 1450 °C and germanium at 1000 °C; at the melting point, the transport properties of those materials are highly modified:

(i) Melting produces a sudden increase in the electrical conductivity, from 580 Ω^{-1} cm^{-1} to 12,000 Ω^{-1} cm^{-1} for silicium, from 1,250 to 14,000 Ω^{-1} cm^{-1} for germanium. In the liquid state the conductivity of these two semiconductors is close to the conductivity of liquid metals such as tin or mercury (10^4 Ω^{-1} cm^{-1}).

(ii) At high temperature, before melting, the electrical conductivity of solid silicon and germanium increases with temperature, indicating an intrinsic conduction. After melting, the electrical conductivity decreases monotonically with temperature.

(iii) At the melting point there is a sudden fall in the thermoelectric power up to very low values, typical of metals. This is due to a strong increase in the carrier density.

(iv) At the melting point, the absolute of the magnetic susceptibility abruptly decreases.

(v) There is a considerable fall in the mobility of charge carriers in germanium and in silicon upon melting. At the melting temperatures, the solid state electron mobilities are 16 $cm^2/V \cdot s$ and 370 $cm^2/V \cdot s$ for Si and Ge, respectively; they decrease to 0.3 and 0.4 $cm^2/V \cdot s$ after melting. This effect can be attributed to a strong increase in the efficiency of the electron scattering by thermal vibrations and/or structure defects corresponding to a drastic change in the long-range order.

Silicon and germanium, upon melting, are transformed from semiconductors to metals. A change in the nature of the chemical bonding and in the long-range and short-range orders is associated with this transition. Molecular materials should behave differently; molecular subunits are only loosely bound to each other and the nature of the cohesion forces involved does not require a strict geometry. Liquid crystalline molecular semiconductors can therefore be envisaged; the corresponding mobility will be, however, limited to about 1 cm^2/V · s.

d Chlorophylls and Porphyrins

Chlorophyll [1169–1177] and porphyrins [1178–1183] have been extensively used for making solar cells. Chlorophyll a (Chl$_a$) yields fairly different results depending on the thermal treatment effected. Films obtained from electroplating absorb at 745 nm; this probably corresponds to Chl$_a$-H$_2$O aggregates [1173]. After treatment at 100 °C, the absorption peak is shifted to 670 nm; this maximum wavelength is also found when Chl$_a$ is studied in solution, it therefore probably corresponds to monomeric Chl$_a$ [1173].

Power conversion efficiencies of the order of $10^{-2}\%$ have been reported for Cr/Chl$_a$/Hg cells (745 nm form, monochromatic light, 1 mW/cm^2) [1174]. From cells to cells, orders-of-magnitude differences in efficiencies have been noticed [1174]. Huge differences in photovoltages result from different methods of fabrication; it corresponds mainly to variations in crystallinities [1179]. The photoconductivity of Chl$_a$ is enhanced in the presence of water [1184]. Magnetic circular dichroism measurements indicate that O$_2$ or H$_2$O doping is not a surface effect, and that they affect the axial ligand sensitive region of the spectra throughout the films [1181]. Other gases (O$_2$, N$_2$, air, Ar, N$_2$O, SF$_6$, H$_2$) also influence the characteristics of M$_1$/Chl$_a$/M$_2$ cells [1175].

Using the Langmuir-Blodgett technique, monolayers of Chl$_a$ have been deposited on metallic electrodes [1169, 1171, 1177]. Lamellar composites of Chl$_a$ with electron acceptors such as o-chloranil, vitamin K$_1$ or other quinones have been realized [1169, 1171, 1177]. However, most of the measurements have been carried out in air making difficult the interpretation of the results. The power-conversion efficiencies found are extremely low, of the order of $10^{-2}\%$.

Various para-substituted derivatives of tetraphenylporphyrins have been synthesized and studied [1180]. The quantum yield of generation of charge carriers is dependent upon the electron-donating ability of the substituents. The more easily oxidized porphyrins exhibit the higher photocurrent-quantum yields [1180].

e Polymeric Systems

Devices based on poly(2-vinylpyridine)-I$_2$ [1185, 1186] and poly(vinylcarbazole)-I$_2$ [1187] have been realized. White light illumination hardly yields energy-conversion efficiencies higher than 0.05%. Poled poly(vinylidene)fluoride — which is a widely used ferroelectric polymer [1188] — shows an anomalous photovoltaic effect. Open-circuit photovoltages as high as $4 \cdot 10^4$ V have been reported, orders of magnitude higher than the estimated band gap of the polymer [1189]. The photocurrent action spectrum does not seem, in this example, to depend on the irradiation wavelength from 400 to 700 nm. The real nature of this "photovoltaic effect" may consequently be questioned.

Conclusion

Are molecular semiconductors a reality? The previous chapters have been partly devoted to answering this question. It has been seen that beside the semantic problem of their definition, the denomination "molecular semiconductor" raises several fundamental points. A molecular material is considered to be a semiconductor if it possesses a reasonable mobility of charge carriers and if it demonstrates the existence of a band gap of the order of 1–2 eV. The molecular material must also be able to form typical semiconductor-based devices: junctions and solar cells. With these criteria in mind, no example of molecular semiconductor has been found, up to the present time, throughout the literature. Polymeric systems may potentially lead to high-mobility charge carriers, but either they cannot be doped and are insulators, or they are too impure and too inhomogeneous to exhibit experimentally accessible high mobilities. The purity of molecular crystals is satisfactory but the corresponding charge carrier mobilities are limited by the poor intersubunit overlap. Charge transfer complexes and radical ion salts presumably form semiconducting materials but at very low temperatures ($T < 60$ K) and with very small band gaps ($E_g \sim 0.1$ eV).

Metallo-organic derivatives seem promising for forming future semiconductors. The variety of orbital overlaps possible increases the chemical flexibility of the system. Intermediate phases between solid crystals and liquid phases (mesophases or liquid crystals) are probably more appropriate to obtain molecular materials whose structure may be easily predicted from simple structural considerations. The key to the problem is still in the chemist's hands; no physical or theoretical considerations limit the possibility of obtaining molecular semiconductors.

Molecular insulators, metals and supraconductors have all been reported. Molecular semiconductors are undoubtedly at least as difficult to prepare. Because of the low density of charge carriers, semiconductors are more sensitive to defects and chemical impurities increasing the chemical difficulties of synthesis. Molecular materials are, moreover, highly anisotropic, enhancing the localizing effect of the defects. Two-dimensional molecular systems would remove these kinds of difficulties.

Once the existence of molecular semiconductors has been proven, the realization of well-behaved devices will be reasonably envisaged.

Paris, le 13 Mars 1984   J. Simon   J.-J. André

References

1. Kittel, C.: Introduction à la Physique de l'Etat Solide, Paris, Dunod 1969
2. Ashcroft, N. W., Mermin, N. D.: Solid State Physics, New York, Holt, Rinehart and Winston 1976
3. Ziman, J. M.: Principles of the Theory of Solids, Cambridge, University Press 1969
4. Karplus, M., Porter, R. N.: Atoms and Molecules, Menlo Park, California, W. A. Benjamin Inc. 1970
5. Pauling, L.: The Nature of the Chemical Bond, 3rd edition, New York, Cornell University Press 1962
6. Silinsh, E. A.: Organic Molecular Crystals, Berlin, Springer Verlag 1980
7. Salem, L.: The Molecular Orbital Theory of Conjugated Systems, New York, W. A. Benjamin Inc. 1966
8. André, J.-J., Bieber, A., Gautier, F., Ann. Phys. *1*, 145 (1976)
9. Epstein, A. J., Conwell, E. M., Miller, J. S.: Annals of New York Acad. of Sciences *313*, 183 (1978).
10. Schein, L. B.: Phys. Rev. *B 15*, 1024 (1977).
11. Friedman, L.: Phys. Rev. *A 133*, 1668 (1964)
12. Marcus, R. A.: J. Chem. Phys. *24*, 966 (1956); *26*, 871 and 876 (1957)
13. Marcus, R. A.: Ann. Rev. of Phys. Chem. *15*, 155 (1964)
14. Halpern, J., Orgel, L. E.: Discussion of the Faraday Soc. *29*, 32 (1960)
15. Meier, H.: Organic Semiconductors, Berlin, Verlag Chemie 1974
16. Kearns, D.: J. Chem. Phys. *35*, 2269 (1961)
17. Eley, D. D., Parfitt, G. D.: Trans. Faraday Soc. *51*, 1529 (1955)
18. Eley, D. D.: J. of Polymer Sci. *C 17*, 73 (1967)
19. Keller, R. A., Mast, H. E.: J. Chem. Phys. *36*, 2640 (1962)
20. Glaeser, R. M., Berry, R. S.: J. Chem. Phys. *44*, 3797 (1966)
21. Sumi, H.: J. Chem. Phys. *70*, 3775 (1979)
22. Schein, L. B., Duke, C. B., McGhie, A. R.: Phys. Rev. Lett. *40*, 197 (1978)
23. Gutmann, F., Lyons, L. E.: Organic Semiconductors, New York, J. Wiley & Sons 1967
24. Elermann, R., Hofberger, W., Baessler, H.: J. Non-Cryst. Solids *28*, 415 (1978)
25. Lyons, L. E.: J. Chem. Soc. 5001 (1957)
26. Loutfy, R. O., Cheng, Y. C.: J. Chem. Phys. *73*, 2902 (1980)
27. Karl, N.: Organic Semiconductors, Braunschweig, Vieweg 1974
28. Pohl, H. A., Rosen, S. L.: Proceedings on the Fifth Carbon Conference, Vol. II, Oxford, Pergamon Press 1963
29. Seanor, D. A.: Adv. Polymer Sci. *4*, 317 (1965)
30. Rembaum, A.: J. Polym. Sci. *C* 157 (1970)
31. Goodings, E. P.: Endeavour *34*, 123 (1975)
32. Paushkin, Ya. M. et al.: J. Polym. Sci. *A 5*, 1203 (1967)
33. Kossmehl, G.: Ber. Bunsenges Phys. Chem. *83*, 417 (1979)
34. Naarmann, H.: Ber. Bunsenges Phys. Chem. *83*, 427 (1979)
35. Schopov, I., Kirowa, P.: Makromol. Chem. *181*, 1405 (1980)
36. Dresselhaus, M. S., Dresselhaus, G.: Advances in Physics *30*, 139 (1981)
37. Hennig, G. R.: Prog. Inorg. Chem. *1*, 125 (1959)
38. Davies, J. E. D.: J. Chem. Ed. *54*, 536 (1977)

39. Baughman, R. H.: Contemporary Topics in Polymer Science Pearce, E. M., Schaefgen, J. R. (ed.), Vol. 2, New York, Plenum Publishing Corp. 1977
40. Kaiser, J., Wegner, G., Fischer, E. W.: Israel J. of Chemistry *10*, 157 (1972)
41. Spannring, W., Bässler, H.: Ber. Bunsenges. Phys. Chem. *83*, 433 (1979)
42. Tieke, B., Lieser, G., Wegner, G.: J. Polym. Sci. Polym. Chem. Ed. *17*, 1631 (1979)
43. Mondong, R., Bässler, H.: Chem. Phys. Lett. *78*, 371 (1981)
44. Chance, R. R., Baughmann, R. H., Reucroft, P. J., Takahashi, K.: Chem. Phys. *13*, 181 (1976)
45. Reimer, B., Bässler, H.: Chem. Phys. Lett. *43*, 81 (1976)
46. Spannring, W., Bässler, H.: Chem. Phys. Lett. *84*, 54 (1981)
47. Duke, C. B., Fabish, T. J., Paton, A.: Chem. Phys. Lett. *49*, 133 (1977)
48. Duke, C. B. et al.: Phys. Rev. *B 18*, 5717 (1978)
49. Duke, C. B.: Mol. Cryst. Liquid Cryst. *50*, 63 (1979)
50. Anderson, P. W.: Phys. Rev. *109*, 1492 (1958)
51. Elermann, E., Parkinson, G. M., Bässler, H., Thomas, J. M.: J. Phys. Chem. *86*, 313 (1982)
52. Windsor, M. W.: Physics and Chemistry of the Organic Solid State, Vol. II, Fox, D., Labes, M. M., Weissberger, A. (eds.), New York, Interscience Pub. 1965
53. Nudelman, S., Mitra, S. S.: Optical Properties of Solids, New York, Plenum Press 1969
54. Willardson, R. K., Beer, A. C.: Semiconductors and Metals, Vol. 3, New York, Academic Press 1967
55. Calvert, J. G., Pitts, Jr., J. N.: Photochemistry, New York, J. Wiley & Sons 1967
56. Lewis, G. N., Kasha, M.: J. Amer. Chem. Soc. *67*, 994 (1945)
57. Lamola, A. A., Turro, N. J.: Energy Transfer and Organic Chemistry, Technique of Organic Chemistry Vol. 5, Leermakers, P. A., Weissberger, A. (ed.), New York, Interscience Pub. 1969
58. Turro, N. J.: Pure Appl. Chem. *49*, 405 (1977)
59. Craig, D. P., Walmsley, S. H.: Excitons in Molecular Crystals, New York, W. A. Benjamin Inc. 1968
60. Dexter, D. L., Knox, R. S.: Excitons, Interscience Tracts on Physics and Astronomy, Marshak, R. E. (ed.), New York, Interscience Pub. 1965
61. Cho, K.: Excitons, Topics in Current Chemistry, Vol. 14, Berlin, Springer-Verlag 1979
62. Harris, C. B., Zwemer, D. A.: Ann. Rev. Phys. Chem. *29*, 473 (1978)
63. Voltz, R.: Radiation Res. Rev. *1*, 301 (1968)
64. Weill, G. in Problémes Physiques dans les Systèmes Biologiques, C. de Witt, Matricon: J. ed., New York, Gordon and Breach Sci. Pub. 1969, p. 369
65. Frenkel, J.: Phys. Rev. *37*, 1276 (1931)
66. Peierls, R.: Ann. Physik. *13*, 905 (1932)
67. Davydov, A. S.: Theory of Molecular Excitons (Eng. Transl.) Kasha, M., Oppenheimer Jr., M. (ed.), New York, McGraw Hill 1962
68. Förster, Th.: Dis. Farad. Soc. *27*, 7 (1959)
69. Förster, Th.: Modern Quantum Chemistry, Vol. III, Shinanoglu, O. (ed.), New York, Academic Press 1965
70. Dexter, D. L.: J. Chem. Phys. *21*, 836 (1953)
71. Schipper, P. E.: Mol. Cryst. Liquid Cryst. *28*, 401 (1974)
72. Crisp, G. M., Walmsley, S. H.: Mol. Cryst. Liquid Cryst. *58*, 71 (1980)
73. Goode, D., Lupien, Y., Siebrand, W., Williams, D. F., Thomas, J. M., Williams, J. O.: Chem. Phys. Lett. *25*, 308 (1974)
74. Hochstrasser, R. M.: Radiation Res. *20*, 107 (1963)
75. Rice, S. A., Jortner, J.: J. Chem. Phys. *44*, 4470 (1966)
76. Hong, H.-K., Kopelman, R.: J. Chem. Phys. *55*, 5380 (1971)
77. Kopelman, R.: J. Phys. Chem. *80*, 2191 (1976)
78. Abram, I. I., Hochstrasser, R. M.: J. Chem. Phys. *72*, 3617 (1980)
79. Kopelman, R., Argyrakis, P.: J. Chem. Phys. *72*, 3053 (1980)
80. Onsager, L.: J. Chem. Phys. *2*, 599 (1934)
81. Onsager, L.: Phys. Rev. *54*, 554 (1938)
82. Pai, D. M., Enck, R. C.: Phys. Rev. *B 11*, 5163 (1975)
83. Pai, D. M.: J. Appl. Phys. *46*, 5122 (1975)
84. Melz, P. J.: J. Chem. Phys. *57*, 1694 (1972)

85. Haberkorn, R., Michel-Beyerle, M. E.: Chem. Phys. Lett. *23*, 128 (1973); Phys. Status Solidi *B67*, K61 (1975)
86. Singh, J., Baessler, H.: Phys. Status Solidi *B63*, 425 (1974)
87. Noolandi, J., Hong, K. M.: J. Chem. Phys. *70*, 3230 (1979)
88. Bullot, J., Cordier, P., Gauthier, M.: Chem. Phys. Lett. *54*, 77 (1975)
89. Yokoyama, M., Endo, Y., Mikawa, H.: Chem. Phys. Lett. *34*, 597 (1975)
90. Chapin, D. M., Fuller, C. S., Pearson, G. L.: J. Appl. Phys. *25*, 676 (1954)
91. Linmeyer, J., Wrigley, C. Y.: "Development of a 20% Efficient Solar Cell" (NSF-43090), National Science Foundation, Washington D.C. 1975
92. Morel, D. L.: Mol. Cryst. Liquid Cryst. *50*, 127 (1979)
93. Morel, D. L., Ghosh, A. K., Feng, T., Stogryn, E. L., Purwin, P. E., Shaw, R. F., Fishman, C.: Appl. Phys. Lett. *32*, 495 (1978)
94. Ghosh, A. K., Feng, T.: J. Appl. Phys. *49*, 5982 (1978)
95. Sze, S. M.: Physics of Semiconductor Devices, New York, Wiley-Interscience 1969
96. Dalven, R.: Introduction to Applied Solid State Physics, New York, Plenum Press 1980
97. Hovel, H. J.: Semiconductors and Semimetals, Vol. 11, Solar Cells, New York, Academic Press 1975
98. Sitting, M.: Solar Cells for Photovoltaic Generation of Electricity, Park Ridge, Noyes Data Corp. 1979
99. Seraphin, B. O.: Solar Energy Conversion, Berlin, Springer-Verlag 1979
100. Mott, N. F., Gurney, R. W.: Electronic Processes in Ionic Crystals, Oxford, Clarendon Press 1940
101. Rose, A.: Phys. Rev. *97*, 1538 (1955)
102. Lampert, M. A.: Injection Currents in Solids, New York, Academic Press 1965
103. Lampert, M. A., Rose, A., Smith, R. W.: J. Phys. Chem. Solids *8*, 464 (1958)
104. Mark, P., Helfrich, W.: J. Appl. Phys. *33*, 205 (1962)
105. De Levie, R., Moreira, H.: J. Membr. Biol. *10*, 171 (1972)
106. Sinharay, N., Meltzer, B.: Solid State Electron *7*, 125 (1964)
107. Page, D. J.: Solid State Electron *9*, 255 (1966)
108. Ormancey, G., Godefroy, G.: J. Phys. Paris *35*, 135 (1974)
109. O'Reilly, T. J., DeLucia, J.: Solid State Electron *18*, 965 (1975)
110. Bonham, J. S., Jarvis, D. H.: Aust. J. Chem. *30*, 705 (1977)
111. Nicolet, M. A.: J. Appl. Phys. *37*, 4224 (1966)
112. Delannoy, P., Schott, M., Berrehar, J.: Phys. Status Solidi *A32*, 577 (1975)
113. Sworakowski, J.: J. Appl. Phys. *41*, 292 (1970)
114. Hwang, W., Kao, K. C.: Solid State Electron *15*, 523 (1972)
115. Bonham, J. S.: Aust. J. Chem. *31*, 2117 (1978)
116. Delannoy, P.: Mater. Sci. *7*, 13 (1981)
117. Henisch, H. K., Manifacier, J.-C., Callarotti, R. C., Schmidt, P. E.: J. Appl. Phys. *51*, 3790 (1980)
118. Killesreiter, H., Baessler, H.: Chem. Phys. Lett. *11*, 411 (1971)
119. Vaubel, G., Baessler, H., Möbius, D.: Chem. Phys. Lett. *10*, 334 (1971)
120. Kallmann, H., Vaulbe, G., Baessler, H.: Phys. Status Solidi *44*, 813 (1971)
121. Singh, H., Baessler, H.: Phys. Status Solidi *B62*, 147 (1974)
122. Bonham, J. S.: Aust. J. Chem. *29*, 2123 (1976)
123. Levinson, J., Burshtein, Z., Many, A.: Mol. Cryst. Liquid Cryst. *26*, 329 (1974)
124. Pope, M., Weston, W.: Mol. Cryst. Liquid Cryst. *25*, 205 (1974)
125. Huggins, C. M., Sharbaugh, A. H.: J. Chem. Phys. *38*, 393 (1963)
126. Goodman, A. M.: J. Appl. Phys. *34*, 329 (1963)
127. Vincent, G., Bois, D., Pinard, P.: J. Appl. Phys. *46*, 5173 (1975)
128. Simmons, J. G., Nadkarni, G. S., Lancaster, M. C.: J. Appl. Phys. *41*, 538 (1970)
129. Braun, A., Tcherniac, J.: Ber. Dtsch. Chem. Ges. *40*, 2709 (1907)
130. De Diesbach, H., Von der Weid, E.: Helv. *10*, 886 (1927)
131. Linstead, R. P.: J. Chem. Soc. 1016 and 1031 (1934)
132. Lawton, E. A.: J. Phys. Chem. *62*, 384 (1958)
133. Linstead, R. P., Weiss, F. T.: J. Chem. Soc. 2981 (1950)
134. Linstead, R. P., Noble, E. G., Wright, J. M.: J. Chem. Soc. 911, 933 (1937)

135. Barrett, P. A., Bent, C. E., Linstead, R. P.: J. Chem. Soc. 1719 (1937)
136. Lever, A. B. P.: Adv. Inorg. and Radiochem. *7*, 27 (1965)
137. Moser, F. H., Thomas, A. L.: Phthalocyanines, A. C. S. Monograph 157, New York, Reinhold Publishing Corp. 1963
138. Berezin, B. D.: Coordination Compounds of Porphyrins and Phthalocyanines, New York, J. Wiley & Sons 1981
139. Kasuga, K., Tsutsui, M.: Coord. Chem. Rev. *32*, 67 (1980)
140. Barrett, P. A., Frye, D. A., Linstead, R. P.: J. Chem. Soc. 1157 (1938)
141. Wöhrle, D., Meyer, G.: Makromol. Chem. *181*, 2127 (1980)
142. Gouterman, M., Sayer, P., Shankland, E., Smith, J. P.: Inorg. Chem. *20*, 87 (1981)
143. Kirin, I. S., Moskalev, P. N., Moskalev, Yu. A.: Russ. J. Inorg. Chem. *10*, 1065 (1965)
144. Buslaev, Yu. A., Kuznetsova, A. A., Goryanov, L. F.: Izv. Akad. Nauk. SSSR, Neorg. Matter. *3*, 1701 (1967)
145. Day, V. W., Marks, T. J., Wachter, W. A.: J. Amer. Chem. Soc. *97*, 4519 (1975)
146. Marks, T. J., Stojakovic, D. R.: J. Amer. Chem. Soc. *100*, 1695 (1978)
147. Lux, F., Dempf, D., Graw, D.: Angew. Chem. Int. Ed. *7*, 819 (1968)
148. Lux, F., Brown, D., Dempf, D., Fischer, R. D., Hagenberg, W.: Angew. Chem. Int. Ed. *8*, 894 (1969).
149. Barnhart, G.: U. S. Pat. 2602800.
150. Brach, P. J., Grammatica, S. J., Ossanna, O. A., Weinberger, L.: J. Hetero. Chem. *7*, 1403 (1970).
151. Yang, C. H., Lin, S. F., Chen, H. L., Chang, C. T.: Inorg. Chem. *19*, 3541 (1980)
152. Yoshihara, K., Kishimoto, M., Takahashi, M., Suzuki, S., Shiokaura, T.: Radiochim. Acta *21*, 148 (1974)
153. Yoshihara, K., Wolf, G. K., Baumgartner, F.: Radiochim. Acta *21*, 96 (1974)
154. Mikhalenko, S. A., Barkanova, S. V., Lebedev, O. L., Lukyanets, E. A.: J. Gen. Chem. USSR *41*, 2770 (1971)
155. Oksengendler, I. G., Kondratenko, N. V., Lukyanets, E. A., Yogupolskii, L. M.: Zh. Org. Khim. *13*, 1554 (1977); C. A. *87*, 137296 (1977)
156. Mikhalenko, S. A., Yagodina, L. A., Lukyanets, E. A.: J. Gen. Chem. USSR *46*, 1557 (1976)
157. Pawlowski, G., Hanack, M.: Synthesis 287 (1980)
158. Pawlowski, G., Hanack, M.: Synthetic Comm. *11*, 351 (1981)
159. Weber, J. H., Busch, D. H.: Inorg. Chem. *4*, 469 (1965)
160. Piechocki, C., Simon, J., Skoulios, A., Guillon, D., Weber, P.: J. Amer. Chem. Soc. *104*, 5245 (1982)
161. Loutfy, R. O., Hsiao, C.-K.: Can. J. Chem. *57*, 2546 (1979)
162. Vogt Jr., L. H., Zalkin, A., Templeton, D. H.: Science *151*, 569 (1966)
163. Vogt Jr., L. H., Zalkin, A., Templeton, D. H.: Inorg. Chem. *6*, 1725 (1967)
164. Elvidge, J. A., Lever, A. B. P.: Proc. Chem. Soc. 195 (1959)
165. Janson, T. R., Kane, A. R., Sullivan, J. F., Knox, K., Kenny, M. E.: J. Amer. Chem. Soc. *91*, 5210 (1969)
166. Owen, J. E., Kenney, M. E.: Inorg. Chem. *1*, 334 (1962)
167. Mooney, J. R., Choy, C. K., Knox, K., Kenney, M. E.: J. Amer. Chem. Soc. *97*, 3033 (1975)
168. Schoch Jr., K. F., Kundalkar, B. R., Marks, T. J.: J. Amer. Chem. Soc. *101*, 7071 (1979)
169. Swift, D. R.: Ph. D. Dissertation, Case Western Reserve University, Cleveland, Ohio (1970)
170. Joyner, R. D., Kenney, M. E.: Inorg. Chem. *1*, 717 (1962)
171. Meyer, G., Wöhrle, D.: Makromol. Chem. *175*, 714 (1974)
172. Meyer, G., Hartmann, M., Wöhrle, D.: Makromol. Chem. *176*, 1919 (1975)
173. Berezin, B. D., Akopov, A. S., Lapshina, O. B.: Vysokomol. Soedin Ser. A *16*, 450 (1974)
174. Joyner, R. D., Kenney, M. E.: J. Amer. Chem. Soc. *82*, 5790 (1960)
175. Kroenke, W. J., Sutton, L. E., Joyner, R. D., Kenney, M. E.: Inorg. Chem. *2*, 1064 (1963)
176. Dirk, C. W., Mintz, E. A., Schoch Jr., K. F., Marks, T. J.: J. Macromol. Sci. Chem. *A16*, 275 (1981)
177. Kuznesof, P. M., Nohr, R. S., Wynne, K. J., Kenney, M. E.: J. Macromol. Sci. Chem. *A16*, 299 (1981)
178. Linsky, J. P., Paul, T. R., Nohr, R. S., Kenney, M. E.: Inorg. Chem. *19*, 3131 (1980)
179. Schneider, O., Hanack, M.: Angew. Chem. Int. Ed. *19*, 392 (1980)
180. Metz, J., Hanack, M.: Nouveau J. de Chimie *5*, 541 (1981)

181. Nohr, R. S., Wynne, K. J.: J. C. S. Chem. Comm. 1210 (1981)
182. Hanack, M., Mitulla, K., Pawlowski, G., Subramanian, L. R.: Angew. Chem. Int. Ed. *18*, 322 (1979)
183. Seelig, F. F.: Z. Naturforsch. *34 A*, 986 (1979)
184. Marvel, C. S., Rassweiler, J. H.: J. Amer. Chem. Soc. *80*, 1197 (1958)
185. Epstein, A., Wildi, B. S.: J. Chem. Phys. *32*, 324 (1960)
186. Wildi, B. S., Katon, J. E.: J. Polym. Sci. *A 2*, 4709 (1964)
187. Berlin, A. A., Sherle, A. I.: Inorg. Macromol. Rev. *1*, 235 (1971)
188. Norrell, C. J., Pohl, H. A., Thomas, M., Berlin, K. D.: J. Polym. Sci. Physics *12*, 913 (1974)
189. Wöhrle, D.: Advances in Polymer Sci. *10*, 35 (1972)
190. Boston, D. R., Bailar Jr., J. C.: Inorg. Chem. *11*, 1578 (1972)
191. Wöhrle, D., Wahl, B.: Tetrahedron Lett. 227 (1979)
192. Bannehr, R., Meyer, G., Wöhrle, D.: Polymer Bulletin *2*, 841 (1980)
193. Bannehr, R., Haeger, N., Meyer, G., Wöhrle, D.: Makromol. Chem. *182*, 1633 (1981)
194. Baumann, F., Binert, B.: German Pat. 83939; U. S. Pat. 2768867; British Pat. 687655
195. Barnhart, G., Skiles, B. F.: U. S. Pat. 2772284; British Pat. 745359
196. Friedel, M. K., Hoskins, B. F., Martin, R. L., Mason, S. A.: J. C. S. Chem. Comm. 400 (1970)
197. Magner, G., Savy, M., Scarbeck, G.: J. Electrochem. Soc. *127*, 1076 (1980)
198. Linstead, R. P., Robertson, J. M.: J. Chem. Soc. 1195 and 1636 (1936)
199. Robertson, J. M., Linstead, R. P., Dent, C. E.: Nature *135*, 506 (1935)
200. Robertson, J. M.: J. Chem. Soc. *615* (1935)
201. Robertson, J. M., Woodward, I. J.: J. Chem. Soc. 219 (1937); 36 (1940)
202. Suito, E., Uyeda, N.: J. Phys. Chem. *84*, 3223 (1980)
203. Suito, E., Uyeda, N.: Kolloid Z. u. Z. Polym. *193*, 7 (1963)
204. Sharp, J. H., Lardon, M.: J. Phys. Chem. *72*, 3230 (1968)
205. Mason, R., Williams, G. A., Fielding, P. E.: J. C. S. Dalton 676 (1979)
206. Ercolani, C., Neri, C., Porta, P.: Inorg. Chim. Acta *1*, 415 (1967)
207. Brown, C. J.: J. Chem. Soc. *A* 2494 (1968)
208. Susich, G.: Anal. Chem. *22*, 425 (1950)
209. Tarantino, F. R., Stubbs, D. H., Cooke, T. F., Melsheimer, L. A.: Am. Ink. Maker *29*, 35 and 425 (1950)
210. Ebert Jr., A. A., Gottlieb, H. B.: J. Amer. Chem. Soc. *74*, 2806 (1952)
211. Karasek, F. W., Decius, J. C.: J. Amer. Chem. Soc. *74*, 4716 (1952)
212. Shigemitsu, M.: Bull. Chem. Soc. Japan *32*, 607 (1959)
213. Abkowitz, M., Chen, I.: J. Chem. Phys. *54*, 811 (1971)
214. Fustoss-Wegner, M.: Thermochim. Acta *23*, 93 (1978)
215. Mindorff, M. S., Brodie, D. E.: Can. J. Phys. *59*, 249 (1981)
216. Sharp, J. H., Miller, R. L.: J. Phys. Chem. *72*, 3335 (1968)
217. Wihksne, K., Newkirk, A. E.: J. Chem. Phys. *34*, 2184 (1961)
218. Harrison, S. E., Ludewig, K. H.: J. Chem. Phys. *45*, 343 (1966)
219. Uyeda, N., Ashida, M., Suito, E.: J. Appl. Phys. *36*, 1453 (1965)
220. Buchholz, J. C., Somorjai, G. A.: J. Chem. Phys. *66*, 573 (1977)
221. Boucher, L. J.: Coordination Chemistry of Macrocyclic Compounds, G. A. Melson (ed.), New York, Plenum Press 1979
222. Rogers, D., Osborn, R. S.: J. C. S. Chem. Comm. 840 (1971)
223. Cariati, F., Morazzoni, F., Zocchi, M.: Inorg. Chim. Acta *14*, L31 (1975)
224. Kobayashi, T., Kurokawa, F., Ushida, T., Yeda, W. A. U., Suito, E.: J. C. S. Chem. Comm. 1631 (1971)
225. Omiya, S., Tsutsui, M., Meyer Jr., E. F., Bernal, I., Cullen, D. L.: Inorg. Chem. *19*, 134 (1980)
226. Ukei, K.: Acta. Cryst. *B 29*, 2290 (1973)
227. Iyechika, Y., Yakushi, K., Ikemoto, I., Kuroda, H.: Acta. Cryst. *B 38*, 766 (1982)
228. Williams, G. A., Figgis, B. N., Mason, R., Mason, S. A., Fielding, P. E.: J. C. S. Dalton Trans. 1688 (1980)
229. Bennett, W. E., Broberg, D. E., Baenziger, N. C.: Inorg. Chem. *12*, 930 (1973)
230. Gieren, A., Hoppe, W.: J. C. S. Chem. Comm. 413 (1971)

231. Kasuga, K., Tsutsui, M., Pettersen, R. C., Tatsumi, K., Van Opdenbosh, N., Meyer, E. F.: J. Amer. Chem. Soc. *102*, 4835 (1980)
232. Boyd, P. D. W., Smith, T. D.: J. C. S. Dalton 839 (1972)
233. Cookson, D. J., Smith, T. D., Boas, J. F., Pilbrow, J. R.: J. C. S. Dalton Trans, 1791 (1976)
234. Schelly, Z. A., Harward, D. J., Hemmes, P., Eyring, E. M.: J. Phys. Chem. *74*, 3040 (1970)
235. Farina, F. D., Halko, D. J., Swinehart, J. H.: J. Phys. Chem. *76*, 2343 (1972)
236. Kratky, O., Oelschlaeger, H.: J. Colloid. Int. Sci. *31*, 490 (1969)
237. Bernauer, K., Fallab, S.: Helv. *44*, 1287 (1961)
238. Gruen, L. C., Blagrove, R. J.: Aust. J. Chem. *26*, 319 (1973)
239. De Bolfo, J. A., Smith, T. D., Boas, J. F., Pilbrow, J. R.: J. C. S. Faraday Trans II *72*, 481 (1976)
240. Abkowitz, M., Monahan, A. R.: J. Chem. Phys. *58*, 2281 (1973)
241. Monahan, A. E., Brado, J. A., De Luca, A. F.: J. Phys. Chem. *76*, 446 (1972)
242. Hughes, A.: Proc. Roy. Soc. *B 155*, 710 (1936)
243. Kuhn, H.: J. Chem. Phys. *17*, 1198 (1949)
244. Kuhn, H.: Angew. Chem. *71*, 93 (1959)
245. Basu, S.: Indian J. Phys. *28*, 511 (1954)
246. Schaffer, A. M., Gouterman, M., Davidson, E. R.: Theor. Chim. Acta *30*, 9 (1973)
247. Edwards, L., Gouterman, M.: J. Mol. Spectrosc. *33*, 292 (1970)
248. Almlöf, J.: Int. J. Quantum Chem. *3*, 915 (1974)
249. Case, D. A., Karplus, M.: J. Amer. Chem. Soc. *99*, 6182 (1977)
250. Zerner, M., Gouterman, M.: Theor. Chim. Acta *4*, 44 (1966)
251. Ross, B., Sundbom, M.: J. Mol. Spect. *36*, 8 (1970)
252. Weiss, C., Kobayashi, H., Gouterman, M.: J. Mol. Spect. *16*, 415 (1965)
253. Mc Hugh, A., Gouterman, M., Weiss, C.: Theor. Chim. Acta *24*, 346 (1972)
254. Berkovitch-Yellin, Z., Ellis, D. E.: J. Amer. Chem. Soc. *103*, 6066 (1981)
255. Huang, T.-H., Rieckhoff, K. E., Voigt, E. M.: J. Phys. Chem. *85*, 3322 (1981)
256. Lever, A. B. P., Pickens, S. R., Minor, P. C., Licoccia, S., Ramaswamy, B. S., Magnell, K.: J. Amer. Chem. Soc. *103*, 6800 (1981)
257. Harrison, S. E., Assour, J. M.: J. Chem. Phys. *40*, 365 (1964)
258. Guzy, C. M., Raynor, J. B., Symons, M. C. R.: J. Chem. Soc. A 2299 (1969)
259. Henriksson, A., Ross, B., Sundbom, M.: Theor. Chim. Acta *27*, 303 (1972)
260. Hollebone, B. R., Stillman, M. J.: J. Chem. Soc. Faraday Trans. II *74*, 2107 (1978)
261. Stillman, M. J., Thomson, A. J.: J. C. S. Faraday Trans. II *70*, 805 (1974)
262. Stillman, M. J., Thomson, A. J.: J. C. S. Faraday Trans. II *70*, 790 (1974)
263. Schott, M.: J. Chem. Phys. *44*, 429 (1966)
264. Fielding, P. E., McKay, A. G.: Aust. J. Chem. *17*, 750 (1964) and *28*, 1445 (1975)
265. Yoshino, K., Kaneto, K., Inuishi, Y.: Energy and Charge Transfer in Organic Semiconductors, p. 37, Masuda, K., Silver, M. (ed.), New York, Plenum Press 1974
266. Yoshino, K., Hikida, M., Tatsuno, K., Kaneto, K., Inuishi, Y.: J. Phys. Soc. Jpn *34*, 441 (1973)
267. Roll, U., Ewert, S., Lüth, H.: Chem. Phys. Lett. *58*, 91 (1978)
268. Day, P., Williams, R. J. P.: J. Chem. Phys. *37*, 567 (1962)
269. Lucia, E. A., Verderame, F. D.: J. Chem. Phys. *48*, 2674 (1968)
270. Chadderton, L. T.: J. Phys. Chem. Solids *24*, 751 (1963)
271. Dmietrierskii, D., Ermolaev, V. L., Terenin, A. N.: Dokl. Akad. Nauk. SSSR *114*, 751 (1957)
272. Evstigneev, V. B., Krasnoskii, A. A., Gavrilova, V. A.: Dokl. Akad. Nauk. SSSR *70*, 261 (1950)
273. Gachkovskii, V. F.: Dokl. Akad. Nauk. SSSR *82*, 739 (1953)
274. Litvin, F. F., Personov, R. I.: Fiz. Prob. Spektroskopii *1*, 229 (1963)
275. Lyalin, G. N., Kobyshev, G. I.: Opt. i. Spectroskopiya *15*, 253 (1963)
276. Huang, T.-H., Rieckhoff, K. E., Voigt, E.-M.: Chem. Phys. *19*, 25 (1977)
277. Menzel, E. R., Rieckhoff, K. E., Voigt, E.-M: J. Chem. Phys. *58*, 5726 (1973)
278. Vincett, P. S., Voigt, E.-M., Rieckhoff, K. E.: J. Chem. Phys. *55*, 4131 (1971)
279. Brannon, J. H., Magde, D.: J. Amer. Chem. Soc. *102*, 62 (1980)
280. McVie, J., Sinclair, R. S., Truscott, T. G.: J. C. S. Faraday Trans. II 1870 (1978)
281. Strickler, S. J., Berg, R. A.: J. Chem. Phys. *37*, 814 (1962)
282. Sevchenko, A. N.: Izv. Akad. Nauk. SSSR Ser. Fiz. *26*, 53 (1962)

283. Kosonocky, W. F., Harrison, S. E.: J. Appl. Phys. *37*, 4789 (1966)
284. Gradyushko, A. T., Sevchenko, A. N., Solovyov, K. N., Tsvirko, M. P.: Photochem. Photobiol. *11*, 387 (1970)
285. Seybold, P. G., Gouterman, M.: J. Mol. Spectrosc. *31*, 1 (1969)
286. Robinson, G. W., Frosch, R. P.: J. Chem. Phys. *37*, 1962 (1962) and *38*, 1187 (1963)
287. Bryne, J. P., McCoy, E. F., Ross, I. G.: Aust. J. Chem. *18*, 1589 (1965)
288. Kosonocky, W. F., Harrison, S. E., Stander, R.: J. Chem. Phys. *43*, 831 (1965)
289. Jacques, P., Braun, A.: Helv. Chim. Acta *64*, 1800 (1981)
290. Kraut, E. A., Grant, R. W., Waldrop, J. R., Kowalczyk, S. P.: Phys. Rev. Lett. *44*, 1620 (1980)
291. Maroie, S., Savy, M., Verbist, J. J.: Inorg. Chem. *18*, 2560 (1979)
292. Vilesov, F. I., Zagrubskii, A. A., Garbuzov, D. F.: Fiz. Tverd. Tela *5*, 2000 (1963); Sov. Phys. Sol. State *5*, 1460 (1964)
293. Zeller, M. V., Hayes, R. G.: J. Amer. Chem. Soc. *95*, 3855 (1973)
294. Tatsumi, K., Kasuga, K., Tsutsui, M.: J. Amer. Chem. Soc. *101*, 484 (1979)
295. Niwa, Y., Kobayashi, H., Tsuchiya, T.: J. Chem. Phys. *60*, 799 (1974)
296. Höchst, H., Goldmann, A., Hüfner, S., Malter, H.: Phys. Status Solidi *B76*, 559 (1976)
297. Ouedrago, G. V., Benlian, D., Porte, L.: J. Chem. Phys. *73*, 642 (1980)
298. Koch, E. E., Grobman, W. D.: J. Chem. Phys. *67*, 837 (1977)
299. Battye, F. L., Goldmann, A., Kasper, L.: Phys. Status Solidi *B80*, 42 (1977)
300. Berkowitz, J.: J. Chem. Phys. *70*, 2819 (1979)
301. Iwan, M., Koch, E. E., Chiang, T. C., Eastman, D. E., Himpsel, F.-J.: Solid State Commun. *34*, 57 (1980)
302. Eley, D. D., Hazeldine, D. J., Palmer, T. F.: J. C. S. Faraday Trans. II *12*, 1808 (1973)
303. Pope, M.: J. Chem. Phys. *36*, 2810 (1962)
304. Wolberg, A., Manassen, J.: J. Amer. Chem. Soc. *92*, 2982 (1970)
305. Giraudeau, A., Bard, A. J.: Private Communication
306. Giraudeau, A., Fan, F.-R. F., Bard, A. J.: J. Amer. Chem. Soc. *102*, 5137 (1980)
307. Manasse, J., Bar-Ilan, A.: J. Catal. *17*, 86 (1970)
308. Lever, A. B. P., Wilshire, J. P.: Can. J. Chem. *54*, 2514 (1976)
309. Lever, A. B. P., Wilshire, J. P.: Inorg. Chem. *17*, 1145 (1978)
310. Clack, D. W., Hush, N. S., Woolsey, I. S.: Inorg. Chim. Acta *19*, 129 (1976)
311. Rollman, L. D., Iwamoto, R. I.: J. Amer. Chem. Soc. *90*, 1455 (1968)
312. Taube, R.: Pure Appl. Chem. *38*, 427 (1974)
313. Felton, R. H., Linschitz, H.: J. Amer. Chem. Soc. *88*, 1113 (1966)
314. Clack, D. W., Yandle, J. R.: Inorg. Chem. *11*, 1738 (1972)
315. Dodd, J. W., Hush, N. S.: J. Chem. Soc. 4607 (1964)
316. Day, P., Hill, H. A. O., Price, M. G.: J. Chem. Soc. A 91 (1968)
317. Whitlock, N. W., Bower, B. K.: Tet. letters 4837 (1965)
318. Myers, J. F., Canham, G. W. R., Lever, A. B. P.: Inorg. Chem. *14*, 461 (1975)
319. Taube, R.: Z. Chem. *6*, 8 (1966)
320. Clack, D. W., Hush, N. S., Yandle, J. R.: Chem. Phys. Lett. *1*, 157 (1967)
321. Linder, R. E., Rowlands, J. R., Hush, N. S.: Mol. Phys. *21*, 417 (1971)
322. Guzy, C. M., Raynor, J. B., Stodulski, L. P., Symons, M. C. R.: J. Chem. Soc. A 997 (1969)
323. Canham, G. W. R., Myers, J., Lever, A. B. P.: J. C. S. Chem. Comm. 483 (1973)
324. Ercolani, C., Neri, C.: J. Chem. Soc. A 1715 (1967)
325. Ingram, D. J. E., Bennett, J. E.: J. Chem. Phys. *22*, 1136 (1954)
326. Ingram, D. J. E., Bennett, J. E.: Disc. Faraday Soc. *19*, 140 (1955)
327. Kivelson, D., Neiman, R.: J. Chem. Phys. *35*, 149 (1961)
328. Neiman, R., Kivelson, D.: J. Chem. Phys. *35*, 156 (1961)
329. Assour, J. M., Harrison, S. E.: Phys. Rev. *A136*, 1368 (1964)
330. Klemm, L., Klemm, W.: J. Prakt. Chem. *143*, 82 (1935)
331. Ray, P., Sen, D. N.: J. Indian, Chem. Soc. *25*, 473 (1948)
332. Rollman, L. D., Chan, S. I.: Inorg. Chem. *10*, 1978 (1971)
333. Assour, J. M.: J. Amer. Chem. Soc. *87*, 4701 (1965)
334. Martin, R. L., Mitra, S.: Chem. Phys. Lett. *3*, 183 (1969)
335. Lever, A. B. P.: J. Chem. Soc. 1821 (1965)
336. Barraclough, C. G., Martin, R. L., Mitra, S., Sherwood, R. C.: J. Chem. Phys. *53*, 1638 (1970)

337. Miyoshi, H., Ohya-Nishiguchi, H., Deguchi, Y.: Bull. Chem. Soc. Jpn. *46*, 2724 (1973)
338. George, P., Ingram, D. J. E., Bennett, J. E.: J. Amer. Chem. Soc. *79*, 1870 (1957)
339. Bobrovskii, A. P., Sidorov, A. N.: Zhur-Strukt. Khimii *17*, 63 (1976); J. Struct. Chem. *17*, 50 (1976)
340. Lexa, R., Reix, M.: J. Chimie Physique *71*, 517 (1974)
341. Aoyagi, Y., Masuda, K., Namba, S.: J. Phys. Soc. Jpn. *31*, 524 (1971)
342. Schramm, C. J., Scaringe, R. P., Stojakovic, D. R., Hoffman, B. M., Ibers, J. A., Marks, T. J.: J. Amer. Chem. Soc. *102*, 6702 (1980)
343. Phillips, T. E., Hoffman, B. M., Soos, Z. G.: Solid State Commun *33*, 51 (1980)
344. Phillips, T. E., Scaringe, R. P., Hoffman, B. M., Ibers, J. A.: J. Amer. Chem. Soc. *102*, 3435 (1980)
345. Kholmogrov, V. E.: Optics & Spectroscopy *17*, 155 (1964)
346. Singh, J.: Phys. Status Solidi *B82*, 263 (1977)
347. Sukigara, M., Nelson, R. C.: Mol. Phys. *17*, 387 (1969)
348. Chen, I.: J. Chem. Phys. *51*, 3241 (1969)
349. Mathur, S. C., Singh, J., Singh, D. C.: J. Phys. C Solid State Phys. *4*, 3122 (1971)
350. Devaux, P., Delacote, G.: Chem. Phys. Lett. *2*, 337 (1968); J. Chem. Phys. *52*, 4922 (1970)
351. Rodot, M., Barbé, M., Dixmier, J.: Revue de Phys. App. *12*, 1223 (1977)
352. Mathur, S. C., Ramesh, N.: Chem. Phys. Lett. *37*, 276 (1976)
353. Musser, M. E., Dahlberg, S. C.: Surface Sci. *100*, 605 (1980)
354. Eley, D. D.: Nature *162*, 819 (1948)
355. Eley, D. D., Parfitt, G. D., Perry, M. J., Taysum, D. H.: Trans. Faraday Soc. *49*, 79 (1953)
356. Vartanyan, A. T.: Zh. Fiz. Khim. *22*, 769 (1948)
357. Fielding, P. E., Gutmann, F. J.: J. Chem. Phys. *26*, 411 (1957)
358. Livingston, R., Fujimori, E.: J. Amer. Chem. Soc. *80*, 5610 (1958)
359. Vartanyan, A. T., Karpovich, I. A.: Zh. Fiz. Khim. *32*, 178 and 274 (1958)
360. Cox, G. A., Knight, P. C.: J. Phys. Chem. Solids *34*, 1655 (1973)
361. Usov, N. N., Benderskii, V. A.: Phys. Status Solidi *B37*, 535 (1970)
362. Hamann, C.: Phys. Status Solidi *20*, 481 (1967)
363. Heilmeier, G. H., Harrison, S. E.: Phys. Rev. *132*, 2010 (1963)
364. Heilmeier, G. H., Warfield, G., Harrison, S. E.: Phys. Rev. Lett. *8*, 309 (1962)
365. Delacote, G., Schott, M.: Phys. Status Solidi *2*, 1460 (1962)
366. Cox, G. A., Knight, P. C.: Phys. Status Solidi *B50*, K135 (1972)
367. Cox, G. A., Knight, P. C.: J. Phys. *C7*, 146 (1974)
368. Devaux, P., Quedec, P.: Phys. Lett. *28 A*, 537 (1969)
369. Westgate, C. R., Warfield, G.: J. Chem. Phys. *46*, 94 (1967)
370. Kearns, D. R., Calvin, M.: J. Chem. Phys. *34*, 2022 (1961)
371. Lehmann, G., Hamann, C.: Phys. Status Solidi *B55*, 585 (1973)
372. Munn, R. W., Siebrand, W.: Disc. Farad. Soc. *51*, 17 (1971)
373. Sussman, A.: J. Appl. Phys. *38*, 2738 and 2748 (1967)
374. Delacote, G. M., Fillard, J. P., Marco, F. J.: Solid State Commun. *2*, 373 (1964)
375. Fillard, F.: Thesis, University of Montpellier (France) 1968
376. Hamann, C.: Phys. Status Solidi *26*, 311 (1968)
377. Brunschwig, B. S., Logan, J., Newton, M. D., Sutin, N.: J. Amer. Chem. Soc. *102*, 5798 (1980)
378. Martinsen, J., Pace, L. J., Phillips, T. E., Hoffman, B. M., Ibers, J. A.: J. Amer. Chem. Soc. *104*, 83 (1982)
379. Barbe, D. F., Westgate, C. R.: J. Chem. Phys. *52*, 4046 (1970)
380. Heilmeier, G. H., Warfield, G.: J. Chem. Phys. *38*, 163 (1963)
381. Morel, D. L., Berger, H.: J. Appl. Phys. *46*, 863 (1975)
382. Westgate, C. R., Warfield, G.: J. Chem. Phys. *46*, 537 (1967)
383. Hoshino, Y.: J. Appl. Phys. *52*, 5655 (1981)
384. Bradley, R. S., Grace, J. D., Munro, D. C.: Trans. Faraday Soc. *58*, 776 (1962)
385. Onodera, A., Kawai, N., Kobayashi, T.: Solid State Commun. *17*, 775 (1975)
386. Sawyer, D. T., Gibian, M. J., Morrison, M. M., Seo, E. T.: J. Amer. Chem. Soc. *100*, 627 (1978)
387. Wilshire, J., Sawyer, D. T.: Acc. Chem. Res. *12*, 105 (1979)
388. Rosengerg, B., Camiscoli, J. F.: J. Chem. Phys. *35*, 982 (1961)

389. Ingram, D. J. E., Bennett, J. E.: Phil. Mag. *45*, 545 (1954)
390. Raynor, J. B., Robson, M., Torrens-Burton, A. S. M.: J. Chem. Soc. Dalton 2360 (1977)
391. Boas, J. F., Fielding, P. E., McKay, A. G.: Aust. J. Chem. *27*, 7 (1974)
392. Neiman, R., Kivelson, D.: J. Chem. Phys. *35*, 162 (1961)
393. Assour, J. M., Harrison, S. E.: J. Phys. Chem. *68*, 872 (1964)
394. Winslow, F. H., Baker, W. O., Yager, W. A.: J. Amer. Chem. Soc. *77*, 4751 (1955)
395. Fu Yen, T., Erdman, J. G., Saraceno, A. J.: Anal. Chem. *34*, 694 (1962)
396. Sharoyan, E. G., Tikhomirova, N. N., Blyumenfel's, L. A.: Zhur. Strukt. Khim. *6*, 843 (1965)
397. Yasunaga, H., Kojima, K., Yohda, H., Takeya, K.: J. Phys. Soc. Jpn. *37*, 1024 (1974)
398. Contour, J. P., Lenfant, P., Vijh, A. K.: J. Catal. *29*, 8 (1973)
399. Briegleb, G.: Angew. Chem. Int. Ed. *3*, 617 (1964)
400. Milazzo, G., Caroli, S.: Tables of Standard Electrode Potentials, New York, J. Wiley & Sons 1978
401. Curry, J., Cassidy, E. P.: J. Chem. Phys. *57*, 2154 (1962)
402. Kearns, D., Calvin, M.: J. Amer. Chem. Soc. *83*, 2110 (1961)
403. Balzani, V.: Topics in Current Chemistry, Vol. 75, p. 30, Berlin, Springer Verlag 1978
404. Petersen, J. L., Schramm, C. S., Stojakovic, D. R., Hoffman, B. M., Marks, T. J.: J. Amer. Chem. Soc. *99*, 286 (1977)
405. Orr, W. A., Dahlberg, S. C.: J. Amer. Chem. Soc. *101*, 2875 (1979)
406. Schramm, C. J., Stojakovic, D. R., Hoffman, B. M., Marks, T. J.: Science *200*, 47 (1978)
407. Barzaghi, M., Beringhelli, T., Morazzoni, F.: J. Mol. Cat. *14*, 357 (1982)
408. Etemad, S., Penney, T., Engler, E. M., Scott, B. A., Seiden, P. E.: Phys. Rev. Lett. *34*, 741 (1975)
409. Copper, J. R.: Phys. Rev. *B19*, 2404 (1979)
410. Kuznesof, P. M., Wynne, K. J., Nohr, E. S., Kenney, M. E.: J. C. S. Chem. Comm. 121 (1980)
411. Nohr, R. S., Kuznesof, P. M., Wynne, K. J., Kenney, M. E., Siebenman, P. G.: J. Amer. Chem. Soc. *103*, 4371 (1981)
412. Schoch, Jr., K. F., Marks, T. J.: Unpublished results (Northwestern University 1981)
413. Bourdon, J., Schnuriger, B.: Physics and Chemistry of the Organic Solid State, Vol. 3, Fox, D., Labes, M., Weissberger, A., (ed.), New York, Interscience Pub. 1967
414. Meier, H.: Photochem. Photobiol. *16*, 219 (1972)
415. Kearns, D.: Radiation Res. Sup. *2*, 407 (1956)
416. Meier, H.: Angew. Chem. Int. Ed. *4*, 619 (1965)
417. Uth, H.-J., Wöhrle, D.: Bremer Briefe zur Chemie 5 (1978)
418. Kearns, D. R., Tollin, G., Calvin, M.: J. Chem. Phys. *32*, 1020 (1960)
419. Cheng, Y. C., Loutfy, R. O.: J. Chem. Phys. *73*, 2911 (1980)
420. Hamann, C., Starke, M., Wagner, H.: Phys. Status Solidi *A16*, 463 (1973)
421. Evstigneev, V. B., Gavrilova, V. A.: Dokl. Akad. Nauk. SSSR *92*, 381 (1953)
422. Kleitman, D., Goldsmith, G. J.: Phys. Rev. *98*, 1544 (1955)
423. Putseiko, E., Terenin, A. N.: Dokl. Akad. Nauk. SSSR *90*, 1005 (1953)
424. Liang, C. Y., Scalco, E. G.: J. Electrochem. Soc. *110*, 779 (1963)
425. Bornmann, J. A.: J. Chem. Phys. *27*, 604 (1954)
426. Harrison, S. E.: J. Chem. Phys. *50*, 4739 (1969)
427. Yasunaga, H., Kasai, K., Takeya, K.: J. Phys. Soc. Jpn. *46*, 839 (1979)
428. Day, P., Price, M. G.: J. Chem. Soc. A 236 (1969)
429. Gamo, K., Masuda, K., Yamaguchi, J.: J. Phys. Soc. Jpn. *25*, 431 (1968)
430. Day, P., Williams, R. J. P.: J. Chem. Phys. *42*, 4049 (1965)
431. Bergkamp, M. A., Dalton, J., Netzel, T. L.: J. Amer. Chem. Soc. *104*, 253 (1982)
432. Holten, D., Gouterman, M., Parson, W. W., Windsor, M. W., Rockley, M. G.: Photochem. Photobiol. *23*, 415 (1976)
433. Gouterman, M., Holten, D.: Photochem. Photobiol. *25*, 85 (1977)
434. Seeley, G. R.: Photochem. Photobiol. *27*, 639 (1978)
435. Calvin, M., Cockbain, E. G., Polanyi, M.: Trans. Farad. Soc. *32*, 1436 (1936)
436. Sakaguchi, M., Ohta, M., Nozawa, T., Takada, M.: J. Electrochem. Soc. *127*, 1832 (1980)
437. Day, P., Scregg, G., Williams, R. J. P.: Nature *197*, 589 (1963)
438. Dahlberg, S. C., Musser, M. E.: J. Chem. Phys. *72*, 6706 (1980)
439. Dahlberg, S. C., Musser, M. E.: Surface Sci. *90*, 1 (1979)

440. Popovic, Z. D., Sharp, J. H.: J. Chem. Phys. *66*, 5076 (1977)
441. Menzel, E. R., Popovic, Z. D.: Chem. Phys. Lett. *55*, 177 (1978)
442. Menzel, E. R., Loutfy, R. O.: Chem. Phys. Lett. *72*, 522 (1980)
443. Popovic, Z. D., Menzel, E. R.: J. Chem. Phys. *71*, 5090 (1979)
444. Menzel, E. R., Sharp, J. H.: J. Chem. Phys. *66*, 67 (1977)
445. Putseĭko, E.: Dokl. Akad. Nauk. SSSR *59*, 471 (1948)
446. Calvin, M., Kearns, D. R.: US Patent 3,057,947 CA 2990a (1963)
447. Kearns, D., Calvin, M.: J. Chem. Phys. *29*, 950 (1958)
448. Meier, H.: Umschau. Wiss. Tech. *66*, 438 (1966) CA *63*, 14202g (1966)
449. Morgan, D. V., Frey, J.: J. Appl. Phys. *52*, 5702 (1981)
450. Freeouf, J. L., Woodall, J. M.: Appl. Phys. Lett. *39*, 727 (1981)
451. Haak, F. A., Nolta, J. P.: J. Chem. Phys. *38*, 2648 (1963)
452. Martin, M., André, J.-J., Simon, J.: J. Appl. Phys. *54*, 2792 (1983)
453. Michaelson, H. B.: J. Appl. Phys. *48*, 4730 (1977)
454. Hall, K. J., Bonham, J. S., Lyons, L. E.: Aust. J. Chem. *31*, 1661 (1978)
455. Ashwell, G. J., Bonham, J. S., Lyons, L. E.: Aust. J. Chem. *33*, 1619 (1980)
456. Fedorov, M. I., Benderskii, V. A.: Sov. Phys. Semicond. *4*, 1198 and 1720 (1971)
457. Ghosh, A. K., Morel, D. L., Feng, T., Shaw, R. F., Rowe Jr., C. A.: J. Appl. Phys. *45*, 230 (1974)
458. Barkhalov, B. Sh., Vidadi, Yu. A.: Thin Solid Films *40*, L5 (1977)
459. Tantzscher, C., Hamann, C.: Phys. Status Solidi *A 26*, 443 (1974)
460. Hamann, C., Wagner, H.: Kristall und Technik *6*, 307 (1971)
461. Chamberlain, G. A., Cooney, P. J.: Chem. Phys. Lett. *66*, 88 (1979)
462. Flannery, W. E., Pollack, S. R.: J. Appl. Phys. *37*, 4417 (1966)
463. Simmons, J. G.: J. Appl. Phys. *34*, 2581 (1963)
464. Fisher, J. C., Giaver, I.: J. Appl. Phys. *32*, 172 (1961)
465. Frenkel, J.: Phys. Rev. *54*, 647 (1938)
466. Loutfy, R. O., Sharp, J. H.: J. Chem. Phys. *71*, 1211 (1979)
467. Fan, F.-R., Faulkner, L. R.: J. Chem. Phys. *69*, 3334 (1978)
468. Loutfy, R. O., Sharp, J. H., Hsiao, C. K., Ho, R.: J. Appl. Phys. *52*, 5218 (1981)
469. Katz, W., Evans Jr., C. A., Eaton, D. R., Faulkner, L. R.: J. Vac. Sci. Technol. *15*, 1561 (1978)
470. Twarowski, A. J., Albrecht, A. C.: J. Chem. Phys. *72*, 1797 (1980)
471. Shing, Y. H., Loutfy, R. O.: J. Appl. Phys. *52*, 6961 (1981)
472. Vidadi, Yu. A., Kocharli, K. Sh., Barkhalov, B. Sh., Sadreddinov, S. A.: Phys. Status Solidi *A 34*, K77 (1976)
473. Yasunaga, H., Shintaku, H.: J. Appl. Phys. *51*, 2149 (1980)
474. Bigorgne, J. P., Jourdan, M., Polignac, A., De Ropars, F., Despujols, J.: Thin Solid Films *23*, 239 (1974)
475. Henish, H. F.: "Rectifying Semiconductor Contacts" New York, Oxford University Press 1957
476. Popovic, Z.: Appl. Phys. Lett. *34*, 694 (1979)
477. Popovic, Z.: J. Appl. Phys. *52*, 4871 (1981)
478. Isett, L. C.: Appl. Phys. Lett. *36*, 376 (1980)
479. Isett, L. C.: J. Appl. Phys. *52*, 4873 (1981)
480. Ghosh, A. K., Feng, T.: J. Appl. Phys. *44*, 2871 (1973)
481. Kotani, M., Akamatu, H.: Disc. Faraday Soc. *51*, 94 (1971)
482. Bube, R. H.: Photoconductivity of Solids, New York, J. Wiley & Son, 1960
483. Görlich, P.: Advances in Electronics and Electron Physics, Marton, L., (ed.), Vol. 14, New York, Academic Press 1961
484. Mort, J., Pai, D. M.: Photoconductivity and Related Phenomena, Amsterdam, Elsevier Pub. Comp. 1976
485. Reucroft, P. J., Takahashi, K., Ullal, H.: Appl. Phys. Lett. *25*, 664 (1974)
486. Reucroft, P. J., Takahashi, K., Ullal, H.: J. Appl. Phys. *46*, 5218 (1975)
487. Reucroft, P. J., Ullal, H.: Solar Energy Mat. *2*, 217 (1979/80)
488. Rose, A.: RCA Rev. *12*, 362 (1951)
489. Heilmeier, G. H., Harrison, S. E.: J. Appl. Phys. *34*, 2732 (1963)

490. Fan, F.-R., Faulkner, L. R.: J. Chem. Phys. *69*, 3341 (1978)
491. Terman, L. M.: Solid State Electron *2*, 1 (1961)
492. Hackett, C. F.: J. Chem. Phys. *55*, 3178 (1971)
493. Komissarov, G. G., Shumov, Yu. S., Borisevich, Yu. E.: Dokl. Akad. Nauk. SSSR 187, 670 (1969)
494. Fedorov, M. E., Benderskii, V. A.: Fiz. Tekh. Poluprov. 4, 1403 (1970); CA *73*, 114, 311 q.
495. Tang, C. W.: Research Disclosure *162*, 71 (1977)
496. Fedorov, M. I., Zinov'eva, E. P., Shorin, V. A., Nutrikhina, L. I.: Izv. Vyssh. Uch. Zav. Fiz. *20*, 159 (1977); CA *86*, 149, 485 v.
497. Fedorov, M. I., Zinov'eva, E. P., Shashaurov, V. N., Nutikhina, L. I.: Izv. Vyssh. Uch. Zav. Fiz. *20*, 157 (1977); CA *87*, 110, 291 z.
498. Stepanova, G. A., Volkova, L. S., Gainullina, M. A., Yumakulova, F. F.: Zh. Fiz. Khim. *51*, 1771 (1977)
499. Popovic, Z. D., Zamin, J.: Chem. Phys. Lett. *80*, 135 (1981)
500. Popovic, Z. D., Loutfy, R. O.: J. Appl. Phys. *52*, 6190 (1981)
501. Martin, M., André, J.-J., Simon, J.: Nouv. J. Chim. *5*, 485 (1981)
502. Benderskii, V. A., Al'yanov, M. I., Fedorov, M. I., Fedorov, L. M.: Dokl. Akad. Nauk. SSSR *239*, 856 (1978)
503. Ilatovshii, V. A., Komissarov, G. G.: Zh. Fiz. Khim. *49*, 1352 and 1353 (1975)
504. Kampas, F. J., Yamashita, K., Fajer, J.: Nature *284*, 40 (1980)
505. Fabian, J., Hartmann, H.: Light Absorption of Organic Colorants, Berlin, Springer Verlag 1980
506. Solymar, L., Walsh, D.: Lectures on the Electrical Properties of Materials, Oxford University Press 1975²
507. Sayer, P., Gouterman, M., Connell, C. R.: Acc. Chem. Res. *15*, 73 (1982)
508. Sugimoto, H., Higashi, T., Mori, M.: Chem. Lett. 801 (1982)
509. Collins, K. E., Jardin, I. C. S. F., Collins, C. H.: Radiochem. Radioanal. Lett. *52*, 193 (1982)
510. Achar, B. N., Fohlen, G. M., Parker, J. A.: J. Polym. Sci. Polym. Chem. Ed. *20*, 1785 (1982)
511. Kobayashi, T., Fujiyoshi, Y., Iwatsu, F., Uyeda, N.: Acta Crystallogr. Sect. *A* 37, 692 (1981)
512. Leicknam, J. P., Anitoff, O. E., Gallice, M. J., Henry, M., Tayeb, A. E. K.: J. de Chimie Physique *78*, 587 (1981)
513. Przyborowski, F., Hamann, C., Müller, M., Reinhardt, C., Starke, M., Vollmann, W.: Wiss. Z. Techn. Hochschule Karl-Marx-Stadt *7*, 709 (1980)
514. Przyborowski, F., Hamann, C.: Cryst. Res. and Technol. *17*, 1041 (1982)
515. Figgis, B. N., Williams, G. A., Forsyth, J. B., Mason, R.: J. C. S. Dalton 1837 (1981)
516. Davidson, A. T.: J. Chem. Phys. *77*, 168 (1982)
517. Lever, A. B. P., Minor, P. C.: Adv. Mol. Rel. Inter. Process. *18*, 115 (1980)
518. Lever, A. B. P., Minor, P. C.: Inorg. Chem. *20*, 4015 (1981)
519. Lever, A. B. P., Licoccia, S., Magnell, K., Minor, P. C., Ramaswany, B. S.: A. C. S. Symposium series 1 (1982)
520. Kobayashi, T., Fujiyoshi, Y., Uyeda, N.: Acta Crystallogr. Sect. *A* 38, 356 (1982)
521. Abdel-Malik, T. G., Abdeen, A. M., Aly, A. A.: Phys. Status Solidi *A* 70, 703 (1982)
522. Harbour, J. R., Loutfy, R. O.: J. Phys. Chem. Solids *43*, 513 (1982)
523. Mizugushi, J.: Jpn. J. Appl. Phys. *20*, 1855, 2065 and 2073 (1981)
524. Twarowski, A. J.: J. Chem. Phys. *76*, 2640 (1982)
525. Popovic, Z. D.: J. Chem. Phys. *76*, 2714 (1982)
526. Pananakakis, G., Viktorovitch, P., Ponpon, J. P.: Revue de Phys. Appl. *13*, 449 (1978)
527. Ponpon, J. P., Grob, J. J., Grob, A., Stuck, R., Siffert, P.: Nucl. Instrum. Methods *149*, 647 (1978)
528. Ponpon, J. P., Siffert, P.: J. Appl. Phys. *50*, 5050 (1979)
529. Twarowski, A.: J. Chem. Phys. *77*, 5840 (1982)
530. Champetier, G., Job, A.: C. R. Acad. Sci. *189*, 1089 (1929)
531. Champetier, G., Job, A.: Bull. Soc. Chim. *47*, 279 (1930)
532. Champetier, G., Martynoff, M.: Bull. Soc. Chim. 2083 (1961); C. R. Acad. Sci. *252*, 633 (1961)
533. Natta, G., Mazzanti, G., Corradini, P.: Atti. Acad. Nazl. Lincei, Rend. Classe Sci. Fis. Mat. Nat. *25*, 3 (1958)

534. Natta, G., Mazzanti, G., Pino, P.: Angew. Chem. *69*, 685 (1957)
535. Luttinger, L. B.: Chem. and Ind. 1135 (1960); J. Org. Chem. *27*, 1591 (1962)
536. Luttinger, L. B., Colthup, E. C.: J. Org. Chem. *27*, 3752 (1962)
537. Green, M. L. H., Nehme, N., Wilkonson, G.: Chem. and Ind. 1136 (1960)
538. Hatano, M., Kambara, S., Okamato, S.: J. Polym. Sci. *51*, S26 (1961)
539. Watson, W. H., McMordie, W. C., Lands, L. G.: J. Polym. Sci. *55*, 137 (1961)
540. Pople, J. A., Walmsley, S. H.: Mol. Phys. *5*, 15 (1962)
541. Topchiev, A. Y., Heyderich, A. V., Davydov, B. E., Kargin, V. A., Krentsel, B. A., Kustanovich, I. M., Polak, L. S.: Dokl. Akad. Nauk. SSSR *128*, 312 (1959)
542. Nechtschein, M.: J. Polym. Sci. Part C *4*, 1367 (1964)
543. Holob, G. M., Ehrlich, P., Allendoerfer, R. D.: Macromolecules *5*, 569 (1972)
544. Berlin, A. A.: Polymer Sci. USSR *13*, 2727 (1971)
545. Berlin, A. A., Vinogradov, G. A., Ovchinnikov, A. A.: Int. J. Quantum Chem. *6*, 263 (1972)
546. Ovchinnikov, A. A., Ukrainskii, I. I., Krentsel, G. V.: Sov. Phys. Uspekhi *15*, 575 (1973); Usp. Fiz. Nauk. *108*, 81 (1972)
547. Shirakawa, H., Ikeda, S.: Polym. J. *2*, 231 (1971)
548. Shirakawa, H., Ito, T., Ikeda, S.: Polym. J. *4*, 460 (1973)
549. Ito, T., Shirakawa, H., Ikeda, S.: J. Polym. Sci. Polym. Chem. Ed. *12*, 11 (1974); *13*, 1943 (1975)
550. Shirakawa, H., Louis, E. J., McDiarmid, A. G., Chiang, C. K., Heeger, A. J.: J. Chem. Soc. Chem. Comm. 578 (1977)
551. Chiang, C. K., Fincher, C. R. Jr., Park, Y. W., Heeger, A. J., Shirakawa, H., Louis, E. J., Gau, S. C., McDiarmid, A. G.: Phys. Rev. Lett. *39*, 1098 (1977)
552. Chiang, C. K., Druy, M. A., Gau, S. C., Heeger, A. J., Louis, E. J., McDiarmid, A. G., Park, Y. W., Shirakawa, H.: J. Amer. Chem. Soc. *100*, 1013 (1978)
553. Enkelmann, V., Muller, W., Wegner, G.: Synth. Metals *1*, 185 (1979/80)
554. Daniels, W. E.: J. Org. Chem. *29*, 2936 (1964)
555. Merriwether, L. S.: J. Org. Chem. *26*, 5163 (1961)
556. Nicolescu, I. V., Angelescu, E. M.: J. Polym. Sci. A1 *4*, 2963 (1966)
557. Angelescu, E., Nicolescu, I. V.: J. Polym. Sci. C, *22*, 203 (1968)
558. Aldissi, M., Linaya, C., Sledz, J., Schué, F., Giral, L., Fabre, J. M., Rolland, M.: Polymer *23*, 243 (1980)
559. Hsu, S. L., Signorelli, A. J., Pez, G. P., Baughman, R. H.: J. Chem. Phys. *69*, 106 (1978)
560. Aldissi, M., Schué, F., Giral, L., Rolland, M.: Polymer *23*, 246 (1982)
561. Nicolescu, I. V., Angelescu, E. M.: J. Polym. Sci. *A 3*, 1227 (1965)
562. Chien, J. C. W., Karasz, F. E., Wnek, G. E., McDiarmid, A. G., Heeger, A. J.: J. Polymer Sci. Polym. Lett. Ed. *18*, 45 (1980)
563. Ciardelli, F., Benedetti, E., Pieroni, O.: Makromol. Chemie *103*, 1 (1967)
564. Noguchi, H., Kambara, S.: J. Polym. Sci. Polym. Lett. Ed. *1*, 553 (1963)
565. Voronkov, M. G., Pukhnarevich, V. B., Sushchinskaya, S. P., Annenkova, V. Z., Annenkova, V. M., Andreeva, N. J.: J. Polym. Sci. Polym. Chem. Ed. *18*, 53 (1980)
566. Woon, P. S., Farona, M. F.: J. Polym. Sci. Polym. Chem. Ed. *12*, 1749 (1974)
567. Lutz, E. F.: J. Amer. Chem. Soc. *83*, 2551 (1961)
568. Ikeda, S., Tamaki, A.: Polym. Lett. *4*, 605 (1966)
569. Reikhsfeld, V. O., Makovetskii, K. L.: Dokl. Akad. Nauk. SSSR *155*, 414 (1964)
570. Schrauzer, G. N., Glockner, P., Eichler, S.: Angew. Chem. *76*, 28 (1964)
571. Teyssié, Ph., Dawans, F.: Theory of Coordination Catalysts in the Stereo Rubbers, Saltman, W. M. (ed.), New York, J. Wiley & Sons Inc. 1977
572. Lieser, G., Wegner, G., Müller, W., Enkelmann, V.: Makro. Chem. Rapid Comm. *1*, 621 (1980)
573. Wegner, G.: Angew. Chem. Int. Ed. *20*, 361 (1981)
574. Chien, J. C. W., Yamashita, Y., Hirsch, J. A., Fan, J. L., Schen, M. A., Karasz, F. E.: Nature *299*, 608 (1982)
575. Cotton, F. A., Wilkinson, G.: Advanced Inorganic Chemistry New York, Interscience Pub. 1972
576. Takeda, M., Iimura, K., Nozawa, Y., Hisatome, M., Koide, N.: J. Polym. Sci. C *23*, 741 (1968)
577. Hirai, H., Hiraki, K., Noguchi, I., Makishima, S.: J. Polym. Sci. A1 *8*, 147 (1970)

578. Hiraki, K., Kaneko, S., Hirai, H.: J. Polym. Sci. Polym. Lett. Ed. *10*, 199 (1972)
579. Bailey, G. C.: Catalyst Rev. *3*, 37 (1970)
580. Maricq, M. M., Waugh, J. S., McDiarmid, A. G., Shirakawa, H., Heeger, A. J.: J. Amer. Chem. Soc. *100*, 7729 (1978)
581. Kambara, S., Hatano, M., Hosoe, T.: Koggyo Kagaku, Zasshi *65*, 720 (1962)
582. Hatano, M.: Koggyo Kagaku Zasshi *65*, 723 (1962)
583. Shirakawa, H., Ikeda, S.: Synth. Metals *1*, 175 (1979/80)
584. Hatano, M., Shimamura, K., Ikeda, S., Kambara, S.: Rep. Prog. Polym. Phys. Japan *11*, 123 (1968)
585. Karasz, F. E., Chien, J. C. W., Galkiewicz, R., Wnek, G. E., Heeger, A. J., McDiarmid, A. G.: Nature *282*, 286 (1979)
586. McDiarmid, A. G., Heeger, A. J.: Synth. Metals *1*, 101 (1979/80)
587. Edwards, J. H., Feast, W. J.: Polymer *21*, 595 (1980)
588. Marvel, C. S., Sample, J. H., Roy, M. F.: J. Amer. Chem. Soc. *61*, 3241 (1939)
589. Roth, J.-P., Rempp, P., Parrod, J.: J. Polym. Sci. C *4*, 1347 (1964)
590. Berlin, A. A., Tcherkashin, M. I., Selskaya, O. G., Limanov, V. E.: Vysokomol. Soedin. *1*, 1817 (1959)
591. Korshak, V. V., Polyakova, A. M., Suchkova, D. M.: Vysokomol. Soedin *2*, 1246 (1960)
592. Chantarovitch, P. S., Chliapnikova, I. A.: Vysokomol. Soedin *3*, 363 and 1495 (1961)
593. Barkalov, I. M., Berlin, A. A., Goldanski, V. I., Dzantiev, B. G.: Vysokomol. Soedin *2*, 1103 (1960)
594. Tabata, Y., Saito, B., Shibano, H., Sobue, H., Oshima, K.: Makromol. Chem. *76*, 89 (1964)
595. Wnek, G. E., Chien, J. C. W., Karasz, F. E., Druy, M. A., Park, Y. W., McDiarmid, A. G., Heeger, A. J.: J. Polym. Sci. Polym. Lett. Ed. *17*, 779 (1979)
596. Baughman, R. H., Hsu, S. L., Pez, G. P., Signorelli, A. J.: J. Chem. Phys. *68*, 5405 (1978)
597. Ito, T., Shirakawa, H., Ikeda, S.: Kobunshi Ronbunshu *33*, 339 (1976); english ed. *5*, 470 (1976); CA *85*, 78449c (1976)
598. Ruland, W.: Acta Crystallogr. *14*, 1180 (1961)
599. Akaishi, T., Miyasaka, K., Ishikawa, K., Shirakawa, H., Ikeda, S.: J. Polym. Sci. Polym. Phys. Ed. *18*, 745 (1980)
600. Fincher, Jr., C. R., Moses, D., Heeger, A. J., McDiarmid, A. G.: Synth. Metals *6*, 243 (1983)
601. Shimamura, K., Karasz, F. E., Hirsch, J. A., Chien, J. C. W.: Makromol. Chem. Rap. Comm. *2*, 473 (1981)
602. Meyer, W. H.: Synth. Metals *4*, 81 (1981)
603. Lieser, G., Wegner, G., Müller, W., Enkelmann, V., Meyer, W. H.: Makromol. Chem. Rap. Comm. *1*, 627 (1980)
604. Druy, M. A., Tsang, C. H., Brown, N., Heeger, A. J., McDiarmid, A. G.: J. Polym. Sci. Polym. Phys. Ed. *18*, 429 (1980)
605. Shirakawa, H., Sato, M., Hamano, A., Kawakami, S., Soga, K., Ikeda, S.: Macromolecules *13*, 457 (1980)
606. Wnek, G. W., Capistran, J., Chien, J. C. W., Dickinson, L. C., Gable, R., Gooding, R., Gourley, K., Karasz, F. E., Lillya, C. P., Yao, K. D.: Polym. Sci. Technol. *15*, 183 (1981); CA *95*, 204461x (1981)
607. Chien, J. C. W., Capistran, J., Karasz, F. E., Schen, M., Fan, J.-L.: Las Vegas Meeting of the Amer. Chem. Soc. *23*, 76 (1982)
608. Bernier, P., Schué, F., Sledz, J., Rolland, M., Giral, L.: Chem. Scripta *17*, 151 (1981)
609. Bernier, P., Linaya, C., Rolland, M., Aldissi, M.: J. Physique Lett. *42*, L295 (1981)
610. Lefrant, S., Rzepka, E., Bernier, P., Rolland, M., Aldissi, M.: Polymer *21*, 1235 (1980)
611. Yen, S. P. S., Somoano, R., Khanna, S. K., Rembaum, A.: Solid State Comm. *36*, 339 (1980)
612. Yoshino, K., Sakai, T., Yamamoto, Y., Inuishi, Y.: Jpn. J. Appl. Phys. *20*, 867 (1981)
613. Deits, W., Cukor, P., Rubner, M., Jopson, H.: Synth. Metals *4*, 199 (1982)
614. Pochan, J. M., Pochan, D. F., Rommelmann, H., Gibson, H. W.: Macromolecules *14*, 110 (1981)
615. Naarmann, H.: Ecole d'hiver de Font Romeu, CNRS, Montpellier (France), 1982
616. Kanicki, J.: Meeting "Polymères électroactifs" RCP 457, Autrans 9–11 June 1982
617. Berets, D. J., Smith, D. S.: Trans. Farad. Soc. *64*, 823 (1968)
618. Snow, A., Brant, P., Weber, D., Yang, N. L.: J. Polym. Sci. Polym. Lett. Ed. *17*, 263 (1979)

619. Nechtschein, M., Devreux, F., Genoud, F., Guglielmi, M., Holczer, K.: Phys. Rev. B *27*, 61 (1983)
620. Bernier, P., Rolland, M., Linaya, C., Disi, M.: Polymer *21*, 7 (1980)
621. Meurer, B.: private communication
622. Pochan, J. M., Gibson, H. W., Bailey, F. C.: J. Polym. Sci. Polym. Lett. Ed. *18*, 447 (1980)
623. Pochan, J. M., Gibson, H. W., Bailey, F. C., Pochan, D. F.: Polymer Commun. *21*, 250 (1980)
624. Pochan, J. M., Hinman, D. F.: J. Polym. Sci. Polym. Phys. Ed. *14*, 1871 (1976)
625. Pochan, J. M., Hinman, D. F., Nash, R.: J. Appl. Phys. *46*, 4115 (1975)
626. Tsuji, K., Seiki, T.: J. Polym. Sci. A1 *9*, 3063 (1971); J. Polym. Sci. B *8*, 817 (1970)
627. François, B., Bernard, M.. André, J.-J.: J. Chem. Phys. *75*, 4142 (1981)
628. Devreux, F., Döry, I., Mihaly, L., Pekker, S., Janossy, A., Kertesz, M.: J. Polym. Sci. Polym. Phys. Ed. *19*, 743 (1981)
629. Karpfen, A., Höller, R.: Solid. State Commun. *37*, 179 (1981)
630. Haberkorn, H., Naarmann, H., Penzien, K., Schlag, J., Simak, P.: Synth. Metals *5*, 51 (1982)
631. Suhai, S.: J. Chem. Phys. *73*, 3843 (1980)
632. Brédas, J. L.: Ph. D. Thesis, Namur, Belgium 1979
633. Brédas, J. L., André, J. M., Delhalle, J.: J. Mol. Struct. *87*, 237 (1982)
634. Yamabe, T., Tanaka, K., Terama-e, H., Fukui, K., Imamura, A., Shirakawa, H., Ikeda, S.: Sol. State Commun. *29*, 329 (1979); J. Phys. C *12*, L257 (1979)
635. Karpfen, A., Petkov, J.: Solid State Commun. *29*, 251 (1979); Theor. Chim. Acta *53*, 65 (1979)
636. Carreira, L. A.: J. Chem. Phys. *62*, 3851 (1975)
637. Salem, L.: "The Molecular Orbital Theory of Conjugated Systems", New York, W. A. Benjamin Inc., 1966
638. Longuet-Higgins, H. C., Salem, L.: Proc. Roy. Soc. A *251*, 172 (1959)
639. Fincher, Jr., C. R., Chen, C.-E., Heeger, A. J., McDiarmid, A. G., Hastings, J. B.: Phys. Rev. Lett. *48*, 100 (1982)
640. Cram, D. J., Hammond, G. S.: "Organic Chemistry", 2nd ed., New York, Mc Graw Hill, 1964
641. Shirakawa, H., Ito, T., Ikeda, S.: Makromol. Chem. *179*, 1565 (1978)
642. Montaner, A., Galtier, M., Benoit, C., Aldissi, M.: Solid State Commun. *39*, 99 (1981)
643. Fincher, J., C. R., Ozaki, M., Tanaka, M., Peebles, D. L., Lauchlan, L., Heeger, A. J., McDiarmid, A. G.: Phys. Rev. B *20*, 1589 (1979)
644. Harada, I., Tasumi, M., Shirakawa, H., Ikeda, S.: Chem. Lett. of Chem. Soc. Jpn. 1411 (1978)
645. Lichtmann, L. S., Sarhangi, A., Fitchen, D. B.: Solid State Commun. *36*, 869 (1980)
646. Lichtmann, L. S., Fitchen, D. B., Temkin, H.: Synth. Metals *1*, 139 (1979/80)
647. Kuzmany, H.: Phys. Status Solidi B *97*, 521 (1980)
648. Lefrant, S., Lichtmann, L. S., Temkin, H., Fitchen, D. B., Miller, D. C., Whitwell II, G. E., Burlitch, J. M.: Solid State Commun. *29*, 191 (1979)
649. Vancsó, G., Pekker, S., Egyed, O., Jánossy, A.: to be published
650. Kleist, F. D., Byrd, N. R.: J. Polym. Sci. A1 *7*, 3419 (1969)
651. Mathis, C., François, B.: Ecole d'hiver de Font Romeu, CNRS. Montpellier 1982
652. Chien, J. C. W., Karasz, F. E., Wnek, G. E.: Nature *285*, 390 (1980)
653. Chien, J. C. W., Karasz, F. E., Shimamura, K.: J. Polym. Sci. Polym. Lett. Ed. *20*, 97 (1982)
654. Baughman, R. H., Hsu, S. L.: J. Polym. Sci. Polym. Lett. Ed. *17*, 185 (1979)
655. Meurer, B., Spegt, P., Weill, G., Mathis, C., François, B.: Solid State Commun. *44*, 201 (1982)
656. Mihály, L., Pekker, S., Jánossy, A.: Synth. Metals *1*, 349 (1979/80)
657. Baughman, R. H., Hsu, S. L., Anderson, L. R., Pez, G. P., Signorelli, A. J.: "Molecular Metals", Hatfield, W. E., (ed.), New York, Plenum Press 1970
658. Davenas, J.: personal communication
659. Kuhn, H.: Helv. Chim. Acta *31*, 1441 (1948); J. Chem. Phys. *17*, 1198 (1949)
660. Kertész, M., Koller, J., Ažman, A.: J. Chem. Phys. *67*, 1180 (1977); Phys. Rev. B *19*, 2034 (1979)
661. Horsch, P.: Phys. Rev. B *24*, 7351 (1981)
662. Kertész, M.: Chem. Phys. *44*, 349 (1979)
663. Grant, P. M., Batra, I. P.: Solid State Commun. *29*, 225 (1979)

664. Bredas, J. L., Chance, R. R., Baughmann, R. H., Silbey, R.: J. Chem. Phys. *76*, 3673 (1982)
665. Szabó, A., Langlet, J., Malrieu, J. P.: Chem. Phys. *13*, 173 (1976)
666. Duke, C. B., Paton, A., Salaneck, W. R., Thomas, H. R., Plummer, E. W. Heeger, A. J., McDiarmid, A. G.: Chem. Phys. Lett. *59*, 146 (1978)
667. Peierls, R. E.: "Quantum Theory of Solids" Oxford, Clarendon Press 1955, p. 108
668. Su, W. P., Schrieffer, J. R., Heeger, A. J.: Phys. Rev. B *22*, 2099 (1980)
669. Bredas, J. L., Chance, R. R., Silbey, R., Nicolas, G., Durand, Ph.: J. Chem. Phys. *75*, 255 (1981)
670. André, J.-M., Lerov, G.: Int. J. Quantum Chem. *5*, 557 (1971)
671. Karpfen, A., Petkov, J.: Theor. Chim. Acta (Berlin) *53*, 65 (1979)
672. Young, V.: Solid State Commun. *35*, 715 (1980)
673. Fincher, Jr., C. R., Peebles, D. L., Heeger, A. J., Druy, M. A., Matsumura, Y., McDiarmid, A. G., Shirakawa, H., Ikeda, S.: Solid State Commun. *27*, 489 (1978)
674. Tanaka, M., Watanabe, A., Tanaka, J.: Bull. Chem. Soc. Jpn. *53*, 3430 (1980)
675. Park, Y.-W., Heeger, A. J., Druy, M. A., McDiarmid, A. G.: J. Chem. Phys. *73*, 946 (1980)
676. Salanek, W. R., Thomas, H. R., Duke, C. B., Paton, A., Plummer, E. W., Heeger, A. J., McDiarmid, A. G.: J. Chem. Phys. *71*, 2044 (1979)
677. Tani, T., Grant, P. M., Gill, W. D., Street, G. B., Clarke, T. C.: Solid State Commun. *33*, 499 (1980)
678. Young, V., Suck, S. H., Hellmuth, E. W.: J. Appl. Phys. *50*, 6088 (1979)
679. Yamabe, T., Akagi, K., Matsui, T., Fukui, K., Shirakawa, H.: J. Phys. Chem. *86*, 2365 (1982)
680. Chiang, C. K., Fincher, Jr., C. R., Park, Y. W., Heeger, A. J., Shirakawa, H., Louis, E. J., Gan, S. C., McDiarmid, A. G.: Phys. Rev. Lett. *39*, 1098 (1977)
681. Mihály, G., Vancsó, G., Pekker, S., Janóssy, A.: Synth. Metals *1*, 357 (1979/80)
682. Grant, P. M., Krounbi, M.: Solid State Commun. *36*, 291 (1980)
683. Feldblum, A., Park, Y. W., Heeger, A. J., McDiarmid, A. G., Wnek, G., Karasz, F., Chien, J. C. W.: J. Polym. Sci., Polym. Phys. Ed. *19*, 173 (1981)
684. Bychkov, Y. A., Gorkov, L. P., Dzyaloshinskii, I. E.: Sov. Phys. JETP *23*, 489 (1966)
685. Lauchlan, L., Etemad, S., Chung, T.-C., Heeger, A. J., McDiarmid, A. G.: Phys. Rev. B *24*, 3701 (1981)
686. Su, W. P.: Solid State Commun. *35*, 899 (1980)
687. Su, W. P., Schrieffer, J. R., Heeger, A. J.: Phys. Rev. Lett. *42*, 1698 (1979)
688. Rice, M. J.: Phys. Lett. *71* A, 152 (1979)
689. Bishop, A. R. and Schneider, T. in: Solitons and Condensed Matter Physics. Springer Verlag Berlin 1978
690. Scott, A. C., Chu, F. Y. F., Mc Laughlin, D. W.: Proc. IEEE *61*, 1443 (1973)
691. Rice, M. J., Mele, E. J.: Solid State Commun. *35*, 487 (1980)
692. Takayama, H., Lin-Liu, Y. R., Maki, K.: Phys. Rev. B *21*, 2388 (1980)
693. Allen, W. N., Brant, P., Carosella, C. A., Decorpo, J. J., Ewing, C. T., Saalfeld, F. E., Weber, D. C.: Synth. Metals *1*, 151 (1979/80)
694. Feldblum, A., Kaufman, J. H., Etemad, S., Heeger, A. J., Chung, T.-C., McDiarmid, A. G.: Phys. Rev. B *26*, 815 (1982)
695. McInnes, Jr., D., Druy, M. A., Nigrey, P. J., Nairns, D. P., McDiarmid, A. G., Heeger, A. J.: J. Chem. Soc. Chem. Commun. 317 (1981)
696. Clarke, T. C., Krounbi, M. T., Lee, V. Y., Street, G. B.: J. Chem. Soc. Chem. Commun. 384 (1981)
697. Weber, D. C., Brant, P., Carosella, C., Banks, L. G.: J. Chem. Soc. Chem. Commun. 522 (1981)
698. Clarke, T. C., Street, G. B.: Synth. Metals *1*, 119 (1979/80)
699. Selig, H., Pron, A., Druy, M. A., McDiarmid, A. G., Heeger, A. J.: J. Chem. Soc. Chem. Commun. 1288 (1981)
700. Clarke, T. C., Geiss, R. H., Gill, W. D., Grant, P. M., Morawitz, H., Street, G. B.: Synth. Metals *1*, 21 (1979/80)
701. Allen, W. N., Decorpo, J. J., Saalfeld, F. E., Wyatt, J. R., Weber, D. C.: Synth. Metals *1*, 371 (1979/80)
702. Kaindl, G., Wortmann, G., Roth, S., Menke, K.: Solid State Commun. *41*, 75 (1982)

703. Salaneck, W. R., Thomas, H. R., Bigelow, R. W., Duke, C. B., Plummer, E. W., Heeger, A. J., McDiarmid, A. G.: J. Chem. Phys. 72, 3674 (1980)
704. Mathis, C., François, B.: Synth. Metals 9, 347 (1984)
705. Thomas, H. R., Salaneck, W. R., Duke, C. B., Plummer, E. W., Heeger, A. J., McDiarmid, A. G.: Polymer 21, 1328 (1980)
706. Epstein, A. J., Rommelmann, H., Druy, M. A., Heeger, A. J., McDiarmid, A. G.: Solid State Commun. 38, 683 (1981)
707. Ritsko, J. J.: Phys. Rev. Lett. 46, 849 (1981)
708. Fincher, Jr., C. R., Peebles, D. L., Heeger, A. J., Druy, M. A., Matsumura, Y., McDiarmid, A. G., Shirakawa, H., Ikeda, S.: Solid State Commun. 27, 489 (1978)
709. Moses, D., Denestein, A., Chen, J., Heeger, A. J., McAndrew, P., Woerner, T., McDiarmid, A. G., Park, Y. W.: Phys. Rev. B 25, 7652 (1982)
710. Jánossy, A., Pogány, L., Pekker, S., Swietlik, R.: Hung. Acad. of Science HFKI 29 (1981)
711. Kiess, H., Meyer, W., Baeriswyl, D., Harbeke, G.: J. Electron. Mat. 9, 763 (1980)
712. Ikemoto, I., Sakairi, M., Tsutsumi, T., Kuroda, H., Harada, I., Tasumi, M., Shirakawa, H., Ikeda, S.: Chem. Lett. 1189 (1979)
713. Seeger, K., Gill, W. D., Clarke, T. C., Street, G. B.: Solid State Commun. 28, 873 (1978)
714. Inoue, T., Osterholm, J. E., Yasuda, H. K., Levenson, L. L.: Appl. Phys. Lett. 36, 101 (1980)
715. Chien, J. C. W., Karasz, F. E., Shimamura, K.: Macromolecules 15, 1012 (1982)
716. François, B., Mermilliod, N., Zuppiroli, L.: Synth. Metals 4, 131 (1981)
717. Baughman, R. H., Brédas, J. L., Chance, R. R., Eckhardt, H., Elsenbaumer, R. L., Ivory, D. M., Miller, G. G., Preiziosi, A. F., Shacklette, L. W.: "Macromolecular Metals and Semi-conductors" in "Conducting Polymers", R. B. Seymours (ed.), New York, Plenum Press 1981
718. Monkenbusch, M., Morra, B. S., Wegner, S.: Makromol. Chem., Rapid Commun. 3, 69 (1982)
719. Suzuki, N., Ozaki, M., Etemad, S., Heeger, A. J., McDiarmid, A. G.: Phys. Rev. Lett. 45, 1209 (1980)
720. Harbeke, G., Kiess, H., Meyer, W., Baeriswyl, D.: J. Phys. Soc. Jpn. 49, Sup A, 865 (1980)
721. Chung, T. C., Feldblum, A., Heeger, A. J., McDiarmid, A. G.: J. Chem. Phys. 74, 5504 (1981)
722. Heeger, A. J., McDiarmid, A. G.: Chemica Scripta 17, 115 (1981)
723. Maki, K., Nakahara, M.: Phys. Rev. B 23, 5005 (1981)
724. Kivelson, S., Lee, T. K., Lin-Liu, Y. R., Peschal, I., Yu, L.: Phys. Rev. B 25, 4173 (1982)
725. Glick, A. J.: Phys. Rev. Lett. 49, 804 (1982)
726. Tomkiewicz, Y., Schultz, T. D., Brom, H. B., Taranko, A. R., Clarke, T. C., Street, G. B.: Phys. Rev. B 24, 4348 (1981)
727. Rice, M. J., Timonen, J.: Phys. Lett. 73 A, 368 (1979)
728. Mele, E. J., Rice, M. J.: Phys. Rev. B 23, 5397 (1981)
729. Bredas, J. L., Chance, R. R., Silbey, R.: J. Phys. Chem. 85, 756 (1981)
730. Fincher, Jr., C. R., Ozaki, M., Heeger, A. J., McDiarmid, A. G.: Phys. Rev. B 19, 4140 (1979)
731. Mele, E. J., Rice, M. J.: Phys. Rev. Lett. 45, 926 (1980)
732. Etemad, S., Pron, A., Heeger, A. J., McDiarmid, A. G., Mele, E. J., Rice, M. J.: Phys. Rev. B 23, 5137 (1981)
733. Horowitz, B.: Solid State Commun. 41, 729 (1982)
734. Chien, J. C. W.: J. Polym. Sci. Polym. Lett. Ed. 19, 249 (1981)
735. Goldberg, I. B., Crowe, H. R., Newman, P. R., Heeger, A. J., McDiarmid, A. G.: J. Chem. Phys. 70, 1132 (1979)
736. Bernier, P., Rolland, M., Galtier, M., Montaner, A., Regis, M., Candille, M., Benoit, C., Aldissi, M., Linaya, C., Schué, F., Sledz, J., Fabre, J. M., Giral, L.: J. Phys. lett. 40, L297 (1979)
737. Tomkiewicz, Y., Schutz, T. D., Broom, H. B., Clarke, T. C., Street, G. B.: Phys. Rev. Lett. 43, 1532 (1979)
738. Weinberger, B. R., Ehrenfreund, E., Pron, A., Heeger, A. J., McDiarmid, A. G.: J. Chem. Phys. 72, 4749 (1980)
739. Weinberger, B. R., Kaufer, J., Heeger, A. J., Pron, A., McDiarmid, A. G.: Phys. Rev. B 20, 223 (1979)

740. Ikehata, S., Kaufer, J., Woerner, T., Pron, A., Druy, M. A., Sivak, A., Heeger, A. J., McDiarmid, A. G.: Phys. Rev. Lett. *45*, 1123 (1980)
741. Holczer, K., Devreux, F., Nechtschein, M., Travers, J. P.: Solid State Commun. *39*, 881 (1981)
742. Kinoshita, N., Tokumoto, M.: J. Phys. Soc. Jpn. *50*, 2779 (1981)
743. Nechtschein, M., Devreux, F., Greene, R. L., Clarke, T. C., Street, G. B.: Phys. Rev. Lett. *44*, 356 (1980)
744. Peo, M., Förster, H., Menke, K., Hocker, J., Gardner, J. A., Roth, S., Dransfeld, K.: Solid State Commun. *38*, 467 (1981)
745. Clarke, T. C., Scott, J. C.: Solid State Commun. *41*, 389 (1982)
746. Rachdi, F., Bernier, P., Faulques, E., Lefrant, S., Schué, F.: Polymer *23*, 173 (1982)
747. Chiang, C. K., Heeger, A. J., McDiarmid, A. G.: Ber. Bunsen. Phys. Chem. *83*, 407 (1979)
748. Epstein, A. J., Rommelmann, H., Abkowitz, M., Gibson, H. W.: Phys. Rev. Lett. *47*, 1549 (1981)
749. Gamoudi, M., André, J. J., François, B., Maitrot, M.: J. Physique *43*, 953 (1982)
750. Park, Y. W., Denenstein, A., Chiang, C. K., Heeger, A. J., McDiarmid, A. G.: Solid State Commun. *29*, 747 (1979)
751. Kwak, J. F., Clarke, T. C., Greene, R. L., Street, G. B.: Bull. Amer. Chem. Soc. *23*, 56 (1978)
752. Ozaki, M., Peebles, D., Weinberger, B. R., Heeger, A. J., McDiarmid, A. G.: J. Appl. Phys. *51*, 4252 (1980)
753. Chiang, C. K., Franklin, A. D.: Solid State Commun. *40*, 775 (1981)
754. Weinberger, B. R., Akhtar, M., Gau, S. C.: Synth. Metals 4, 187 (1982)
755. Moses, D., Chen, J., Denenstein, A., Kaveh, M., Chung, T. C., Heeger, A. J., McDiarmid, A. G., Park, Y. W.: Solid State Commun. *40*, 1007 (1981)
756. Kivelson, S.: Phys. Rev. Lett. *46*, 1344 (1981); Phys. Rev. B *25*, 3798 (1982)
757. André, J. J., Bernard, M., François, B., Mathis, C.: J. de Physique C3 *44*, 199 (1983)
758. Hoffman, D. M., Tanner, D. B., Epstein, A. J., Gibson, H. W.: Mol. Cryst. Liq. Cryst. *86*, 1175 (1982)
759. Bak, P., Pokrovsky, V. L.: Phys. Rev. Lett. *47*, 958 (1981)
760. Moses, D., Denenstein, A., Pron, A., Heeger, A. J., McDiarmid, A. G.: Solid State Commun. *36*, 219 (1980)
761. Audendaert, M., Gusman, G., Deltour, R.: Phys. Rev. B *24*, 7380 (1981)
762. Mermilliod, N., Zuppirolli, L., François, B.: J. Phys. *41*, 1453 (1980)
763. Sheng, P., Sichel, E. K., Gittleman, J. L.: Phys. Rev. Lett. *40*, 1197 (1978)
764. Sichel, E. K., Gittleman, J., Sheng, P.: Phys. Rev. B *18*, 5712 (1978)
675. Sheng, P.: Phys. Rev. B *21*, 2180 (1980)
766. Gould, C. M., Bates, D. M., Bozler, H. M., Heeger, A. J., Druy, M. A., McDiarmid, A. G.: Phys. Rev. B *23*, 6820 (1981)
767. Epstein, A. J., Rommelmann, H., Abkowitz, M., Gibson, H. W.: Mol. Cryst. Liq. Cryst. *77*, 81 (1981)
768. Chiang, C. K., Park, Y. W., Heeger, A. J., Shirakawa, H., Louis, E. J., McDiarmid, A. G.: J. Chem. Phys. *69*, 5098 (1978)
769. Mortensen, K., Thewalt, M. L. W., Tomkiewicz, Y., Clarke, T. C., Street, G. B.: Phys. Rev. Lett. *45*, 490 (1980)
770. Maitrot, M., François, B.: personal communication
771. Etemad, S., Mitani, T., Ozaki, M., Chung, T. C., Heeger, A. J., McDiarmid, A. G.: Solid State Commun. *40*, 75 (1981)
772. Brazovskii, S. A.: JETP Lett. *28*, 606 (1979)
773. Brazovskii, S. A., Kirova, N. N.: JETP Lett. *33*, 5 (1981)
774. Su, W. P., Schrieffer, J. R.: Proc. Nat. Acad. Sci. 77, 5626 (1980)
775. Flood, J. D., Ehrenfreund, E., Heeger, A. J., McDiarmid, A. G.: Solid State Commun. *44*, 1055 (1982)
776. Salem, L.: Account Chem. Res. *12*, 87 (1979)
777. Sethna, J. P., Kivelson, S.: Phys. Rev. B *26*, 3513 (1982)
778. Tani, T., Gill, W. D., Grant, P. M., Clarke, T. C., Street, G. B.: Synth. Metals *1*, 301 (1979/80)
779. Tsukamoto, J., Ohigashi, H., Matsumura, K., Takahashi, A.: Synth. Metals 4, 177 (1982)
780. Cade, N. A.: Chem. Phys. Lett. *53*, 45 (1978)

781. Ozaki, M., Peebles, D. L., Weinberger, B. R., Chiang, C. K., Gau, S. C., Heeger, A. J., McDiarmid, A. G.: Appl. Phys. Lett. *35*, 83 (1979)
782. Waldrop, J. R., Cohen, M. J., Heeger, A. J., McDiarmid, A. G.: Appl. Phys. Lett. *38*, 53 (1981)
783. Weinberger, B. R., Gau, S. C., Kiss, Z.: Appl. Phys. Lett. *38*, 555 (1981)
784. Tsukamoto, J., Ohigashi, H., Matsumura, K., Takahashi, A.: Jpn. J. Appl. Phys. *20*, L127 (1981)
785. Grant, P. M., Tani, T., Gill, W. D., Krounbi, M., Clarke, T. C.: J. Appl. Phys. *52*, 869 (1981)
786. Vander Donckt, E., Kanicki, J.: Eur. Polym. J. *16*, 677 (1980)
787. Kanicki, J., Boué, S., Vander Donckt, E.: Mol. Cryst. Liq. Cryst. *83*, 1351 (1982)
788. Kanicki, J., Fedorko, P., Boué, S., Van der Donckt, E.: Proceedings of Fourth E. C. Photovoltaic Solar Energy Conf., Stresa 10–14 May 1982, D. Reidel, Publish. Co. p. 562
789. Chiang, C. K., Gau, S. C., Fincher, Jr., C. R., Park, Y. W., McDiarmid, A. G., Heeger, A. J.: Appl. Phys. Lett. *33*, 18 (1978)
790. Yamase, T., Harada, H., Ikawa, T., Ikeda, S., Shirakawa, H.: Bull. Chem. Soc. Jpn. *54*, 2817 (1981)
791. Shirakawa, S., Ikeda, S., Aizawa, M., Yoshitake, J., Suzuki, S.: Synth. Metals *4*, 43 (1981)
792. Belkind, A. I., Grechov, V. V.: Phys. Status Solidi *A 26*, 377 (1974)
793. Gallegos, E. J.: J. Phys. Chem. *72*, 3452 (1968)
794. Schein, L. B.: Electrical & Related Properties of Organic Solids; Papers of Int. Conf., Karpacz Poland, Sept. 18–23 (1978)
795. Probst, K. H., Karl, N.: Phys. Status Solidi *A 27*, 499 (1975)
796. Armington, A. F.: Organic Semiconductors, Okamoto, Y., Brenner, W. (ed.), New York, Reinhold Pub. Co. 1964
797. Hatfield, W. E.: Molecular Metals, (ed.), New York, Plenum Press 1979
798. Chen, R., Trucano, P., Stewart, R. F.: Acta Crystallogr. *A 33*, 823 (1977)
799. Moore, A. W.: Chemistry and Physics of Carbon, Vol. 11, Walker Jr., P. C., Thrower, P. A., (ed.), New York, Dekker 1973
800. An overview of the properties of graphite may be found in the Proceedings of the second conference on intercalation compounds of graphite; Synth. Metals *2* (1980) and *3* (1981)
801. Wright, S. K., Schramm, C. J., Phillips, T. E., Scholler, D. M., Hoffman, B. M.: Synth. Metals *1*, 43 (1979)
802. Pace, L. J., Martinsen, J., Ulman, A., Hofman, B. M., Ibers, J. A.: J. Amer. Chem. Soc. *105*, 2612 (1983)
803. Konishi, S., Hoshino, M., Imamura, M.: J. Phys. Chem. *84*, 3437 (1980)
804. Fulton, G. P., La Mar, G. N.: J. Amer. Chem. Soc. *98*, 2119 (1980)
805. Ulman, A., Manassen, J., Frolow, F., Rabinovich, D.: Inorg. Chem. *20*, 1987 (1981)
806. Gleizes, A., Marks, T. J., Ibers, J. A.: J. Amer. Chem. Soc. *97*, 3545 (1975)
807. Kalina, D. W., Lyding, J. W., Ratajack, M. T., Kannewurf, C. R., Marks, T. J.: J. Amer. Chem. Soc. *102*, 7854 (1980)
808. Bowman Mertes, K., Ferraro, J. R.: J. Chem. Phys. *70*, 646 (1979)
809. Wuu, Yee-Min, Peng, Shie-Ming: J. Inorg. Nucl. Chem. *42*, 205 (1980)
810. Cassoux, P., Gleizes, A.: Inorg. Chem. *19*, 665 (1980)
811. Cassoux, P., Interrante, L., Kasper, J.: Mol. Cryst. Liq. Cryst. *81*, 293 (1982)
812. Thomas, T. W., Underhill, A. E.: Chem. Soc. Rev. *1*, 99 (1972)
813. Miller, J. S., Eptstein, A. J.: Prog. Inorg. Chem. *20*, 1 (1976)
814. Ferraro, J. R.: Coord. Chem. Rev. *43*, 205 (1982)
815. Miller, J. S. (ed.): Extended Linear Chain Compounds, Vol. I–IV, New York, Plenum Press 1982 and 1983
816. Whangbo, M.-H., Hoffman, R.: J. Amer. Chem. Soc. *100*, 6093 (1978)
817. Wood, D. J., Underhill, A. E., Schultz, A. J., Williams, J. M.: Solid State Comm. *30*, 501 (1979)
818. Krogman, K.: Angew. Chem. Int. ed. *8*, 35 (1969)
819. Osborn, R. S., Rogers, D.: J. C. S. Dalton 1002 (1974)
820. Little, W. A.: Low Dimensional Cooperative Phenomena, Keller, H. J. (ed.), New York, Plenum Press, 1975

821. Little, W. A.: J. de Physique Colloque *C3 44*, 819 (1983)
822. Winkler, T., Mayer, C.: Helv. Chim. Acta *55*, 2351 (1972)
823. Torrance, J. B.: Account. Chem. Res. *12*, 79 (1979)
824. Garito, A. F., Heeger, A. J.: Account Chem. Res. *7*, 232 (1974)
825. Scott, B. A., Kaufman, F. B., Engler, E. M.: J. Amer. Chem. Soc. *98*, 4342 (1976)
826. Peover, M. E.: Trans. Faraday Soc. *58*, 2370 (1962)
827. Peover, M. E.: Trans. Faraday Soc. *58*, 1956 (1962)
828. André, J.-J.: in "Recent Advances in the Quantum Theory of Polymers", Lect. Notes in Physics *113*, 35 (1980)
829. Shchegolev, I. F.: Phys. Status Solidi (a) *12*, 9 (1972)
830. Tombs, G. A.: Physics Rep. C, *40*, 181–240 (1978)
831. Bulaevskii, L. N.: Soviet Phys. Usp. *18*, 131 (1975)
832. Berlinsky, A. J.: Contemp. Phys. *17*, 331–354 (1976)
833. Friedel, J., Jérome, D.: Contemp. Phys. *23* (1982)
834. Engler, E. M.: Chem. Tech. p. 274 (1976)
835. Khidekel, M. L., Zhilyaeva, E. I.: Synth. Metals *4*, 1–34 (1981)
836. Epstein, A. J., Miller, J. S.: Sci. Amer., October p. 52 (1979)
837. Bechgaard, K., Jérome, D.: Sci. Amer. July, p. 52 (1982)
838. Narita, M., Pittman, C. H. Jr.: Synthesis, pp. 489–514
839. Little, W. A.: J. Polym. Sci. part 2, *29* (1969)
840. Schuster, H. G.: Lecture Notes in Physics N° 34, 1975
841. Pal, L., Grüner, G., Janossy, A., Solyom, J., (Siofok Conference), Lecture Notes in Physics N° 65, 1977
842. Miller, J. S., Epstein, A. J.: Ann. N. Y. Acad. Sci. *313*, 1–828 (1978)
843. Barisic, S., Bjelis, A., Cooper, J. R., Leontic, B.: (Dubrovnik Conference), Lecture Notes in Physics N° 95 and 96, 1979
844. Helsingor Conference: Chem. Scripta *17*, 1–230 (1981)
845. Boulder Conference: Epstein, A. J., Conwell, E. M.: Mol. Cryst. Liquid Cryst. *77, 79, 81, 83, 85, 86* (1981–1982)
846. Conférences des Arcs: ed. by Comes, R., Bernier, P., Rouxel, J.: J. de Physique Colloque 3, *44* (1983)
847. Jerome, D., Schutz, H. J.: Adv. Phys. *31*, 299 (1982)
848. Low-Dimensional Cooperative Phenomena, edited by Keller, H. H., 1975, New York, Plenum Press
849. Chemistry and Physics of One-Dimensional Metals, ed. Keller, H. J., 1977, New York, Plenum Press
850. The Physics and Chemistry of Low-Dimensional Solid, ed. Alcacer, L., 1980, Dordrecht D. Reidel
851. Physics in One Dimension, ed. Bernasconi, J., Schneider, T., 1981, Springer-Verlag
852. Highly Conducting One-Dimensional Solids, ed. Devreese, J. T., Van Doren, V. E., 1979, Plenum Press
853. Herbstein, F. H., in Perspective in Structural Chemistry, vol. IV (eds. Dunitz, J. D., Ibers, J. A.), Wiley, New York 1971
854. Soos, Z. G.: Ann. Rev. Phys. Chem. *25*, 121 (1974)
855. Nordio, P. L., Soos, Z. G., McConnell, H. M.: Ann. Rev. Phys. Chem. *17*, 237 (1966)
856. Mülliken, R. S., Pearson, W. B.: "Molecular Complexes", Wiley Interscience, New York 1969
857. Ikemoto, I., Kuroda, H.: Acta Cryst. *B24*, 383 (1968)
858. Kistenmacher, T. J., Phillips, T. E., Cowan, D. O.: Acta Cryst. *B30*, 763 (1974)
859. Hanson, A. W.: Acta Cryst. *B24*, 768 (1968)
860. Sundaresan, T., Wallwork, S. C.: Acta Cryst. *B28*, 491 (1972)
861. Chasseau, D., Gaultier, J., Hauw, C., Schoerer, M.: C. R. Acad. Sci. (Paris) *C 275*, 1491 (1972)
862. De Boer, J. L., Vos, A.: Acta Cryst. *B28*, 835 (1972)
863. Bechgaard, K.: Mol. Cryst. Liq. Cryst. *79*, 1 (1982)
864. Phillips, T. E., Kistenmacher, T. J., Bloch, A. N., Ferraris, J. P., Cowan, D. O.: Acta Cryst. *B33*, 422 (1977)
865. Torrance, J. B., Vasquez, J. E., Mayerlé, J. J., Lee, V. Y.: Phys. Rev. Lett. *46*, 253 (1981)

866. Khanna, J. K., Pouget, J. P., Comes, R., Garito, A. F., Heeger, A. J.: Phys. Rev. *B 16*, 1468 (1977)
867. Weyl, C., Engler, E. M., Bechgaard, K., Jehanno, G., Etemad, S.: Solid State Commun. *19*, 925 (1976)
868. Megtert, S., Pouget, J. P., Comes, R., Garito, A. F., Bechgaard, K., Fabre, J. M., Giral, L.: J. Phys. Lett. *39*, L118 (1978)
869. Inabe, T., Matsunaga, Y.: Bull. Chem. Soc. Jpn. *51*, 2813 (1978)
870. Girlando, A., Marzola, F., Pecile, C., Torrance, J. B.: J. Chem. Phys. *79*, 1075 (1983)
871. Tomkiewicz, Y., Torrance, J. B., Scott, B. A., Green, D. C.: J. Chem. Phys. *60*, 5111 (1974)
872. Torrance, J. B., Silverman, B. D.: Phys. Rev. *B 15*, 788 (1977)
873. Silverman, B. D., Grobman, W. D., Torrance, J. B.: Chem. Phys. Lett. *50*, 152 (1977)
874. Klymenko, V. E., Ya. Krivnov, V., Dchinnikov, A. A., Ukrainskii, I. I.: J. Phys. Chem. Solids *39*, 359 (1978)
875. Scott, B. A., Laplaca, S. J., Torrance, J. B., Silverman, B. D., Welber, B.: J. Amer. Chem. Soc. *99*, 6631 (1977)
876. Silverman, B. D., Laplaca, S. J.: J. Chem. Phys. *69*, 2585 (1978)
877. Ciobani, G., Kimun, S., Lee, T., Das, T. P.: Chem. Phys. Lett. *38*, 500 (1976)
878. Wheland, R. C.: J. Amer. Chem. Soc. *98*, 3926 (1976)
879. Epstein, A. J., Conwell, E. M., Miller, J. S.: ref. 842, p. 183
880. Weger, M.: in reference 797, p. 123
881. Coleman, L. B., Cohen, M. J., Sandman, D. J., Yamagishi, F. G., Garito, A. F., Heeger, A. J.: Solid. State Commun. *12*, 1125 (1973)
882. Cohen, M. J., Coleman, L. B., Garito, M. F., Heeger, A. J.: Phys. Rev. *B 13*, 5111 (1976)
883. Bloch, A. W., Carruthers, T. F., Poehler, T. O., Cowan, D. O.: in reference 849, p. 47
884. Thomas, G. A., Schafer, D. E., Wudl, F., Horn, P. M., Rimai, D., Cook, J. W., Glocker, D. A., Skove, M. J., Chu, C. W., Groff, R. P., Gillson, J. L., Wheland, R. C., Melby, L. R., Salamon, M. B., Craven, R. A., de Pasquali, G., Bloch, A. N., Cowan, D. O., Walatka, V. V., Pyle, R. E., Gemmer, R., Poehler, T. O., Johnson, G. R., Miles, M. G., Wilson, J. D., Ferraris, J. P., Finnegan, T. F., Warmack, R. J., Ramen, V. F., Jerome, D.: Phys. Rev. *B 13*, 5105 (1976)
885. Ferraris, J. P., Finnegan, T. F.: Solid State Commun. *18*, 1169 (1976)
886. Wheland, R. C., Gillson, J. L.: J. Amer. Chem. Soc. *98*, 3916 (1976)
887. Warmack, R. J., Callcott, T. A., Watson, C. R.: Phys. Rev. *B 12*, 3336 (1975)
888. Laplaca, S. J., Corfield, P. W. R., Thomas, R., Scott, B. A.: Solid State Commun. *17*, 635 (1975)
889. Greene, R. L., Mayerlé, J. J., Schumaker, R., Castro, G., Chaikin, P. M., Etemad, S., Laplaca, S. J.: Solid State Commun. *20*, 943 (1976)
890. Delhaes, P., Flandrois, D., Amiell, J., Keryer, G., Toreilles, E., Fabre, J. M., Giral, L., Jacobsen, C. S., Bechgaard, K.: J. Phys. Lett. *38*, 233 (1977)
891. Friend, R. M., Jerome, D., Fabre, J. M., Giral, L., Bechgaard, K.: J. Phys. *C 11*, 263 (1978)
892. Engler, E. M., Patel, V. V.: J. Amer. Chem. Soc. *96*, 7376 (1974)
893. Jacobsen, C. S., Mortensen, K., Andersen, J. R., Bechgaard, K.: Phys. Rev. *B 18*, 905 (1978)
894. Conwell, E. M., Banik, N. C.: Mol. Cryst. Liq. Cryst. *79*, 95 (1982)
895. Bechgaard, K., Jacobsen, C. S., Mortensen, K., Pedersen, H. J., Thorup, N.: Solid State Commun. *33*, 1189 (1980)
896. Bechgaard, K., Jacobsen, C. S., Andersen, N. H.: Solid State Commun. *25*, 875 (1978)
897. Mogensen, B., Friend, R. H., Jerome, D., Bechgaard, K., Carneiro, K.: to be published
898. Buravov, L. I., Eremenko, O. N., Lyubovskii, R. B., Rozenberg, L. P., Khidekel, M. L., Shibaeva, R. P., Shchegolev, I. F., Yagubskii, E. B.: J. E. T. P. Lett. *20*, 208 (1974) (Zh ETF Pis. Red. *20*, 457 (1974))
899. Somoana, R. B., Yen, S. P. S., Hadek, V., Khanna, S. K., Novotny, M., Datta, T., Hermann, A. M., Woollam, J. A.: Phys. Rev. *B 17*, 2853 (1978)
900. Buravov, L. I., Zvereva, G. I., Kaminskii, V. F., Rozenberg, L. P., Khidekel, M. L., Shibaeva, R. P., Shchegolev, I. F., Yagubskii, E. B.: J. Chem. Soc. Chem. Commun. 720 (1976)
901. Cooper, J. R., Weger, M., Delplanque, G., Jerome, D., Bechgaard, K.: J. Phys. Lett. *37*, L349 (1976)
902. Korin, B., Cooper, J. R., Miljak, M., Hamzic, A., Bechgaard, K.: Chem. Scripta *17*, 45 (1981)

903. Blakemore, J. S., Lane, J. E., Woodbury, D. A.: Phys. Rev. *B 18*, 6797 (1978)
904. Shirotani, I., Sakai, N.: J. Solid State Chem. *18*, 17 (1976)
905. Hurditch, R. J., Vincent, V. M., Wright, J. D.: J. C. S. Faraday Trans. I *68*, 465 (1972)
906. Coleman, L. B., Cohen, J. A., Garito, A. F., Heeger, A. J.: Phys. Rev. *B 7*, 2122 (1973)
907. Buravov, L. I., Fedutin, D. N., Shchegolev, I. F.: Sov. Phys. JETP *32*, 612 (1971)
908. Sakai, N., Shirotani, I., Minomura, S.: Bull. Chem. Soc. Jpn. *45*, 3314 (1972)
909. Ferraris, J., Cowan, D. O., Walat Ka, Jr., V., Perlstein, J. M.: J. Amer. Chem. Soc. *95*, 948 (1973)
910. Zuppiroli, L., Mutka, H., Bouffard, S.: Mol. Cryst. Liq. Cryst. *85*, 1 (1982)
911. Copper, J. R., Miljak, M., Delplanque, G., Jerome, D., Wegner, M., Fabre, J. M., Giral, L.: J. de Phys. *38*, 1097 (1977)
912. Heeger, A. J.: in reference 849, p. 87
913. Denoyer, F., Comes, R., Garito, A. F., Heeger, A. J.: Phys. Rev. Lett. 35, *445* (1975)
914. Kagoshima, S., Anzai, H., Kajimura, K., Ishiguro, T.: J. Phys. Soc. Jpn. *39*, 1143 (1975)
915. Pouget, J. P., Khanna, S. K., Denoyer, F., Comes, R., Garito, A. F., Heeger, A. J.: Phys. Rev. Lett. *37*, 437 (1976)
916. Kagoshima, S., Ishiguro, T., Anzai, H.: Phys. Soc. Jpn. *41*, 2061 (1976)
917. Megtert, S., Pouget, J. P., Comes, R.: in reference 797, p. 87
918. Mook, H. A., Watson, Jr., G. R.: Phys. Rev. Lett. *36*, 801 (1976)
919. Comes, R., Shapiro, S. M., Shirane, G., Garito, A. F., Heeger, A. J.: Phys. Rev. Lett. *35*, 1518 (1975)
920. Shirane, G., Shapiro, S. M., Comes, R., Garito, A. F., Heeger, A. J.: Phys. Rev. *B 14*, 2325 (1976)
921. Comes, R., Shirane, G., Garito, S. M., Garito, A. F., Heeger, A. J.: Phys. Rev. *B 14*, 2376 (1976)
922. Schultz, T. D., Etemad, S.: Phys. Rev. *B 13*, 4928 (1976)
923. Horn, P. M., Rimai, D.: Phys. Rev. Lett. *36*, 809 (1976)
924. Etemad, S.: Phys. Rev. *B 13*, 2254 (1976)
925. Cooper, J. R., Lukatela, J., Miljak, M., Fabre, J. M., Giral, L., Eharo-Shalom, E.: Solid State Commun. *25*, 949 (1978)
926. Ishiguro, T., Kagoshima, S., Anzai, H.: J. Phys. Soc. Jpn. *41*, 351 (1976)
927. Khanna, S. K., Ehrenfreund, E., Garito, A. F., Heeger, A. J.: Phys. Rev. *B 10*, 2205 (1974)
928. Bak, P., Emery, V. J.: Phys. Rev. Lett. *36*, 978 (1976)
929. Kjurek, D., Franulovic, K., Prester, M., Tomic, S., Giral, L., Fabre, J. M.: Phys. Rev. Lett. *38*, 715 (1977)
930. Chaikin, P. M., Greene, R. L., Etemad, S., Engler, E.: Phys. Rev. *B 13*, 1627 (1976)
931. Rybaczewski, E. F., Smith, L. S., Garito, A. F., Heeger, A. J., Silbernagel, B. G.: Phys. Rev. *B 14*, 2746 (1976)
932. Tomkiewicz, Y., Taranko, A. R., Torrance, J. B.: Phys. Rev. Lett. *36*, 751 (1976)
933. Rybaczewski, E. F., Smith, L. S., Garito, A. F., Heeger, A. J., Silbernagel, B. G.: Phys. Rev. *B 14*, 2746 (1976)
934. Chaikin, P. M., Greene, R. L., Engler, E. M.: Solid State Commun. *25*, 1009 (1978)
935. Tomkiewicz, Y., Craven, R. A., Schultz, T. D., Engler, E. M., Taranko, A. R.: Phys. Rev. *B 15*, 3643 (1977)
936. Engler, E. M., Scott, B. A., Etemad, S., Tenney, T., Patel, V. V.: J. Amer. Chem. Soc. *99*, 5909 (1977)
937. Chaikin, P. M., Kwak, J. M., Greene, R. L., Etemad, S., Engler, E. M.: Solid State Commun. *19*, 1201 (1976)
938. Herman, F.: Phys. Scr. *16*, 303 (1977)
939. Debray, D., Miller, R., Jerome, D., Barisic, S., Giral, L., Fabre, J. M.: J. Phys. Lett. *38*, L227 (1977)
940. Jerome, D., Schultz, H. J.: in reference 815, Vol. II, p. 159
941. Bouffard, S., Chipaux, R., Jerome, D., Bechgaard, K.: Solid State Commun. *37*, 405 (1981)
942. Lee, L. P., Rice, T. M., Anderson, P. W.: Phys. Rev. Lett. *31*, 462 (1973)
943. Frohlich, H.: Proc. Roy. Soc. London *A 233*, 296 (1954)
944. Bardeen, J.: Solid State Commun. *13*, 357 (1973)
945. Bouffard, S., Chipaux, R., Jerome, D., Bechgaard, K.: Solid State Commun. *37*, 407 (1981)

946. Etemad, S., Penney, T., Engler, E. M., Scott, B. A., Seiden, P. E.: Phys. Rev. Lett. *34*, 741 (1975)
947. Chaikin, P. M., Greene, R. L., Etemad, S., Engler, E. M.: Phys. Rev. *B 13*, 1627 (1976)
948. Cooper, J. R., Jerome, D., Etemad, S., Engler, E. M.: Solid State Commun. *22*, 257 (1977)
949. Kagoshima, S., Ishiguro, T., Schultz, T. D., Tomkiewicz, Y.: Solid State Commun. *28*, 485 (1978)
950. Tomkiewicz, Y., Andersen, J. R., Taranko, A. R.: Phys. Rev. *B 17*, 1579 (1978)
951. Tomkiewicz, Y., Taranko, A. R., Engler, E. M.: Phys. Rev. Lett. *37*, 1705 (1976)
952. Thomas, J. F., Jerome, D.: Solid State Commun. *36*, 813 (1980)
953. Scott, J. C., Etemad, S., Engler, E. M.: Phys. Rev. *B 17*, 2269 (1978)
954. Bates, F. E., Eldridge, J. E., Bryce, M. R.: Can. J. Phys. *59*, 339 (1981)
955. Schultz, T. D.: Solid State Commun. *22*, 289 (1977)
956. Tomkiewicz, Y., Engler, E. L., Schultz, T. D.: Phys. Rev. Lett. *35*, 456 (1975)
957. Bloch, A. N., Cowan, D. O., Bechgaard, K., Pyle, R. E., Banks, R. M., Poehler, T. O.: Phys. Rev. Lett. *34*, 1561 (1975)
958. Soda, G., Jerome, D., Weger, M., Bechgaard, K., Pedersen, E.: Solid State Commun. *20*, 107 (1976)
959. Cooper, J. R., Weger, M., Jerome, D., Lefur, L., Bechgaard, K., Bloch, A. N., Cowan, D. O.: Solid State Commun. *19*, 749 (1976)
960. Miljak, M., Andrieux, A., Friend, R. H., Malfait, G., Jerome, D., Bechgaard, K.: Solid State Commun. *26*, 969 (1978)
961. Weger, M.: Solid State Commun. *19*, 1149 (1976)
962. Engler, E. M., Greene, R., Haen, P., Tomkiewicz, Y., Mortensen, K., Berendzen, J.: Mol. Cryst. Liq. Cryst. *79*, 15 (1982)
963. Thorup, N., Rindorf, G., Soling, H., Bechgaard, K.: Acta Crystallogr. *B 37*, 1236 (1981)
964. Bechgaard, K., Carneiro, K., Rasmussen, F. B., Olsen, M., Rindorf, G., Jacobsen, C. S., Pedersen, H., Scott, J. C.: J. Amer. Chem. Soc. *103*, 2440 (1981)
965. Williams, J. M., et al.: J. de Physique C3 *44*, 941 (1983)
966. Williams, J. M., Beno, M. A., Appelman, E. H., Capriotti, J. P., Wudl, F., Aharon-Shalom, E., Nalewajek, D.: Mol. Cryst. Liq. Cryst. *79*, 319 (1972)
967. Walsh, W. M., Wudl, F., Thimas, G. A., Nalewajek, D., Hauser, J. J., Lee, P. A., Poehler, T.: Phys. Rev. Lett. *45*, 829 (1980)
968. Zettl, A., Grüner, G., Engler, E. M.: Phys. Rev. *B 25*, 1443 (1982)
969. Pedersen, H. J., Scott, J. C., Bechgaard, K.: Solid State Commun. *35*, 207 (1980)
970. Andrieux, A., Jerome, D., Bechgaard, K.: J. de Phys. Lett. *42*, L 87 (1981)
971. Scott, J. C., Pedersen, H. J., Bechgaard, K.: Phys. Rev. Lett. *45*, 2125 (1980)
972. Jacobsen, C. S., Mortensen, K., Weger, M., Bechgaard, K.: Solid State Commun. *38*, 423 (1981)
973. Chaikin, P. M., Choi, M. Y., Haen, P., Engler, E. M., Greene, R. L.: Mol. Cryst. Liq. Cryst. *79*, 79 (1982)
974. Chaikin, P. M., Grüner, G., Engler, E. M., Greene, R. L.: Phys. Rev. Lett. *45*, 1874 (1980)
975. Brusetti, R., Ribault, M., Jerome, D., Bechgaard, K.: J. de Phys. *43*, 801 (1982)
976. Liautard, B., Peytavin, S., Brun, G., Maurin, M.: J. de Physique C3 *44*, 951 (1983)
977. Pouget, J. P.: Chem. Scripta *17*, 85 (1981)
978. Ribault, M., Pouget, J. P., Jerome, D., Bechgaard, K.: J. de Phys. Lett. *41*, L 607 (1980)
979. Jerome, D.: J. de Physique *C 3*, 775 (1983)
980. Jacobsen, C. S., Tanner, D. B., Bechgaard, K.: Mol. Cryst. Liq. Cryst. *79*, 25 (1982)
981. Azevedo, L. J.: J. de Physique C 3, **44**, 813 (1983)
982. Grant, P. M.: J. de Physique C 3, *44*, 847 (1983)
983. Grant, P. M.: J. de Physique C 3, *44*, 1121 (1983)
984. Schultz, H. J., Jerome, D., Ribault, M., Mazaud, A., Bechgaard, K.: J. de Phys. Lett. *42*, L 51 (1981)
985. Ribault, M., Benedek, G., Jerome, D., Bechgaard, K.: J. Phys. Lett. *41*, L 397 (1980)
986. Parkin, S. S. P., Ribault, M., Jerome, D., Bechgaard, K.: J. Phys. *C 14*, L 445 (1981)
987. Parkin, S. S. P., Jerome, D., Bechgaard, K.: Mol. Cryst. Liq. Cryst. *79*, 213 (1982)
988. Bechgaard, K., Carneiro, K., Olsen, M., Rasmussen, F. B., Jacobsen, C. S.: Phys. Rev. Lett. *46*, 852 (1981)

989. Mailly, D., Ribault, M., Bechgaard, K.: J. de Physique C3, *44*, 1077 (1983)
990. Schwenk, H., Neumaier, K., Andres, K., Wudl, F., Aharon-Shalom, E.: Mol. Cryst. Liq. Cryst. *79*, 277 (1982)
991. Gubser, D. V., Fuller, W. W., Poehler, T. O., Cowan, D. O., Potember, R. S., Chiang, L. Y., Bloch, A. N.: Phys. Rev. *B24*, 478 (1981)
992. Garoche, P., Brusetti, R., Jerome, D., Bechgaard. K.: J. Phys. *C14*, L 445 (1982)
993. Bouffard, S., Rabault, M., Brusetti, R., Jerome, D., Bechgaard, K.: J. de Physique *C15*, 2951 (1982)
994. Greene, R. L., Haen, P., Huang, S. Z., Engler, E. M., Choi, M. Y., Chaikin, P. M.: Mol. Cryst. Liq. Cryst. *79*, 183 (1982)
995. Jerome, D.: Mol. Cryst. Liq. Cryst. *79*, 155 (1982)
996. Machida, K., Matsubara, T.: Mol. Cryst. Liq. Cryst. *79*, 289 (1982)
997. Fenton, E. W., Psaltakis, G. C.: J. de Physique C3, *44*, 1129 (1983)
998. Kwak, J. F., Schirber, J. E., Greene, R. L., Engler, E. M.: Phys. Rev. Lett. *46*, 1296 (1981)
999. Takahashi, T., Jerome, D., Bechgaard, K.: J. de Phys. C3, *44*, 805 (1983)
1000. Azevedo, L. J., Schirber, J. E., Greene, R. L., Engler, E. M.: Physica B *108*, 1183 (1981)
1001. Jerome, D., Mazaud, A., Ribault, M., Bechgaard, K.: J. Phys. Lett. Paris, *41*, L 95 (1980)
1002. Andres, K., Wudl, F., McWhan, D. B., Thomas, G. A., Nalewajek, D., Stevens, A. L.: Phys. Rev. Lett. *45*, 1449 (1980)
1003. Greene, R. L., Engler, E. M.: Phys. Rev. Lett. *45*, 1587 (1980)
1004. Mortensen, K., Tomkiewicz, Y., Schultz, T. D., Engler, E. M.: Phys. Rev. *46*, 1234 (1981)
1005. Mortensen, K., Tomkiewicz, Y., Bechgaard, K.: Phys. Rev. *B25*, 3319 (1982)
1006. Jacobsen, C. S., Pedersen, H. J., Mortensen, K., Rindorf, G., Thorup, N., Torrance, J. B., Bechgaard, K.: J. de Physique *C15*, 2657 (1982)
1007. Parkin, S. S. P., Jerome, D., Bechgaard, K.: Mol. Cryst. Liq. Cryst. *79*, 569 (1982)
1008. Pouget, J. P., Moret, R., Comes, R., Bechgaard, K., Fabre, J. M., Giral, L.: Mol. Cryst. Liq. Cryst. *79*, 129 (1982)
1009. Tomic, S., Pouget, J. P., Jerome, D., Bechgaard, K., Williams, J. M.: J. de Physique *44*, 375 (1983)
1010. Pouget, J. P., Moret, R., Comes, R., Bechgaard, K.: J. Phys. Lett. *42*, L 543 (1981)
1011. Brüinsma, R., Emery, V. J.: J. de Physique C3, *44*, 1115 (1983)
1012. Moret, R., Pouget, V. P., Comes, R., Bechgaard, K.: Phys. Rev. Lett. *49*, 1008 (1982)
1013. Kohlschütter, H. W., Sprengler, L.: Z. Phys. Chem. *B16*, 284 (1932)
1014. Cohen, M. D., Schmidt, G. M. J., Sonntag, F. I.: J. Chem. Soc. 2000 and 2014 (1964)
1015. Magat, M.: Polymer *3*, 449 (1962)
1016. Charlesby, A.: Rep. Prog. Phys. *28*, 464 (1965)
1017. Morawetz, H.: Pure Appl. Chem. *12*, 201 (1966)
1018. Cohen, M. D.: Angew. Chem. Int. ed. *14*, 386 (1975)
1019. Wegner, G.: Pure Appl. Chem. *49*, 443 (1977)
1020. Shklower, V. E., Bokii, N. G., Struchkov, Yu. T.: Russian Chem. Rev. *46*, 706 (1977)
1021. Nakanishi, H., Jones, W., Thomas, J. M., Hasegawa, M., Rees, W. L.: Proc. Roy. Soc. London *A369*, 307 (1980)
1022. Hamilton, W. C.: Mol. Cryst. Liq. Cryst. *9*, 11 (1969)
1023. Aston, J. G.: "Physics and Chemistry of the Organic Solid State", Vol. I, Fox, D., Labes, M. M., Weissberger, A. (ed.), New York, Interscience 1963
1024. Labes, M. M., Love, P., Nichols, L. F.: Chem. Rev. *79*, 1 (1979)
1025. Street, G. B., Greene, R. L.: IBM J. Res. Dev. *21*, 99 (1977)
1026. McDiarmid, A. G., Mikulski, C. M., Saran, M. S., Russo, R. J., Cohen, M. J., Bright, A. A., Garito, A. F., Heeger, A. J.: Adv. Chem. Ser. N° 150, 63 (1976)
1027. Burt, F. P.: J. Chem. Soc. 1171 (1910)
1028. Becke-Goehring, M.: Adv. Inorg. Radiochem. *2*, 169 (1960)
1029. Douillard, A.: Thèse de Doctorat d'Etat, Université Claude Bernard, Lyon (France) 1972
1030. Boudeulle, M.: Cryst. Struct. Commun. *4*, 9 (1975)
1031. Boudeulle, M., Micherl, P.: Acta Cryst. *A28*, 199 (1972)
1032. Mikulski, C. M., Russo, P. J., Saran, M. S., McDiarmid, A. G., Garito, A. F., Heeger, A. J.: J. Amer. Chem. Soc. *97*, 6358 (1975)
1033. Pintschovius, L.: Colloid Polym. Sci. *256*, 883 (1978)

1034. Baughman, R. H.: J. Polym. Sci. Polym. Phys. ed. *12*, 1511 (1974)
1035. Banerjie, A., Lando, J. B.: J. Polym. Sci. Polym. Phys. ed. *17*, 655 (1979)
1036. Yee, K. C.: J. Polym. Sci. Polym. Chem. ed. *17*, 3637 (1979)
1037. Wegner, G.: Z. Naturforsch. *24b*, 824 (1969)
1038. Chance, R. R., Patel, G. N., Turi, E. A., Khanna, Y. P.: J. Amer. Chem. Soc. *100*, 1307 (1978)
1039. Enkelmann, V., Leyer, R. J., Wegner, G.: Makromol. Chem. *180*, 1787 (1979)
1040. Schermann, W., Wegner, G.: Makromol. Chem. *125*, 667 (1974)
1041. Reimer, B., Bässler, H.: Phys. Status Solidi *32a*, 435 (1975)
1042. Chance, R. M., Baughman, R. H.: J. Chem. Phys. *64*, 3889 (1976)
1043. Donovan, K. J., Wilson, E. G.: J. of Phys. *C 12*, 4857 (1979)
1044. Wilson, E. G.: Chem. Phys. Lett. *90*, 221 (1982)
1045. Donovan, K. J., Wilson, E. G.: Phil. Mag. *44*, 9 (1981)
1046. Lochner, K., Reimer, B., Bässler, H.: Chem. Phys. Lett. *41*, 388 (1976)
1047. Siddiqui, A. S., Wilson, E. G.: J. of Phys. *C 12*, 4237 (1979)
1048. Wilson, E. G.: J. Phys. *C 8*, 727 (1975)
1049. Kertèsz, M., Koller, J., Azman, A.: Chem. Phys. *27*, 273 (1978)
1050. Karpfen, A.: J. of Phys. *C 13*, 5673 (1980)
1051. Lochner, K., Bässler, H., Tieke, B., Wegner, G.: Phys. Status Solidi *B 88*, 653 (1978)
1052. Tieke, B., Graf, H.-J., Wegner, G., Naegele, B., Ringsdorf, H., Banerjee, A., Day, D., Lando, J. B.: Colloid Polym. Sci. *225*, 521 (1977)
1053. Tieke, B., Wegner, G., Naegele, D., Ringsdorf, H.: Angew. Chem. Int. ed. *15*, 764 (1976)
1054. Lieser, G., Tieke, B., Wegner, G.: Thin Solid Films *68*, 77 (1980)
1055. Kajzar, F., Messier, J.: Chem. Phys. *63*, 123 (1981)
1056. Day, D., Hub, H. H., Ringsdorf, H.: Isr. J. of Chem. *18*, 325 (1979)
1057. Bader, H., Ringsdorf, H., Skura, J.: Angew. Chem. Int. ed. *20*, 91 (1981)
1058. Akimoto, A., Dorn, K., Gros, L., Ringsdorf, H., Schupp, H.: Angew. Chem. Int. ed. *20*, 90 (1981)
1059. Lopez, E., O'Brien, D. F., Whitesides, T. H.: J. Amer. Chem. Soc. *104*, 305 (1982)
1060. Kiji, J.: Makromol. Chem. *179*, 833 (1978)
1061. Snow, A. W.: Nature *292*, 40 (1981)
1062. Patel, G. N.: Polym. Prepr. *19* (2), 155 (1978)
1063. Patel, G. N., Chance, R. R., Witt, J. D.: J. Polym. Sci. Polym. Lett. ed. *16*, 607 (1978)
1064. Patel, G. N., Walsh, E. K.: J. Polym. Sci. Polym. Lett. ed. *17*, 203 (1979)
1065. Griffiths, J.: "Colour and Constitution of Organic Molecules", London, Academic Press 1976
1066. Malhotra, S. S., Whiting, M. C.: J. Chem. Soc. 3812 (1960)
1067. Dähne, S., Gürtler, O.: J. Prakt. Chem. *315*, 786 (1973)
1068. Tredwell, C. J., Keary, C. M.: Chem. Phys. *43*, 307 (1950)
1069. Hofer, J. E., Grabenstetter, R. J., Wiig, E. O.: J. Amer. Chem. Soc. *72*, 203 (1950)
1070. Scheibe, G.: Kolloid Zeitschr. *82*, 1 (1938)
1071. Jelley, E. E.: Nature *139*, 631 (1937)
1072. Daltrozzo, E., Scheibe, G., Gschwind, K., Haimerl, F.: Photogr. Sci. Eng. *18*, 441 (1974)
1073. Czikkely, V., Försterling, H. D., Kuhn, H.: Chem. Phys. Lett. *6*, 11 (1970)
1074. Kopainsky, B., Hallermeier, J. K., Kaiser, W.: Chem. Phys. Lett. *87*, 7 (1982)
1075. Mizutani, F., Iijima, S.-I., Tsuda, K.: Bull. Chem. Soc. Jpn. *55*, 1295 (1982)
1076. Nelson, R. C.: J. of Opt. Soc. Am. *46*, 10 (1956)
1077. Kuhn, H., Möbius, D.: Angew. Chem. Int. ed. *10*, 620 (1971)
1078. Kuhn, H.: Pure Appl. Chem. *51*, 341 (1979)
1079. Möbius, D.: Account Chem. Res. *14*, 63 (1981)
1080. Baughman, R. H., Bredas, J. L., Chance, R. R., Elsenbaumer, R. L., Shacklette, L. W.: Chem. Rev. *82*, 209 (1982)
1081. Speight, J. G.: J. Macromol. Sci. *C 5*, 295 (1971)
1082. Osa, T., Yildiz, A., Kuwana, T.: J. Amer. Chem. Soc. *91*, 3994 (1969)
1083. Gallegos, E. J.: J. Phys. Chem. *71*, 1647 (1967)
1084. Shacklette, L. W., Eckhardt, H., Chance, R. R., Miller, G. G., Ivory, D. M., Baughman, R. H.: J. Chem. Phys. *73*, 4098 (1980)
1085. Barbarin, F., Berthet, G., Blanc, J. P., Fabre, C., Germain, J. P., Hamdi, M., Robert, H.: Synth. Metals *6*, 53 (1983)

1086. Shacklette, L. W., Chance, R. R., Ivory, D. M., Miller, G. G., Baughman, R. H.: Synth. Metals *1*, 307 (1979)
1087. Ivory, D. M., Miller, G. G., Sowa, J. M., Shacklette, L. W., Baughman, R. H.: J. Chem. Phys. *71*, 1506 (1979)
1088. Chance, R. R., Shacklette, L. W., Miller, G. G., Ivory, D. M., Sowa, J. M., Elsenbaumer, R. L., Baughman, R. H.: J. C. S. Chem. Commun. 348 (1980)
1089. Rabolt, J. F., Clarke, T. C., Kanazawa, K. K., Reynolds, J. R., Street, G. B.: J. C. S. Chem. Commun. 347 (1980)
1090. Shacklette, L. W., Elsenbaumer, R. L., Chance, R. R., Eckhardt, H., Fronmer, J. E., Baughman, R. H.: J. Chem. Phys. *75*, 1919 (1981)
1091. Boscato, J. F., Catala, J. M., Clouet, F., Brossas, J.: Polymer Bulletin *4*, 357 (1981)
1092. Bredas, J. L., Elsenbaumer, R. L., Chance, R. R., Silbey, R.: J. Chem. Phys. *78*, 5656 (1983)
1093. Clarke, T. C., Kanazawa, K. K., Lee, V. Y., Rabolt, J. F., Reynolds, J. R., Street, G. B.: J. Polym. Sci. Polym. Phys. ed. *20*, 117 (1982)
1094. Wnek, G., Chien, J. C. W., Karasz, F. E.: Org. Coat. Plast. Chem. *43*, 882 (1980)
1095. Chien, J. C. W., Wnek, G. E., Karasz, F. E., Hirsch, J. A.: Macromolecules *14*, 479 (1981)
1096. Yamabe, T., Tanaka, K., Terama, E.-H., Fukui, K., Shirakawa, H., Ikeda, S.: Synth. Metals *1*, 321 (1979/80)
1097. McWulty, B. J.: Polymer *7*, 275 (1966)
1098. Deits, W., Cukor, P., Rubner, M., Jopson, H.: Polym. Prep. *22*, 197 (1981)
1099. Hergenrother, P. M.: J. Macromol. Sci. Rev. Macromol. Chem. *C 19*, 1 (1980)
1100. Simionescu, C. R., Dumitrescu, S. V., Persec, V., Negulescu, I., Diaconu, I.: J. Polym. Sci. Symp. *42*, 201 (1973)
1101. Henrici-Olive, G., Olive, S.: Adv. Polym. Sci. *32*, 123 (1979)
1102. Masuda, T., Sasaki, N., Higashimura, T.: Macromolecules *8*, 717 (1975)
1103. Cukor, P., Krugler, J. I., Rubner, M. F.: Makromol. Chem. *182*, 165 (1981)
1104. Kern, R. J.: J. Polym. Sci. *A 1*, 7, 621 (1969)
1105. Masuda, T., Thien, K.-Q., Sasaki, N., Higashimura, T.: Macromolecules *9*, 661 (1976)
1106. Masuda, T., Hasegawa, K., Higashimura, T.: Macromolecules *7*, 728 (1974)
1107. Famili, A., Farona, M. F.: Polym. Bull. *2*, 289 (1980)
1108. Simionescu, C. I., Persec, V., Dimitrescu, S.: J. Polym. Sci. Polym. Chem. ed. *15*, 2497 (1977)
1109. Simionescu, C. I., Persec, V.: J. Polym. Sci. Polym. Chem. ed. *18*, 147 (1980)
1110. Cukor, P., Krugler, J. I., Rubner, M. F.: Makromol. Chem. *182*, 165 (1981)
1111. Ehrlich, P., Kern, R. J., Pierron, E. D., Prouder, T.: J. Polym. Sci. *B 5*, 911 (1967)
1112. Sanford, T. J., Allendoerfer, R. D., Kang, E. T., Ehrlich, P.: J. Polym. Sci. Polym. Phys. ed. *19*, 1151 (1981)
1113. Bhatt, A. P., Anderson, W. A., Kang, E. T., Ehrlich, P.: J. Appl. Phys. *54*, 3973 (1983)
1114. Kuwane, Y., Masuda, T., Higashimura, T.: Polym. J. *12*, 387 (1980)
1115. Kang, E. T., Bhatt, A. P., Villardel, E., Anderson, W. A., Ehrlich, P.: J. Polym. Sci. Polym. Lett. ed. *20*, 143 (1982)
1116. Bloor, D.: Chem. Phys. Lett. *43*, 270 (1976)
1117. Kang, E. T., Bhatt, A. P., Villardel, E., Anderson, W. A., Ehrlich, P.: Mol. Cryst. Liq. Cryst. *83*, 307 (1982)
1118. Bhatt, A. P., Anderson, W. A., Ehrlich, P.: Solid State Commun. *47*, 997 (1983)
1119. Leroy, S.: Thèse de 3ème Cycle, Paris, 1983
1120. Teoh, H., MacInnes Jr., D., Metz, P. D.: J. de Phys. Colloque *C 3*, *44*, 687 (1983)
1121. Teoh, H., Metz, P. D., Wilhelm, W. G.: Mol. Cryst. Liq. Cryst. *83*, 297 (1982)
1122. Fink, J., Ritsko, J. J., Crecelius, G.: J. de Phys. Colloque *C 3*, *44*, 683 (1983)
1123. Reynaud, C., Richard, A., Juret, C., Nuvolone, R., Boiziau, C., Lecayon, G., Le Gressus, C.: Thin Solid Films *92*, 355 (1982)
1124. Burlant, W. J., Parsons, J. L.: J. Polym. Sci. *22*, 249 (1956)
1125. Monahan, A. R.: J. Polym. Sci. *A 4*, 2391 (1966)
1126. Takata, T., Hiroi, J.: J. Polym. Sci. *A 4*, 2391 (1966)
1127. Reynaud, C., Juret, C., Boiziau, C.: Surface Sci. *126*, 733 (1983)
1128. Van Beek, L. K. H.: J. Appl. Polym. Sci. *9*, 553 (1965)
1129. Whangbo, M. H., Hoffmann, R., Woodward, R. B.: Proc. Roy. Soc. London *A 366*, 23 (1979)
1130. Bredas, J. L., Themans, B., André, J. M.: J. Chem. Phys. *78*, 6137 (1983)

1131. Bruck, S. D.: Polymer 6, 319 (1965)
1132. Naarmann, H.: personal communication
1133. Kanazawa, K. K., Diaz, A. F., Geiss, R. H., Gill, W. D., Kwak, J. F., Logan, J. A., Rabolt, J. F., Street, G. B.: J. C. S. Chem. Commun. 854 (1979)
1134. Street, G. B., Clarke, T. C., Krounbi, M., Kanazawa, K. K., Lee, V., Pfluger, P., Scott, J. C., Weiser, G.: Mol. Cryst. Liq. Cryst. 83, 253 (1982)
1135. Kanazawa, K. K., Diaz, A. F., Gardini, G. P., Gill, W. D., Grant, P. M., Kwak, J. F., Street, G. B.: Synth. Metals 1, 329 (1980)
1136. Ford, W. K., Duke, C. B., Salaneck, W. R.: J. Chem. Phys. 77, 5030 (1982)
1137. Diaz, A. F., Crowley, J., Bargon, J., Gardini, G. P., Torrance, J. B.: J. Electroanal. Chem. 121, 355 (1981)
1138. Pfluger, P., Krounbi, M., Street, G. B., Weiser, G.: J. Chem. Phys. 78, 3212 (1983)
1139. Watanabe, A., Tanaka, M., Tanaka, J.: Bull. Chem. Soc. Jpn. 54, 2278 (1981)
1140. Salaneck, W. L., Erlandsson, R., Prejza, J., Lundström, I., Inganäs, O.: Synth. Metals 5, 125 (1982)
1141. Chamberlain, G. A.: Nature 289, 45 (1981)
1142. Meier, H., Albrecht, W.: Berichte der Buns. 64 (1963)
1143. Merritt, V. Y.: IBM J. Res. Develop. 22, 353 (1978)
1144. Chamberlain, G. A.: J. Appl. Phys. 53, 6262 (1982)
1145. Kudo, K., Moriizumin, T.: Jpn. J. Appl. Phys. 19, L 683 (1980); 20, L 553 (1981)
1146. Chamberlain, G. A.: Mol. Cryst. Liq. Cryst. 93, 369 (1983)
1147. Kudo, K., Shinohara, T., Moriizumi, T., Iriyama, K., Sugi, M.: Jpn. J. Appl. Phys. 20, 135 (1981)
1148. Merritt, V. Y., Hovel, H. J.: Appl. Phys. Lett. 29, 414 (1976)
1149. Riblett, S. E., Cowan, D. O., Bloch, A. N., Poehler, T. O.: Mol. Cryst. Liq. Cryst. 85, 69 (1982)
1150. Takagi, S., Kawabe, K.: Technol. Rep. Osaka Univ. 27, 159 (1977); CA 86, 198, 707 C (1977)
1151. Skotheim, T., Yang, J.-M., Otvos, J., Klein, M. P.: J. Chem. Phys. 77, 6144 (1982)
1152. Tavares, A. D.: J. Chem. Phys. 53, 2520 (1970)
1153. Phadke, S. D.: Indian J. Pure Appl. Phys. 17, 261 (1979)
1154. Inokuchi, H., Maruyama, Y., Akamatu, H.: Bull. Chem. Soc. Jpn. 34, 1093 (1961)
1155. Kallmann, H.: J. Chem. Phys. 30, 585 (1959)
1156. Lyons, L. E., Newman, O. M. G.: Aust. J. Chem. 24, 13 (1971)
1157. Sittig, M.: Solar Cells for Photovoltaic generation of electricity, Park Ridge, Noyes Data Corporation 1979
1158. Reucroft, P. J., Kronick, P. L., Hillman, E. E.: Mol. Cryst. Liq. Cryst. 6, 247 (1969)
1159. Silins, E., Belkind, A. I., Balode, D., Biseniece, A., Grechov, V. V., Taure, L., Kurik, M. V., Vertsimakha, Ya., Bok, I.: Phys. Status Solidi A 25, 339 (1974)
1160. Krikorian, E., Sneed, R. J.: J. Appl. Phys. 40, 2306 (1969)
1161. Lipinski, A., Piotrowski, J., Szymanski, A.: Bull. Acad. Pol. Sci. Ser. Sci. Math. Astron. Phys. 15, 833 (1967)
1162. Mukherjee, T. K.: J. Phys. Chem. 74, 3006 (1970)
1163. Kay, R. E., Walwick, E. R.: C. A. 85, 182, 101 c (1976)
1164. Fang, P. H.: Jpn. J. Appl. Phys. 13, 1232 (1974); 11, 1298 (1972)
1165. Matsumura, M., Uohashi, H., Furusawa, M., Yamamoto, N., Tsubomura, H.: Bull. Chem. Soc. Jpn. 48, 1965 (1975)
1166. Sato, S.: Jpn. J. Appl. Phys. 20, 1989 (1981)
1167. Kamei, H., Katayama, Y., Ozawa, T.: Jpn. J. Appl. Phys. 11, 1385 (1972)
1168. Glazov, V. M., Chizhevskaya, S. N., Glagoleva, N. N.: "Liquid Semiconductors", Plenum Press, New York, 1969
1169. Reucroft, P. J., Simpson, W. H.: Disc. Faraday Soc. 51, 202 (1971)
1170. Simpson, W. H., Freeman, R. A., Reucroft, P. J.: Photochem. Photobiol. 11, 319 (1970)
1171. Lawrence, M. F., Dodelet, J.-P., Ringuet, M.: Photochem. Photobiol. 34, 393 (1981)
1172. Tang, C. W., Albrecht, A. C.: J. Chem. Phys. 63, 953 (1975)
1173. Corker, G. A., Lundström, I.: J. Appl. Phys. 49, 686 (1978)
1174. Tang, C. W., Albrecht, A. C.: Nature 254, 507 (1975); J. Chem. Phys. 62, 2139 (1975)
1175. Dodelet, J. P., Le Brech, J., Leblanc, R. M.: Photochem. Photobiol. 29, 1135 (1979)

1176. Janzen, A. F., Bolton, J. R.: J. Amer. Chem. Soc. *101*, 6342 (1979)
1177. Reucroft, P. J., Simpson, W. H.: Photochem. Photobiol. *10*, 79 (1969)
1178. Kampas, F. J., Gouterman, M.: J. Phys. Chem. *81*, 690 (1977)
1179. Musser, M. E., Dahlberg, S. C.: Thin Solid Films *66*, 261 (1980)
1180. Yamashita, K.: Chem. Lett. 627 (1980)
1181. Langford, C. H., Hollebone, B. R., Nadezhdin, D.: Can. J. Chem. *59*, 652 (1981)
1182. Tanimura, K., Kawai, T., Sakata, T.: J. Phys. Chem. *84*, 751 (1980)
1183. Kawai, T., Tanimura, K., Sakata, T.: Chem. Phys. Lett. *56*, 541 (1978)
1184. Terenin, E., Putseiko, E., Akimov, I.: Discuss. Faraday Soc. *27*, 83 (1959)
1185. Van der Donckt, E., Wollast, P., Noirhomme, B., Deltour, R.: Bull. Soc. Chem. Belg. *88*, 263 (1979)
1186. Van der Donckt, E., Noirhomme, B., Kanicki, J.: J. Appl. Polym. Sci. *27*, 1 (1982)
1187. Hermann, A. M., Rembaum, A.: *CA 73*, 40037f (1970); J. Polym. Sci. *B 5*, 445 (1967)
1188. Lovinger, A. J.: Science *220*, 1115 (1983)
1189. Sasabe, H., Nakayama, T., Kumazawa, K., Miyata, S., Fukada, E.: Polym. J. *13*, 967 (1981)
1190. Phillips, T. E., Kistenmacher, T. J., Bloch, A. N., Cowan, D. O.: J. C. S. Chem. Commun. 334 (1976)
1191. Pope, M., Swenberg, C. E.: Electronic Processes in Organic Crystals, Oxford, Clárendon Press 1982

Subject Index

H. Endres

Chemische Aspekte der Festkörper-Physik

Hoschschultext

1984. 117 Abbildungen. IV, 160 Seiten
ISBN 3-540-13604-5

Inhaltsübersicht: Allgemeine Kristallographie und Beugungserscheinungen. – Gitterfehler. – Gitterschwingungen. – Magnetische Eigenschaften. – Elektronen im Festkörper: Das freie Elektronengas. – Das Elektron im periodischen Potential: Energiebänder. – Supraleitung. – Eindimensionale Leiter. – Weiterführende Literatur. – Register.

Das Buch wendet sich an Chemiker, die einen ersten Überblick über die wichtigsten Erscheinungen in Festkörpern suchen. Als Einführung in weiterführende Lehrbücher ist es auch für Physikstudenten interessant. Es zeigt festkörperphysikalische Zusammenhänge erzählerisch auf und verzichtet auf komplizierte formelhafte Darstellung. Dadurch stellt es keine Ansprüche an Mathematikkenntnisse, die über Abiturwissen hinausgehen. Der Band will allen Interessierten den Zugang zur Festkörperphysik eröffnen, ohne ihn durch mathematische und kristallographische Überfrachtung zu erschweren. Dabei werden die wichtigsten Meßverfahren vorgestellt und im letzten Kapitel des Buches ihre Anwendung auf aktuelle chemische Fragen besprochen.

Springer-Verlag
Berlin
Heidelberg
New York
Tokyo

Solar Energy Materials

1982. 86 figures, 15 tables. V, 182 pages
(Structure and Bonding, Volume 49)
ISBN 3-540-11084-4

Contents/Information: The world's conventianal energy
supplies – based mainly on fossil sources – are rapidly
diminuishing. The main short-term alternative to the energy
crisis, nuclear fission energy, has aroused a tremendous
amount of controversy, and the date of the practical realization of nuclear fusion is not yet known. There is no doubt
that solar energy, being a clean and non-hazardous source,
can provide a considerable contribution to the solution of
the energy problem if proper methods are developed to
collect, concentrate, store, and convert solar light, which is
naturally diffuse and intrinsically intermittent. The volume
discusses this problem by three review articles:
1. *R. Reisfeld, C. K. Jørgensen:* **Luminescent Solar Concentra-
tors for Energy Conversion.**
1. *M. Grätzel, K. Kalyanasundaram, J. Kiwi:* **Visible Light
Induced Cleavage of Water into Hydrogen and Oxygen in
Colloidal and Microheterogeneous Systems.**
3. *H. Tributsch:* **Photoelectrochemical Energy Conversion
Involving Transition Metal d-States and Intercalation of
Layer Compounds.**

H. Rickert

Electrochemistry of Solids

An Introduction

1982. 95 figures, 23 tables. XII, 240 pages
(Inorganic Chemistry Concepts, Volume 7)
ISBN 3-540-11116-6

Contents: Introduction. – Disorder in Solids. – Examples of
Disorder in Solids. – Thermodynamic Quantities of Quasi-
Free Electrons and Electron Defects in Semiconductors. –
An Example of Electronic Disorder. Electrons and Electron
Defects in α-Ag_2S. – Mobility, Diffusion and Partial
Conductivity of Ions and Electrons. – Solid Ionic Conduc-
tors, Solid Electrolytes and Solid-Solution Electrodes. –
Galvanic Cells with Solid Electrolytes for Thermodynamic
Investigations. – Technical Applications of Solid Electro-
lytes – Solid-State Ionics. – Solid-State Reactions. –
Galvanic Cells with Solid Electrolytes for Kinetic Investiga-
tions. – Non-Isothermal Systems. Soret Effect, Transport
Processes, and Thermopowers. – Author Index. – Subject
Index.

Springer-Verlag
Berlin
Heidelberg
New York
Tokyo